Calmodulin and Intracellular Ca^{++} Receptors

Calmodulin and Intracellular Ca++ Receptors

Edited by

SHIRO KAKIUCHI
Institute of Higher Nervous Activity
Osaka University Medical School
Osaka, Japan

HIROYOSHI HIDAKA
Mie University School of Medicine
Mie, Japan

and

ANTHONY R. MEANS
Baylor College of Medicine
Texas Medical Center
Houston, Texas

PLENUM PRESS • NEW YORK AND LONDON

Library of Congress Cataloging in Publication Data

Main entry under title:

Calmodulin and intracellular Ca⁺⁺ receptors.

"Papers contributed by participants in the Symposium on Recent Advances in Ca⁺⁺ and Cell Function, Calmodulin and Intracellular Ca⁺⁺ Receptors, held in Kyoto, Japan, July 25-27, 1981. This meeting was one of the satellite symposia to the 8th International Congress of Pharmacology (Tokyo, 1981)"—Pref.
 Includes bibliographical references and indexes.
 1. Calmodulin—Congresses. 2. Calcium—Physiological effect—Congresses. I. Kakiuchi, Shiro, 1929-   . II. Hidaka, Hiroyoshi, 1938-   . III. Means, Anthony R. IV. Symposium on Recent Advances in Ca⁺⁺ and Cell Function, Calmodulin and Intracellular Ca⁺⁺ Receptors (1981: Kyoto, Japan) [DNLM: 1. Calcium—Physiology—Congresses. 2. Calcium-binding proteins—Congresses. 3. Cells—Physiology—Congresses. QU 55 C1464 1981]
QP552.C28C34 1982   615'.7   82-12326
ISBN 0-306-41109-1

Proceedings of a satellite symposium to the Eighth International
Congress of Pharmacology on Recent Advances in Ca⁺⁺ and Cell
Function, Calmodulin and Intracellular Ca⁺⁺ Receptors, held
July 25-27, 1981, in Kyoto, Japan

© 1982 Plenum Press, New York
A Division of Plenum Publishing Corporation
233 Spring Street, New York, N.Y. 10013

All rights reserved

No part of this book may be reproduced, stored in a retrieval system, or transmitted in any form or by any means, electronic, mechanical, photocopying, microfilming, recording, or otherwise, without written permission from the Publisher

Printed in the United States of America

PREFACE

This volume is a collection of papers contributed by participants in the Symposium on Recent Advances in Ca^{++} and Cell Function, Calmodulin and Intracellular Ca^{++} Receptors, held in Kyoto, Japan, July 25-29, 1981. This meeting was one of the satellite symposia to the 8th International Congress of Pharmacology (Tokyo, 1981).

Calcium ion is a well known intracellular messenger. Since troponin and calmodulin were found to be intracellular mediators of a number of actions of Ca^{++}, the functions of this ion, with relevance to its intracellular receptive proteins, have received a great deal of attention from workers in the fields of biochemistry, biology, pharmacology, physiology, endocrinology, clinical medicine, and related fields.

The symposium was designed to illuminate new aspects of this rapidly growing field of research and extensive discussions on various topics were carried out. This volume contains all the papers presented plus the abstracts of the Poster Presentations. Both researchers and students should find the data and references of considerable value in their studies.

We gratefully acknowledge the sponsorship of the Committees of the 8th International Congress of Pharmacology and the Yamada Science Foundation. We also thank Dr. S. Akabori for his encouragement and support.

Special thanks are extented to Ms. S. Ikukawa, Ms. A. Yoshikawa, M. Ohara and all our colleagues who contributed much to the symposium. We also extend our appreciation to Mrs. F. Comes and Patricia M. Vann for the editing and preparation of the manuscript in its final form.

Shiro Kakiuchi
Hiroyoshi Hidaka

November 1981

CONTENTS

MOLECULAR BASIS OF CALMODULIN FUNCTION

The Pharmacology of Calmodulin Antagonism:
A Reappraisal
    F.F. Vincenzi   1

Biopharmacological Assessment of Calmodulin Function:
Utility of Calmodulin Antagonists
    H. Hidaka & T. Tanaka   19

Intracellular Localization of Calmodulin Antagonist, W-7
    S. Ohno, Y. Fujii, N. Usuda, T. Nagata, T. Endo,
    T. Tanaka & H. Hidaka   35

Possible Dual Regulatory Mechanism in Smooth Muscle
Actomyosin
    M. Maruyama & S. Shibata   49

Ion Binding to Calmodulin
    J. Haiech, M.-C. Kilhoffer, D. Gerard &
    J.G. Demaille   55

Structure and Ca^{++} Dependent Conformational Change
of Calmodulin
    K. Yagi, S. Matsuda, H. Nagamoto, T. Mikuni
    & M. Yazawa   75

Guanine Nucleotide Dependence of Calcium-Calmodulin
Stimulation of Adenylate Cyclase in Rat Cerebral
Cortex and Striatum
    K. Seamon & J.W. Daly   93

Patterns of Intracellular Ca^{++} Distribution and
Mobilization in Smooth Muscle
    G.B. Weiss   111

Role of Ca^{++}-Calmodulin in Metabolic Regulation in Plants
    M.J. Cormier, H.W. Jarrett & H. Charbonneau   125

Regulation of Calmodulin in Mammalian Cells
    A.R. Means & J.G. Chafouleas   141

Calcium and Calmodulin Dependent Regulation of Microtubule Assembly
    H. Kumagai, E. Nishida & H. Sakai   153

Calmodulin and Cytoskeleton
    S. Kakiuchi, K. Sobue, M. Fujita, Y. Muramoto, K. Kanda & K. Morimoto   167

Effect of the Calmodulin-Caldesmon System on the Physical State of Actin Filaments
    K. Maruyama, K. Sobue & S. Kakiuchi   183

Mode of Calcium Binding to Smooth Muscle Contractile System
    S. Ebashi, Y. Nonomura & M. Hirata   189

Kinetic Studies of the Activation of Cyclic Nucleotide Phosphodiesterase by Ca^{++} and Calmodulin
    V. Chau, C.Y. Huang, P.B. Chock, J.H. Wang & R.K. Sharma   199

Ca^{++} AND PROTEIN PHOSPHORYLATION

The Regulation of Myosin Light Chain Kinase by Ca^{++} and Calmodulin *in vitro* and *in vivo*
    J.T. Stull, D.K. Blumenthal, B.R. Botterman, G.A. Klug, D.R. Manning & P.J. Silver   219

Phosphorylation and ATPase Activity of Smooth Muscle Myosin
    D.J. Hartshorne, M.P. Walsh & A. Persechini   239

Demonstration of Two Types of Ca^{++} Dependent Protein Kinases in the Brain and Other Tissues
    E. Miyamoto, K. Fukunaga & K. Matsui   255

Calmodulin Dependent Protein Kinases from Rat Brain
    T. Yamauchi & H. Fujisawa   267

CONTENTS

A Highly Active Calcium Ion Stimulated Endogenous Protein Kinase in Myelin: Kinetic and Other Characteristics, Calmodulin Dependence and Comparison with Other Brain Kinase(s)
    P.V. Sulakhe, E.H. Petrali & B.L. Raney   281

Interaction of Calmodulin with cAMP Dependent Protein Kinase in Brain
    C.B. Klee, M.H. Krinks, D.R. Hathaway & D.A. Flockhart   303

Regulation of Contractile Proteins in Smooth Muscle and Platelets by Calmodulin and Cyclic AMP
    R.S. Adelstein, P. de Lanerolle, J.R. Sellers, M.D. Pato & M.A. Conti   313

Two Transmembrane Control Mechanisms for Protein Phosphorylation in Bidirectional Regulation of Cell Functions
    Y. Takai, K. Kaibuchi, T. Matsubara, K. Sano, B. Yu & Y. Nishizuka   333

Ca^{++} REGULATION OF MICROFILAMENT SYSTEM

Calcium Regulation in Amoeboid Movement
    D.L. Taylor & M. Fechheimer   349

Actinogelin: A Calcium Sensitive Regulator of Microfilament System
    A. Asano, N. Mimura & P.F. Kuo   375

Calcium Regulation of Actin Network Structure by Gelsolin
    H.L. Yin & T.P. Stossel   393

Physical Properties of Fragmin, A Ca^{++} Sensitive Regulatory Protein of Actin Polymerization Isolated from Physarum Plasmodium
    S. Hatano, T. Hasegawa, H. Sugino & K. Ozaki   403

CLOSING

Calcium Ion and Calcium Binding Proteins
    S. Ebashi   421

APPENDIX	425
PARTICIPANTS	465
CONTRIBUTOR INDEX	471
SUBJECT INDEX	473

THE PHARMACOLOGY OF CALMODULIN ANTAGONISM: A REAPPRAISAL

F.F. Vincenzi

Department of Pharmacology, University of Washington
Seattle, Washington 98195, USA

INTRODUCTION

Calmodulin (CaM) is now widely accepted as a major intracellular receptor for calcium ion (Ca^{++}). The central role of intracellular Ca^{++} in the regulation of cellular functions has been recognized, if not understood, for decades. Now CaM is implicated in the mediation through multiple cellular effectors of many, if not all, cellular responses to internal Ca^{++} (2,22,32). Pharmacological antagonism of CaM is therefore likely to be a fruitful approach to manipulation of many cellular responses, elucidation of their control, and possibly the treatment of human illnesses. This report is not intended to propose any simple relationships between CaM antagonism in cells and the general systemic pharmacology of anti-CaM drugs. In fact, I want to urge against prematurely accepting any simple models.

I will use the general term "anti-CaM" for drugs which decrease CaM-mediated responses when the mechanism may be unknown. CaM binding drugs (CaM-BDs) is a term which will be reserved for drugs demonstrated to bind to CaM (although that mechanism may not account for all of their actions). "CaM antagonists" should probably be used to signify drugs which, by binding to "CaM receptors," can prevent its effects. Unfortunately, this term has already been used in the literature more or less interchangeably with the other terms.

At the present time we are somewhat uncertain about the Ca^{++} binding properties of CaM (4,13), or the possibly different Ca^{++} stoichiometry of CaM at various effectors (4,38). Therefore I will make no particular assumptions about Ca^{++} stoichiometry as regards effector or drug binding to CaM. We know something about antagonism of CaM, little about mimicking the effects of CaM, less about how CaM activity in cells is regulated (if it is regulated)(22) and probably less about possible relationships between drug induced antagonism of CaM in cells, on the one hand, and tissue, organ and organism responses to such drugs on the other.

I will present data on antagonism of CaM activation of the Ca^{++} pump ATPase of the human red blood cell (RBC) membrane. This CaM dependent enzyme is activated by low levels of CaM (28,37,38), and activation of the enzyme is translated into an increased rate of Ca^{++} transport across the plasma membrane (11,38). In other words, the intracellular receptor for Ca^{++} appears to participate in regulation of the levels of Ca^{++} to which it is exposed. As will be shown, the activation of Ca^{++} pump ATPase by CaM can be antagonized by a variety of pharmacologically unrelated agents. The results will be interpreted to suggest that it is past time to reconsider the model linking CaM antagonism and antipsychotic activity of drugs (43). It will be further suggested that CaM regulation of various effectors may be rather non-specific. CaM-like activation can apparently be accomplished by other agents. Another related observation is that selective antagonism of a process by an anti-CaM drug does not necessarily implicate CaM in that process. Preliminary evidence is offered which suggests that some anti-CaM drugs bind directly to effector enzymes. This raises the possibility that in the future pharmacological antagonists of CaM may be developed which are selective for some, but not all, CaM-mediated cellular responses.

Figure 1 is a simplified and schematic diagram adapted from that presented by Cheung (2). The figure includes generally assumed relationships between Ca^{++}, CaM and CaM effectors in cells. In short, Ca^{++} enters or is released in the cell; it binds to CaM forming a $CaM(Ca^{++})_n$ complex. This complex binds in turn to various effectors and modifies their activities. This sequence initiates the chain of events leading to a cellular response. CaM is usually assumed not to bind to effectors in the absence of Ca^{++}. Some exceptions to this rule exist. For example, phosphorylase kinase binds CaM in the absence of Ca^{++} (3). The known anti-CaM drugs do not interfere with CaM in this state and will not be considered further. Figure 1 includes the generally accepted mechanism by which CaM-BDs such as trifluoperazine (TFP) are thought to antagonize the effects of CaM. The idea is that drugs such as TFP bind to CaM in a Ca^{++} dependent fashion. Drug binding is assumed to prevent CaM from binding to the effector, thereby reducing the effectiveness of CaM in the system being tested. Usually this is the only mechanism assumed to apply for CaM-BDs (or for that matter, for most anti-CaM drugs). The work of Levin and Weiss (17-19) clearly

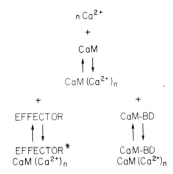

FIGURE 1

Simple model of calmodulin interactions. Calmodulin is visualized as interacting with an undetermined number of intracellular Ca^{++} ions. The calmodulin $(Ca^{++})_n$ complex then undergoes interaction with an effector molecule, of which only one is shown. It is presumed that this interaction produces a change in the effector (for example, activation of an enzyme) which results in a cellular response. In the presence of a calmodulin binding drug (CaM-BD) a competition is set up between the effector molecule(s) and the CaM-BD. If the CaM-BD binds to a significant fraction of calmodulin $(Ca^{++})_n$, then the cellular responses mediated by calmodulin $(Ca^{++})_n$ will be reduced.

demonstrated Ca^{++} dependent binding of TFP and other anti-psychotic drugs to CaM. Weiss and Levin (44), in a diagram functionally equivalent to Fig. 1, used the term "antipsychotic drug." I have chosen the term CaM-BD for two reasons. First, the term is parallel to that of the CaM binding proteins (CaM-BPs) (41,43). There are excellent data available which may be interpreted to suggest that, at least in certain cells, CaM activity (rather than content) may be regulated by CaM-BPs (Vincenzi, submitted for publication). CaM-BPs are presumed not to be cellular effector molecules, but rather "acceptors" or "sinks" for the $CaM(Ca^{++})_n$ complex. Their presence can serve to blunt activation of all CaM dependent effectors. For example, CaM-BP selectively abolished CaM induced activation of the Ca^{++} pump and the associated ATPase (16). CaM-BP did not inhibit basal pump or ATPase activities observed in the absence of added CaM. As will be shown, this is in contrast to typical anti-CaM drugs.

A second reason for the use of the term CaM-BD is as follows: Although there is some controversy in the matter, the relationship between drug binding to CaM and antipsychotic activity (44) cannot be considered to be established. There is excellent correlation between affinity of drugs for CaM and their potency in displacing ^{3}H-TFP from CaM in vitro. However, relationships between anti-CaM activity in vitro and clinical efficacy of antipsychotic drugs in vivo are not clear, if they exist at all. On the basis of previous data from this (11,29) and other (25,30) laboratories, it appears that no simple relationship exists between CaM antagonism and antipsychotic activity of drugs. Beyond this issue I want to raise questions regarding: 1) the specificity of CaM, 2) the possibility that other molecules may mimic CaM effects, either physiologically or pharmacologically, and 3) the apparent binding of anti-CaM drugs directly to effector enzymes. The latter phenomenon may provide a key to future development of anti-CaM drugs which are more selective in antagonizing some, but not all, cellular responses mediated by the ubiquitous protein. Some of these results were presented in a poster at the Eighth International Congress of Pharmacology.

METHODS

Human RBC membranes were isolated in 20 mM imidazole buffer as described by Farrance and Vincenzi (7). The Ca^{++} pump ATPase was isolated (24) and reconstituted (23), as described by Carafoli and co-workers. ATPase activities were determined as outlined by Raess and Vincenzi (28) using a 2 ml assay system and incubation times from 60-120 min. Because the onset of the CaM-induced response of the Ca^{++} pump ATPase (39) and other enzymes (42) may be rather slow, 15 min preincubation of buffer, membranes, drugs and CaM was allowed before the reaction was started with ATP. CaM was isolated from human RBCs as described previously (29). Unless otherwise noted, drugs were dissolved in water as the hydrochloride salts. For drugs with very low water solubility, ethanol or dimethylsulfoxide (DMSO) was used as a solvent. The final volume of these solvents never exceeded 1% in the assay medium. This concentration exerted minimal effects on the various ATPase activities and the activation of Ca^{++} pump ATPase by CaM. Drug solutions were made fresh daily and care was taken to minimize exposure of drugs to light. Protein was determined either by a modification of the Lowry method (21) or by the Peterson assay (26). Except as noted, each figure illustrates results obtained in separate experiments. The different specific activities apparent in these figures are typical of the variability of these preparations observed over the course of approximately one year.

RESULTS AND DISCUSSION

Figure 2 illustrates typical and well known effects of TFP on a CaM dependent enzyme, the (Ca^{++} + Mg^{++})-ATPase, or Ca^{++} pump ATPase of human RBC membranes. As a function of TFP concentration there was pronounced inhibition of the CaM-activated portion of the enzyme activity. This activity was measured in the

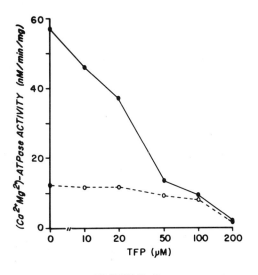

FIGURE 2

Effect of trifluoperazine (TFP) on basal and calmodulin-activated Ca^{++} pump ATPase activities. Ca^{++} pump ATPase activity of isolated human RBC membranes was determined in the absence of (o--o) and presence (•—•) of 160 ng/ml (approximately 10^{-8} M) calmodulin. TFP inhibited both enzyme activities but was selective for the calmodulin-activated portion. Essentially complete abolition of calmodulin activation was produced by 10^{-4} M TFP, a concentration which inhibited the basal activity only slightly.

presence of 160 ng/ml (approximately 10^{-8} M) CaM. By contrast, there was only slight inhibition of the so-called "basal" activity (that measured in the absence of added CaM). Selective inhibition of the CaM-activated portion of enzyme activity has been reported by a variety of authors (8,11,14,20,29,40) for this and other enzymes. Sometimes the antagonism of CaM activation by TFP and other drugs has been shown to exhibit competitive kinetics. Often, on the basis of limited data as in Fig. 2, competitive kinetics are implied or stated without actually measuring activity over a wide range of CaM concentrations.

Figure 3 presents data collected as in Fig. 1, but using the antidepressant drug imipramine. As with trifluoperazine, inhibition of the CaM-activated portion of the activity was quite selective. Imipramine was less potent than TFP, but qualitatively the data are identical, at least as viewed in Fig. 3. Qualitatively identical findings and figures have been widely published. These apparently led some of us to assume that most, if not all, anti-CaM drugs, including butyrophenones (8) and local anesthetics (40), act more

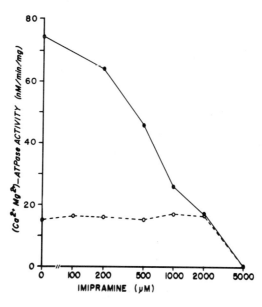

FIGURE 3

Effect of imipramine on basal and calmodulin-activated Ca^{++} pump ATPase activities. Experiment performed and results plotted as in Fig. 2. Imipramine inhibited the calmodulin-activated portion of ATPase activity selectively up to 2×10^{-3} M.

or less like TFP (that is, that they act by virtue of being CaM-BDs). However, as shown in Figs. 4 and 5, respectively, TFP and imipramine display somewhat different apparent kinetics for antagonism of CaM. When various concentrations of these drugs were tested over a wide range of CaM concentrations TFP displayed apparently competitive antagonism, whereas imipramine displayed apparently non-competitive antagonism of CaM. Dibucaine, a local anesthetic agent, also displayed apparent non-competitive antagonism of CaM (Fig. 6). The result in Fig. 6 is similar to that presented by Volpi et al. (40).

Similar measurements were made for a number of drugs of various therapeutic categories. All of the drugs were tested as in Figs. 4-6. With the exception of the phenothiazines, each of the drugs listed in Table 1 gave apparent non-competitive antagonism of CaM activation of the Ca^{++} pump ATPase. Even the cationic detergent cetylpyridinium chloride was a selective antagonist of CaM. Table 1 presents the estimated micromolar concentration of each drug required for 50% inhibition of the $(10^{-8}$ M) CaM activated portion of ATPase activity (IC_{50}). As in Figs. 4-6, there

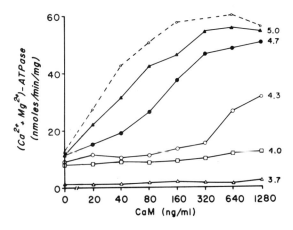

FIGURE 4

Effect of TFP on calmodulin-activation of Ca^{++} pump ATPase of isolated human RBC membranes. Complete dose-response curves for calmodulin-activation of Ca^{++} pump ATPase were performed in the absence (dotted line) and in the presence (various symbols) of various concentrations of TFP. Each curve is labelled with the negative logarithm of the molar concentration of TFP. An apparent competitive antagonism of calmodulin by TFP was obtained. Data at 0 and 160 ng/ml calmodulin were plotted as in Fig. 2.

was little or no inhibition of the basal ATPase activity by these drugs at the IC_{50}. Thus, the antagonism of CaM was selective. At concentrations at or above those needed for complete inhibition of CaM activity, the basal activity usually was also decreased. At such high concentrations there usually was also inhibition of the (Mg^{++})-ATPase and $(Na^+ + K^+ + Mg^{++})$-ATPase activities (data not shown). These effects are probably non-specific and are almost certainly due to membrane disruption.

A wide range of drug potencies is obvious in Table 1. Antipsychotic and psychotropic drugs were generally the most potent. This is probably a reflection of their lipophilic nature rather than a specific characteristic of antipsychotic drugs. In this and similar assay systems anti-CaM activity correlated well with oil-water partition coefficients (25) and both active and inactive analogs of antipsychotic compounds were found to be equipotent (29,30). The general pharmacology of the compounds listed in Table 1 is diverse. Some, but not all of these drugs have been shown to be CaM-BDs. All probably are CaM-BDs, but this almost certainly does not account for either their diverse pharmacology or their apparently non-competitive antagonism of CaM. If the model in Fig. 1 were to apply then one would predict that a pure CaM-BD should be a

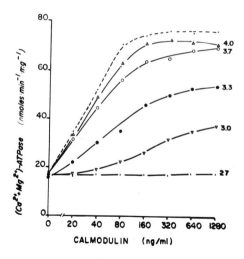

FIGURE 5

Effect of imipramine on calmodulin-activation of Ca^{++} pump ATPase of isolated human RBC membranes. Complete dose-response curves for calmodulin activation of Ca^{++} pump ATPase were performed in the absence (dotted line) and in the presence (various symbols) of various concentrations of imipramine. Each curve is labelled with the negative logarithm of the molar concentration of imipramine. An apparent non-competitive antagonism of calmodulin by imipramine was obtained. Data at 0 and 160 ng/ml calmodulin were plotted as in Fig. 3.

competitive antagonist of CaM. There must be more going on. Because the IC_{50} values of most drugs in Table 1 are so high, it is unlikely that anti-CaM activity forms the basis of their individual specific pharmacological properties. Of course, the non-specific "membrane stabilizing, local anesthetic" properties might be related in part to anti-CaM activity. More likely, the anti-CaM and membrane-perturbing effects of these drugs are simply two properties which result from the amphipathic nature of these compounds (see below).

An obvious question is why selective (even if non-specific) anti-CaM activity can be common to such a diverse group of compounds. The only obvious characteristics common to the compounds in Table 1 are that they are amphipathic and cationic at physiological pH. These are very simple structural requirements. If the characteristics of selective anti-CaM drugs are so simple then perhaps the "CaM receptor" (or rather, the receptor for $CaM(Ca^{++})_n$) is fairly simple. CaM is a highly acidic protein with many anionic groups at physiological pH. There is excellent evidence for Ca^{++} induced

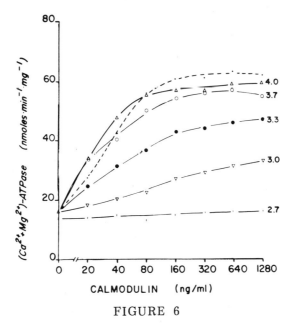

FIGURE 6

Effect of dibucaine on calmodulin-activation of Ca^{++} pump ATPase of isolated human RBC membranes. Complete dose-response curves for calmodulin-activation of Ca^{++} pump ATPase were performed in the absence (dotted line) and in the presence (various symbols) of various concentrations of dibucaine. Each curve is labelled with the negative logarithm of the molar concentration of dibucaine. An apparent non-competitive antagonist of calmodulin by dibucaine was obtained.

hydrophobicity on the surface of CaM (15). A simple interpretation is that the hydrophobic site(s) which appear only when Ca^{++} binds may bind to drugs or to the CaM receptor or to both. That is, some CaM-BDs may not prevent the binding of $CaM(Ca^{++})_n$, but actually accompany the complex to the receptor. Levin and Weiss (17) found two high affinity phenothiazine binding sites on CaM-$(Ca^{++})_n$. It is conceivable that, under some conditions, a CaM-$(Ca^{++})_n$-CaM-BD complex exists which can still bind to the CaM effector molecule. Such drug binding to $CaM(Ca^{++})_n$ might alter its affinity for and/or effectiveness in modulating the effector. Of course, changes in the properties of $CaM(Ca^{++})_n$ might be different for different effectors. I know of no direct evidence for or against this obvious possibility. CaM-BDs with some selectivity against certain effectors might be developed if such a mechanism could be exploited.

TABLE 1: Anti-calmodulin potency of a variety of drugs.

Drug	IC_{50} (μM)
Penfluridol*	6
Pimozide*	25
Cetylpyridinium chloride**	29
Trifluoperazine	30
Thioridazine	38
Chlorprothixene	41
Ellipticine	50
Propranolol	60
Promethazine	84
Cetiedil***	160
Desipramine	480
Imipramine	530
Dibucaine	650
Quinacrine (mepacrine)	750
Quinidine	1000
Daunorubicin	1000
Dimethylpropranolol** (UM-272)	1200
Tetracaine	2500

Calmodulin activation was determined as in Figs. 4-6. IC_{50} was estimated from at least two separate experiments as the concentration of drug required to antagonize by 50% the activation of Ca^{++} pump ATPase caused by 160 ng/ml calmodulin. Drugs added as the hydrochloride salts except as noted.
* free base in DMSO; ** quaternary salt; *** citrate salt

Data accumulating in various laboratories may be interpreted to suggest that the $CaM(Ca^{++})_n$ effect on various effectors is not necessarily entirely specific. For example, acidic phospholipids and free fatty acids activate certain enzymes and the activation resembles that produced by CaM. Oleate activates the Ca^{++} pump ATPase of RBC membranes and exerts no effect on the (Mg^{++})-ATPase or the $(Na^+ + K^+ + Mg^{++})$-ATPase (36). Like any detergent, at high concentrations oleate inhibits all ATPase activities. Figure 7 shows that oleate activation of the Ca^{++} pump ATPase appears to be additive with that produced by CaM. Up to 50 μM oleate activated the ATPase; at 100 μM oleate prevented CaM activation and at 200 μM, oleate abolished all ATPase activity. Qualitatively similar results are apparent in Fig. 8, and were also observed with sodium dodecyl sulfate (data not shown).

One could dismiss these observations as due to an effect of these amphipathic molecules on the environment around the

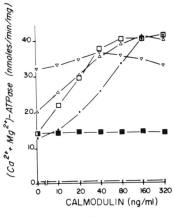

FIGURE 7

Effects of oleate and calmodulin on Ca^{++} pump ATPase activity of isolated human RBC membranes. Calmodulin alone (●—●) produced typical activation of enzyme activity. Increasing concentrations of oleate increased then decreased the activity of Ca^{++} pump ATPase in the absence (points on vertical axis) and in the presence of various concentrations of calmodulin. Concentrations of oleate added were 10 µM (□--□), 20 µM (△—△), 50 µM (▽—▽), 100 µM (■—■) and 200 µM (X—X), respectively. When less than optimal concentrations of both oleate and calmodulin were present there was an apparent additive effect of the two substances. High concentrations of oleate abolished all ATPase activity.

membrane-bound Ca^{++} pump ATPase. Such an influence almost certainly is responsible for inhibition of all ATPase activities at high concentrations. On the other hand, selective activation by oleate of the Ca^{++} pump ATPase at concentrations which exert little or no influence on the (Mg^{++})-ATPase or $(Na^+ + K^+ + Mg^{++})$-ATPase (36) and summation of oleate induced activation of the Ca^{++} pump ATPase with that produced by CaM (Fig. 6) show that there are selective effects on the CaM effector enzyme. Such selectivity is not peculiar to a membrane-bound enzyme. Wolff and Brostrom (45), Pichard and Cheung (27) and Tanaka and Hidaka (35) found that the soluble system, cyclic nucleotide phosphodiesterase was activated by acidic phospholipids and free fatty acids (45).

This discussion is not meant to imply that oleate is an endogenous CaM-like regulator. But an obvious question is whether other biologically relevant molecules might act as regulators of presumed CaM-dependent enzymes. If they exist, these molecules would probably be amphipathic anions. As noted by Wolff and

Brostrom (45) such molecules would not necessarily depend on Ca^{++} binding to mimic the effects of CaM. This idea raises the possibility that at least some supposedly CaM (and Ca^{++})-dependent enzymes can be regulated in other (for example, non-CaM and non Ca^{++}) ways. I am aware of no studies specifically directed to this question. Tanaka and Hidaka (35) reported that sodium dodecyl sulfate activated myosin light chain kinase in a CaM-like fashion in the absence of Ca^{++}. Many CaM mediated effects depend on this and other kinases (33).

Results presented in Fig. 8 show that 10^{-4} M TFP abolished the stimulatory effect of oleate on Ca^{++} pump ATPase. Lower concentrations of TFP antagonized oleate activation (shifted the activation curve "to the right") but did not antagonize oleate inhibition of the ATPase (data not shown) making direct oleate-TFP interaction an unlikely explanation for the results shown in Fig. 8. The results in Figs. 7 and 8 demonstrate that a presumably CaM dependent enzyme (38) can be activated by a molecule other than CaM. What seems to be required to mimic CaM effects is an amphipathic anion. The activation produced by CaM or other amphipathic anions can be selectively antagonized by the classical CaM-BD, TFP. It seems likely that a variety of amphipathic anions may mimic CaM effects at some (possibly all) of its effectors and that the effects could be antagonized by TFP.

As noted before, acidic phospholipids (27,35,45) and fatty acids (45) activate cyclic nucleotide phosphodiesterase. Wolff and Brostrom (45) found that the activation was antagonized by TFP. Taken together, these results may be interpreted to mean that the CaM-like effect, on at least some effectors, may be based on a rather non-specific hydrophobic interaction between an amphipathic anion and some hydrophobic (and probably cationic) region of the effector molecule. Ca^{++} induces surface hydrophobicity on CaM (15). The hydrophobic region(s) on CaM $(Ca^{++})_n$ binds cationic, neutral or anionic compounds (15). In experiments like those in Figs. 4-6 the neutral detergent, Triton X-100, was found to be a potent but inconsistent antagonist of CaM (data not shown). Recently, Triton X-100 was reported to be a CaM-BD and anti-CaM drug (34).

Tanaka and Hidaka (35) suggested that a hydrophobic interaction is important for CaM activation of various enzymes. Unfortunately, at this time it is impossible to be more precise in a description of the necessary and sufficient conditions for "CaM agonists." Other questions arise when the idea of a "$CaM(Ca^{++})_n$ receptor" start to take shape, that is: if there are agonists for $CaM(Ca^{++})_n$ receptors what about competitive antagonists? Are some of the molecules which are known to be anti-CaM actually CaM receptor antagonists? Do some molecules which are CaM-BDs also bind to CaM receptors? If there is hydrophobic complimentarity between

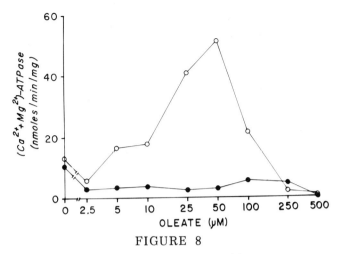

FIGURE 8

TFP antagonism of oleate activation of Ca^{++} pump ATPase of isolated human RBC membranes. Ca^{++} pump ATPase activity was determined at various concentrations of oleate in the absence (o--- o) and in the presence (•—•) of 10^{-4} M TFP.

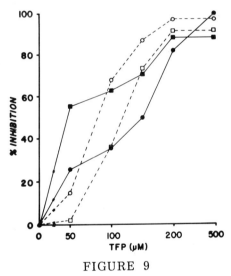

FIGURE 9

TFP inhibition of reconstituted Ca^{++} pump ATPase. Purified Ca^{++} pump ATPase was reconstituted in phosphatidylcholine (o—o); (•—•) or asolectin (□—□ ; ■—■) vesicles. Activity of the ATPase was measured in the absence (solid lines) or presence (dotted lines) of 160 ng/ml calmodulin at various concentrations of TFP, as indicated. Results are expressed as percent inhibition of enzyme activity. The purified Ca^{++} pump ATPase was sensitive to inhibition by TFP in the absence or presence of added calmodulin, both in its low or high activity states.

$CaM(Ca^{++})_n$ and its receptors, isn't it likely that CaM-BDs are also to some extent CaM antagonists in the traditional pharmacologic sense?

What about CaM receptor(s)? It has been considered (1) and rejected (1,12) that there might be a common CaM binding subunit present in various effector proteins on enzymes. Another possibility which has been advanced is that there is a common CaM-binding peptide sequence on various effector enzymes. Mild trypsin treatment activates a number of CaM dependent enzymes (5,6,12,31) and it has been suggested that this sequence may be removed by trypsin. Such a common CaM receptor peptide sequence, if it exists, may be elucidated with application of a recently described photoaffinity derivative of CaM (1,10,37). Because of the widely divergent nature of known CaM effectors, the apparently rather simple requirements for mimicking CaM effects and the selective activation by CaM of adenylate cyclase in an organism which lacks CaM (41), I expect that no such common peptide sequence will be found.

Why does TFP antagonize oleate as well as CaM activation of Ca^{++} pump ATPase? Probably because TFP can bind directly to the ATPase (or to phospholipids near it). For reasons mentioned below I will assume that TFP binds to the Ca^{++} pump ATPase and that this binding prevents oleate induced activation. If it is further assumed that oleate and CaM induced activations are similar then the binding of a variety of drugs (e.g. Table 1) to the enzyme is probably the basis of non-competitive antagonism of CaM. Some drugs, such as TFP, probably bind both to $CaM(Ca^{++})_n$ and the enzyme. Some drugs may bind only to the enzyme. I have recently suggested that effector-directed drugs (rather than CaM-directed) may hold the key to the development of more selective anti-CaM compounds (Vincenzi, submitted for publication).

Another kind of observation provides more direct evidence for interaction of anti-CaM drugs and a CaM effector enzyme. In collaboration with Drs. V. Niggli and E. Carafoli of the ETH (Zurich), the Ca^{++} pump ATPase was isolated by CaM affinity chromatography as they have described (24). The enzyme was reconstituted in phospholipid vesicles (23). In phosphatidylcholine vesicles the enzyme has fairly low activity and in asolectin (acidic phospholipid) vesicles the enzyme has relatively high activity. This is a reflection of the acidic phospholipid (amphipathic anion?) effect on the enzyme which is essentially CaM-free after this isolation procedure (9). In both types of vesicles either in the absence or presence of added CaM the Ca^{++} pump ATPase enzyme was inhibited by TFP (Fig. 9), or other anti-CaM drugs (data not shown). A direct interaction between TFP and the enzyme is a likely but unproven explanation of these results. Details of these results will be presented elsewhere.

SUMMARY

There are some conclusions and some speculation which may be derived from these results. First, there is no simple relationship between selective anti-CaM activity in vitro and general pharmacological properties of drugs. Second, not everything which selectively activates CaM-dependent enzymes in a complex system is necessarily CaM. Third, even if one demonstrates selective drug antagonism of a heat-stable activator of a supposedly CaM-dependent enzyme, that result will not prove that the activator substance is CaM. In more complex systems there is even more uncertainty. In other words, those who, when observing a TFP-induced change in some response in a system proclaim that CaM is, ipso facto, implicated in that response may be leading us down the well traveled "primrose path." The binding of drugs to $CaM(Ca^{++})_n$ is only one possible mechanism of anti-CaM activity. Drugs may also bind directly to CaM effectors and modify the affinity and/or effectiveness of $CaM(Ca^{++})_n$ thereon. The influence may not be equal on all effectors. The history of really significant advancements in pharmacology and therapeutics is rich with examples of receptor differentiation. Differentiation of binding of drugs to CaM effectors appears to offer the greatest hope for selective manipulation of the activity of this important and ubiquitous intracellular receptor protein.

ACKNOWLEDGEMENTS

Supported in part by DHHS grants AM-16436 and GM-24990. Ms. Bonnie Ashleman provided excellent technical assistance.

REFERENCES

1. Andreasen, T.J., Keller, C.H., LaPorte, D.C., Edelman, A.M. & Storm, D.R. (1981): Proc. Nat. Acad. Sci. 78:2782-2785.
2. Cheung, W.Y. (1980): Science 207:19-27.
3. Cohen, P., Klee, C.B., Picton, C. & Shenolikar, S. (1980): Ann. NY Acad. Sci. 356:151-161.
4. Dedman, J.R., Potter, J.D., Jackson, R.L., Johnson, J.D. & Means, A.R. (1977): J. Biol. Chem. 252:8415-8422.
5. Depaoli-Roach, A.A., Gibbs, J.B. & Roach, P.J. (1979): FEBS Lett. 105:321-324.
6. Enyedi, A., Sarkadi, B., Szasz, I., Bot, G. & Gardos, G. (1980): Cell Calcium 1:299-310.
7. Farrance, M.L. & Vincenzi, F.F. (1977): Biochim. Biophys. Acta 471:49-58.
8. Gietzen, K., Mansard, A. & Bader, H. (1980): Biochem. Biophys. Res. Commun. 94:674-681.
9. Gietzen, K., Tejcka, M. & Wolf, H.U. (1980): Biochem. J. 189:81-88.
10. Hinds, T.R. & Andreasen, T.J. (1981): J. Biol. Chem. 256:7877-7882.

11. Hinds, T.R., Raess, B.U. & Vincenzi, F.F. (1981): J. Membrane Biol. 58:57-65.
12. Klee, C.B. (1980): IN Protein Phosphorylation and Bioregulation. (eds) G. Thomas, E.J. Podesta & J. Gordon, Karger, Basel, pp. 61-69.
13. Klee, C.B., Crouch, T.H. & Richman, P.G. (1980): Ann. Rev. Biochem. 49:489-515.
14. Kobayashi, R., Tawata, M. & Hidaka, H. (1979): Biochem. Biophys. Res. Commun. 88:1037-1045.
15. LaPorte, D.C., Wierman, B.M. & Storm, D.R. (1980): Biochemistry 19:3814-3819.
16. Larsen, F.L., Raess, B.U., Hinds, T.R. & Vincenzi, F.F. (1978): J. Supramolec. Struct. 9:269-274.
17. Levin, R.M. & Weiss, B. (1977): Molec. Pharmacol. 13:690-697.
18. Levin, R.M. & Weiss, B. (1978): Biochim. Biophys. Acta 540:197-204.
19. Levin, R.M. & Weiss, B. (1979): J. Pharmacol. Exp. Ther. 208:454-459.
20. Levin, R.M. & Weiss, B. (1980): Neuropharmacology 19:169-174.
21. Lowry, O.H., Rosebrough, N.J., Farr, A.L. & Randall, R.J. (1951): J. Biol. Chem. 193:265-275.
22. Means, A.R. & Dedman, J.R. (1980): Nature 285:73-77.
23. Niggli, V., Adunyah, E.S., Penniston, J.T. & Carafoli, E. (1979): J. Biol. Chem. 256:395-401.
24. Niggli, V., Penniston, J.T. & Carafoli, E. (1979): J. Biol. Chem. 254:9955-9958.
25. Norman, J.A., Drummond, A.H. & Moser, P. (1979): Molec. Pharmacol. 16:1089-1094.
26. Peterson, G.L. (1977): Anal. Biochem. 83:346-356.
27. Pichard, A.L. & Cheung, W.Y. (1977): J. Biol. Chem. 252:4872-4875.
28. Raess, B.U. & Vincenzi, F.F. (1980): J. Pharmacol. Meth. 4:273-283.
29. Raess, B.U. & Vincenzi, F.F. (1980): Molec. Pharmacol. 18:253-258.
30. Roufogalis, B.D. (1981): Biochem. Biophys. Res. Commun. 98:607-613.
31. Sarkadi, B., Enyedi, A. & Gardos, G. (1980): Cell Calcium 1:287-297.
32. Scharff, O. (1981): Cell Calcium 2:1-27.
33. Schulman, H. & Greengard, P. (1978): Proc. Nat. Acad. Sci. 75:5432-5436.
34. Sharma, R.K. & Wang, J.H. (1981): Biochem. Biophys. Res. Commun. 100:710-715.
35. Tanaka, T. & Hidaka, H. (1980): J. Biol. Chem. 255:11078-11080.
36. Vincenzi, F.F. (1981): Proc. West. Pharmacol. Soc. 24:193-196.

37. Vincenzi, F.F., Andreasen, T.J. & Hinds, T.R. (1981): IN Calcium and Phosphate Transport Across Membranes. (eds) F. Bronner & M. Peterlik, Academic Press, New York, pp. 45-50.
38. Vincenzi, F.F. & Hinds, T.R. (1980): IN Calcium and Cell Function, Vol.1. (ed) W.Y. Cheung, Academic Press, New York, pp. 127-165.
39. Vincenzi, F.F., Hinds, T.R. & Raess, B.U. (1980): Ann. NY Acad. Sci. 356:232-244.
40. Volpi, M., Sha'afi, R.I., Epstein, P.M., Andrenyak, D.M. & Feinstein, M.B. (1981): Proc. Nat. Acad. Sci. 78:795-799.
41. Wallace, R.W., Tallant, E.A. & Cheung, W.Y. (1980): Biochemistry 19:1831-1837.
42. Wang, J.H. & Sharma, R.K. (1980): Ann. NY Acad. Sci. 356:190-203.
43. Wang, J.H., Sharma, R.K. & Tam, S. (1980): IN Calcium and Cell Function, Vol.1. (ed) W.Y. Cheung, Academic Press, New York, pp. 305-328.
44. Weiss, B. & Levin, R.M. (1978): Adv. Cyclic Nucleotide Res. 9:285-303.
45. Wolff, D.J. & Brostrom, C.O. (1976): Arch. Biochem. Biophys. 173:720-731.
46. Wolff, J., Cook, G.H., Goldhammer, A.R. & Berkowitz, S.A. (1980): Proc. Nat. Acad. Sci. 77:3841-3844.

BIOPHARMACOLOGICAL ASSESSMENT OF CALMODULIN FUNCTION: UTILITY OF CALMODULIN ANTAGONISTS

H. Hidaka and T. Tanaka

Department of Pharmacology, Mie University School of Medicine, Tsu, Mie 514, Japan

INTRODUCTION

The considerable amount of accumulated data implicates calmodulin as an intracellular Ca^{++}-receptive protein in various species and tissues (2,17,24). The evidence for this implication is based on several lines of experiments: 1) calcium binding properties of purified calmodulin with a high affinity and calcium binding induced conformational changes of calmodulin, 2) calcium-dependent activation of a number of important enzymes, and 3) the calcium-calmodulin complex which prevents the in vitro assembly of microtubule protein. Despite this biochemical evidence which supports the contention that calmodulin is an intracellular Ca^{++}-receptor protein, its physiologic function as a Ca^{++} receptor remains obscure. Pharmacological studies using calmodulin antagonists such as phenothiazines and naphthalenesulfonamides suggest that calmodulin may play an important role in platelet function (25), cell proliferation (10), vascular contraction (12), insulin secretion (29), receptor mediated endocytosis (28), and others.

In 1973 phenothiazine was one of the first drugs to be considered a calmodulin antagonist (15). In 1974 Dr. Weiss and associates (35) did fine and elegant work on the interaction between phenothiazine and Ca^{++}-calmodulin. They demonstrated that trifluoperazine binds to calmodulin in the presence of Ca^{++}, but not in the absence of Ca^{++}, and that calmodulin has two classes of binding sites for the drug, two high affinity sites and 24 low affinity sites, respectively (19). All of these earlier publications were based on findings related to the inhibition of cyclic nucleotide phosphodiesterase and to cyclic nucleotide metabolism.

In 1977 we reported (11) at the USA-Japan Joint Congress in Hawaii that a calmodulin antagonist, naphthalenesulfonamide (W-7), which we synthesized produced relaxation of vascular strips and inhibited the superprecipitation of actomyosin from aortic smooth muscle, thereby suggesting the involvement of calmodulin in muscle contraction. This hypothesis was proved in two different types of experiments, one is a biochemical study in which calmodulin is a subunit of myosin light chain kinase (MLCK) and activates MLCK activity calcium dependently (4) and the other is a pharmacological study in which calmodulin antagonists were found to inhibit Ca^{++}-Mg^{++}-dependent actomyosin ATPase, superprecipitation and produce relaxation of smooth muscle through inhibition of calmodulin dependent myosin light chain phosphorylation (12). These findings led to studies on calcium function and the participation of calmodulin in a number of Ca^{++}-dependent phenomena. While biochemical studies on the functions of calmodulin are feasible, studies using cells and tissues, or even on the whole animal, are relatively difficult to carry out. Thus, to some extent pharmacological techniques are often more appropriate.

We propose that calmodulin antagonists should be defined as agents which bind to calmodulin calcium dependently and inhibit selectively calcium calmodulin dependent enzymes. In this paper, we will describe the character of calmodulin antagonists, the binding sites on calmodulin, the activity-structure relationship of these agents and the useful application of these antagonists to biological functions such as human platelet secretion, vascular smooth muscle contraction and cell proliferation. Particular emphasis is placed on the biological importance of sister compounds which are chlorinated and dechlorinated naphthalenesulfonamide.

Specificity of Naphthalenesulfonamides to Calmodulin

To assess the usefulness of pharmacological methods for studies on calmodulin, we determined the specificity of the binding of N-(6-aminohexyl)-5-chloro-1-naphthalenesulfonamide (W-7) to calmodulin using [^{3}H] labelled W-7. We also synthesized systemically the derivatives of naphthalenesulfonamide and attempted to clarify the structure-activity relationships to delineate the inhibitory effect of these derivatives on several calmodulin dependent enzymes. We also determined the specificity of the binding of these derivatives to calmodulin, using W-7 coupled sepharose 4B affinity chromatography.

Binding to calmodulin: We have demonstrated that [^{3}H]-W-7 binds to calmodulin in the presence of 0.1 mM $CaCl_2$ but not in the absence of calcium ion (1 mM EGTA)(12). In the presence of calcium, the calculated dissociation constant (Kd) for W-7 was 11 μM, and at saturation there were approximately 3 molecules of W-7 specifically found per molecule of calmodulin. In the absence of calcium, only the low affinity, high capacity sites were evident (12).

When the effect of Ca^{++} concentration on the interaction between W-7 and calmodulin was studied using a Ca^{++}-EGTA buffer system, W-7 binding to calmodulin was significantly increased with 1 µM Ca^{++} and almost maximal with 10 µM Ca^{++} (Fig. 1). The binding of [^{3}H]-W-7 to purified calmodulin was investigated using the equilibrium binding technique of Hummel and Dreyer (16) on a Sephadex G-50 gel filtration column, as described previously (12). Calmodulin was purified to homogeneity from bovine brain by the method of Teo et al. (32). The concentration of Ca^{++} required to produce calmodulin-dependent cyclic nucleotide phosphodiesterase activation and the binding of W-7 to calmodulin were indistinguishable, suggesting that in the presence of micromolar concentrations of calcium, W-7 specific binding sites of calmodulin may be exposed by resultant conformational changes as the result of binding with calcium ion. The calcium-dependent binding of W-7 was not observed with other purified proteins tested, except for troponin C and S-100 protein. Troponin C and S-100 protein, however, bound to W-7 Ca^{++} dependently and their affinity to W-7 was less than calmodulin.

Determination of the extent of displacement of [^{3}H]-W-7 from calmodulin by another calmodulin antagonist such as N^α-dansyl-L-arginine-4-t-butylpiperidine amide (No.233) and chlorpromazine (CPZ), showed that these agents can displace [^{3}H]-W-7 competitively from calmodulin.

Activity-Structure Relationship of Naphthalenesulfonamides

We synthesized, systemically, derivatives of naphthalenesulfonamide which were dechlorinated, or in which bromine or fluorine were substituted for chlorine, and also by changing the length of the alkyl chain (8). We then attempted to estimate affinities of these derivatives to calmodulin by determining the concentrations which produce 50% inhibition of [^{3}H]-W-7 binding to calmodulin and Ca^{++}-dependent cyclic nucleotide phosphodiesterase.

The displacement of [^{3}H]-W-7 from [^{3}H]-W-7-calmodulin complex with naphthalenesulfonamide derivatives was studied in the presence of 0.1 mM $CaCl_2$. Calmodulin deficient, Ca^{++}-dependent cyclic nucleotide phosphodiesterase was purified from bovine brain and the activity was measured, as described previously (7,14). Calmodulin, purified to homogeneity from bovine brain (13), was assayed by measuring the extent of activation of a fixed amount of calmodulin-deficient phosphodiesterase under standard conditions (13). One unit of calmodulin was defined as calmodulin-deficient

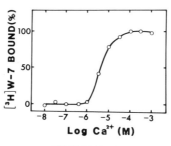

FIGURE 1

[^{3}H]-W-7 binding to calmodulin as a function of Ca^{++} concentration. Sephadex G-50 (0.9 x 28 cm) was pre-equilibrated with the buffer containing 20 mM Tris·HCl, pH 7.5, 20 mM imidazole, 3 mM magnesium acetate, and 0.5 μM [^{3}H]-W-7 at 25°C. Purified calmodulin (210 μg) was used for each experiment. The binding of [^{3}H]-W-7 to calmodulin is given as the percent of maximum binding at the calcium concentrations indicated.

Structure	Displacement of [³H]W-7 from Calmodulin		Inhibition of PDE Activation			
	R=H	R=Cl	R=H	R=F	R=Cl	R=Br
H$_2$N(CH$_2$)$_6$NHSO$_2$-⌬-R 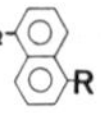	210	31	240	50	26	20
H$_2$N(CH$_2$)$_4$NHSO$_2$-⌬-R 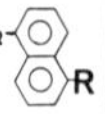	1010	120	1010	–	66	–
H$_2$N(CH$_2$)$_6$NHSO$_2$-⌬-R 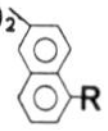	180	16	130	–	14	–

FIGURE 2

Effect of three pairs of naphthalenesulfonamide derivatives with or without chlorine on [^{3}H]-W-7 binding to calmodulin and calmodulin-induced activation of phosphodiesterase. Details were as described elsewhere (12).

phosphodiesterase and was equivalent to 10 ng of protein.

The substitution of hydrogen for the chlorine molecule at 5' position on the naphthalene ring was made in the case of three naphthalenesulfonamides such as N-(6-aminohexyl)-5-chloro-1-naphthalenesulfonamide (W-7), N-(4-aminobutyl)-5-chloro-1-naphthalenesulfonamide (A-5), and N-(6-aminohexyl)-5-chloro-2-naphthalenesulfonamide (W-9). These paired compounds were compared with regard to the affinity to calmodulin. The results are summarized in Fig. 2. The affinity for calmodulin proved to be dependent on the chlorination of the naphthalene ring of these three naphthalenesulfonamide derivatives. We then attempted to introduce other halogen molecules such as bromine or fluorine at the naphthalene ring. Derivatives with bromine were the most potent and those with fluorine the least potent (Fig. 2). The IC_{50} values of naphthalenesulfonamides with the bromine, chlorine and fluorine molecule were 20, 26 and 50 µM, respectively. We synthesized the naphthalenesulfonamides with various lengths of the alkyl chain and found that the action of these compounds as calmodulin antagonists was indeed dependent on the length of the alkyl chain.

Selective inhibition of calmodulin dependent enzymes: We have already demonstrated that these naphthalenesulfonamide derivatives inhibit selectively calmodulin-dependent enzymes such as Ca^{++}-dependent cyclic nucleotide phosphodiesterase, myosin light chain kinase from chicken gizzard and Ca^{++}, Mg^{++}-ATPase of human erythrocyte ghost (9,11,12,18). The concentration of W-7 that almost completely inhibited the activation of Ca^{++}-dependent phosphodiesterase did not affect the activity of Ca^{++}-calmodulin independent phosphodiesterase. Kinetic analysis of W-7-induced inhibition of activation of phosphodiesterase revealed that W-7 inhibits this activity in a competitive fashion with calmodulin (13). As shown in Fig. 3, the selective inhibition of calmodulin-induced activation of myosin light chain kinase from human platelets and myelin phosphorylation by W-7 was also observed (5). Dechlorination of these naphthalenesulfonamide derivatives diminished the potency of the inhibition of this myosin light chain kinase activity, as shown in Fig. 3.

Table 1 summarizes the ratio of the concentration of several calmodulin antagonists producing 50% inhibition of basal activity and Ca^{++}-calmodulin stimulated activity of the phosphodiesterase. W-7 had the highest ratio but the concentration of W-7 inhibiting Ca^{++}-calmodulin stimulated phosphodiesterase activity was higher than that seen with phenothiazine derivatives. IC_{50} values of inactivated Ca^{++}-dependent phosphodiesterase (which may indicate a non-specific action of these agents) were high for this naphthalenesulfonamide (W-7) and relatively low for an antipsychotic agent such as chlorpromazine. Similar data have been reported

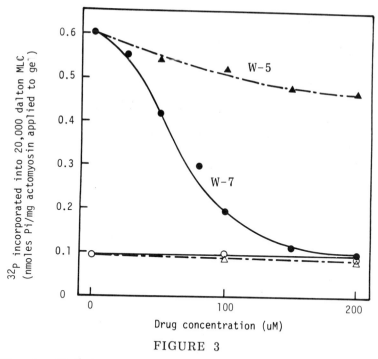

FIGURE 3

Inhibition by W-7 (●—○) or W-5 (▲—△) phosphorylation of 20,000 Mw myosin light chain from human platelets in the presence of 100 µM Ca^{++} (closed symbols) and in the presence of 2 mM EGTA (open symbols).

by Norman, Drummond & Moser (26). These results suggest that W-7 may be useful <u>in vivo</u> or for crude enzyme studies because of a better specificity indicated by such a high ratio.

<u>W-7 coupled affinity chromatography</u>: It is likely that the entire range of pharmacological effects of calmodulin antagonists, such as naphthalenesulfonamide, on living cell function results from a combination of several types of molecular interactions and cannot simply be explained by a single mechanism. We coupled W-7 to cyanogen bromide activated Sepharose 4B and performed Ca-dependent affinity chromatography on W-7 Sepharose 4B of acidic protein solution from bovine brain. Acidic protein solution of bovine brain was prepared as follows: bovine brain was homogenized with 2.5 vol buffer containing 50 mM Tris·HCl (pH 8.0), 5 mM $MgCl_2$, 1 mM EGTA and 4 mM 2-mercaptoethanol. The homogenate was centrifuged at 100,000 x g for 60 min. The supernatant was first brought to 55% $(NH_4)_2SO_4$ saturation, the preparation centrifuged and then brought to pH 4.0 by the slow addition of 1.0 N HCl.

TABLE 1: Effect of several calmodulin antagonists on activated and inactivated phosphodiesterase from bovine brain.

	Phosphodiesterase Inhibition (IC_{50} µM)		Ratio
	Activated	Basal	($\frac{Basal}{Activated}$)
N-(6-aminohexyl)-5-chloro-1-naphthalene-sulfonamide (W-7)	28	1200	42.9
N^{α}-dansyl-L-arginine-4-t-butylpiperidine amide (No.233)	8	60	7.5
Chlorpromazine	16	130	8.1
Prenylamine	18	600	33.3
Trifluoperazine	10	>200	>20.0

Phosphodiesterase activity of a preparation purified from bovine brain was measured in the presence and absence of 20 units of calmodulin and various concentrations of the compounds under study, using 0.4 µM cyclic GMP as substrate. The addition of 20 units of calmodulin produced about an eight-fold increase in phosphodiesterase activity.

The materials precipitated by this procedure were collected by centrifugation and dissolved in a small volume of buffer. The solution was dialyzed overnight against the same buffer and loaded on the W-7 coupled Sepharose 4B. Only calmodulin-like protein bound to W-7-coupled Sepharose 4B in the presence of Ca^{++} and was eluted with buffer containing 4 mM EGTA. This protein activated Ca^{++}-dependent phosphodiesterase and had a very similar amino acid composition and electrophoretical mobility as seen with calmodulin, indicating that this protein is presumably calmodulin itself (Fig. 4). These results indicate that immobilized W-7 interacts with calmodulin selectively. It should be pointed out that W-7 epoxy coupled Sepharose 6B interacted with S-100 protein but not with calmodulin (Fig. 4). This would suggest that the S-100 protein has a molecular structure similar to that of calmodulin.

Determination and Characterization of W-7 Binding Sites on Calmodulin

Octanol-water partition coefficients: Recently we reported that when Ca^{++} binds to the high affinity sites of calmodulin there is a conformational change which exposes the hydrophobic regions (30). These hydrophobic regions of Ca^{++}-calmodulin complex are responsible

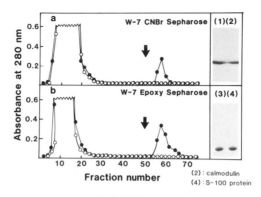

FIGURE 4

Affinity chromatography on W-7 CNBr Sepharose and W-7 epoxy Sepharose. Five ml of acidic protein solution (19.0 mg/ml) were applied to W-7 CNBr Sepharose (a) or W-7 epoxy Sepharose (b). The column was first washed with buffer containing 100 μM Ca^{++} (buffer A). At the point indicated by the arrow, the buffer was changed to that containing 4 mM EGTA (buffer B). Fractions of 1 ml were collected and their absorption at 280 nm was monitored (-●-). EGTA elutable protein peak could not be observed when the acidic protein solution containing 1 mM EGTA was applied to each column equilibrated with buffer B (-○-).

for the activation of Ca^{++}-calmodulin dependent enzymes (30). The octanol-water partition coefficient is considered to be an index of lipid solubility for a molecule and hence of its hydrophobicity. We then determined the octanol-water partition coefficients of naphthalenesulfonamide derivatives with and without chlorine molecule, as an index of their lipid solubility. Octanol-aqueous-buffer partition coefficients were determined experimentally by the method of Leo et al. (20). The aqueous phase was 60 mM phosphate buffer (pH 8.0), saturated with n-octanol. The organic phase was buffer saturated n-octanol. As shown in Table 2, the octanol-water partition coefficients for these drugs correlated very well with their potency in displacement of [^{3}H]-W-7 from calmodulin. The correlation coefficient is r = 0.95, indicating that the affinity of naphthalenesulfonamide to calmodulin is closely related to the lipids solubility of these compounds. Moreover, removal of chlorine from each compound reduced the affinity to calmodulin and also the hydrophobicity of the compounds (Table 2).

TNS fluorescence: Calissano et al. (1) reported that the Ca^{++} binding to S-100 protein induced a conformational change which exposed hydrophobic regions, thus enabling the protein to interact with a hydrophobic probe, 2-p-toluidinyl-naphthalene-6-sulfonate

TABLE 2: Affinity of naphthalenesulfonamide derivatives for calmodulin and their hydrophobicity.

Compounds*	Affinity for Calmodulin** IC_{50} (μM)	Hydrophobicity*** Octanol/Water
W-7 (5-Cl, 1-R, n = 6)	31	8.88
W-5 (5-H, 1-R, n = 6)	210	0.62
A-5 (5-Cl, 1-R, n = 4)	120	2.47
NCM-124 (5-H, 1-R, n = 4)	1010	0.31
W-9 (5-Cl, 2-R, n = 6)	16	11.17
W-6 (5-H, 2-R, n = 6)	180	1.98

* Structure: naphthalene ring numbered 1–8.   R: $SO_2NH(CH_2)_nNH_2 \cdot HCl$

** The affinity of each of these compounds for calmodulin was estimated by determining the displacement of [^{3}H] W-7 from purified calmodulin in the presence of calcium ion by these compounds using the equilibrium binding technique.

*** Octanol/aqueous buffer partition coefficients were determined experimentally by the method of Leo et al. (20).

(TNS). Similar results concerning the calcium-induced exposure of the hydrophobic surface on calmodulin were obtained using TNS (30,31). It is likely that the hydrophobic region of calmodulin and S-100 exposed by Ca^{++} binding is responsible for the binding to W-7. Calmodulin antagonist, W-7, suppressed the TNS fluorescence induced by inhibiting the complex formation with calmodulin and TNS, in the presence of Ca^{++} (Fig. 5). TNS fluorescence was measured as previously reported (30). W-5, a chlorine-deficient analog of W-7 that interacts only weakly with calmodulin (Table 2) had a less inhibitory effect on TNS fluorescence and its IC_{50} value of 78 μM (Fig. 5). These results are explained on the basis of different affinities to calmodulin and imply hydrophobic interactions between the calmodulin-calcium complex and these naphthalenesulfonamides. Similar results are obtained with A-5, NCM-124, W-9 and W-6.

Inhibition of W-7 binding to calmodulin by troponin-I (TN-I): Calmodulin and skeletal muscle troponin C (TN-C) have similar properties and these proteins cross-react in their respective biological systems (2,17,24). Calmodulin as well as TN-C binds to TN-I and inhibits phosphorylation of rabbit skeletal TN-I at one site only, Ser-117, adjacent to the TN-I/actin region of attachment (27). Purified TN-I from rabbit skeletal muscle was found to inhibit [^{3}H] W-7 binding to Ca^{++}-calmodulin complex, in a concentration dependent fashion, and its IC$_{50}$ value for TN-I was 22 μg/ml. The tryptic cleavage peptide of calmodulin E$_2$ (residues of 1-90 of calmodulin) does not possess the ability of peptide E$_1$ (residues 1-106) to interact with TN-I. Furthermore, residues 77-124 have been found to interact with TN-I calcium dependently, but not to activate myosin light chain kinase activity (27). Our results suggest that W-7 binds to the residues 77-124 or to the adjacent regions where TN-I interacts. Experimental conditions were the same as previously reported and purified TN-I was the generous gift of Dr. R. Kobayashi.

Role of methionine residues for W-7 binding to calmodulin: Walsh and Stevens (33) reported that methionine residues of calmodulin molecule have been implicated in the stimulation of Ca^{++}-dependent cyclic nucleotide phosphodiesterase by calmodulin. Treatment of calmodulin with N-chlorosuccinimide in the presence of Ca^{++} was found to result in selective oxidation of methionine residues at positions 71, 72, 76 and possibly 109 in the calmodulin sequence (33). We then achieved selective oxidation of methionine residues of calmodulin. Loss of calmodulin activity in activation of Ca^{++}-dependent phosphodiesterase was concomitant with methionine modification and the linear decrease in [^{3}H]-W-7 binding to calcium calmodulin complex as a function of methionine modification; complete loss of [^{3}H]-W-7 binding to calmodulin after 60-90 min of reaction correlated with the apparent oxidation of four methionine residues per mole of calmodulin. The TN-C-like property of calmodulin such as the Ca^{++}-dependent formation of a complex with troponin I is lost upon oxidation with N-chlorosuccinimide (33). Thus, our results suggest that the W-7 binding domain of calmodulin is located between the second and third Ca^{++} binding loops. It is very interesting that this region, which contains many hydrophobic amino acids, is exposed in the presence of Ca^{++}, as shown in Fig 6 (30).

Pharmacological Action of Calmodulin Antagonists

In this section, we discuss the functions of calmodulin in some biological systems using calmodulin antagonists. We also show that chlorine deficient naphthalenesulfonamide (W-5, NCM-124, W-6) which has a weaker affinity to calmodulin compared with W-7 can serve as an appropriate control drug in vivo or for tissue level experiments.

When a calmodulin antagonist is used for studies on living cells, it is important that the agent penetrate the cell membrane and bind to cytosol calmodulin. In naphthalenesulfonamide, such as W-7, was

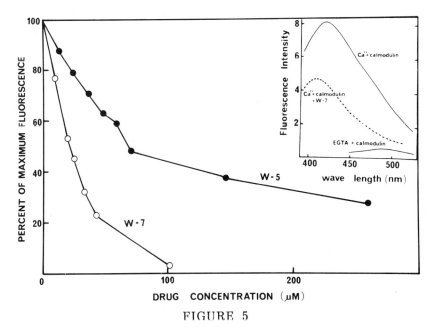

FIGURE 5

Effect of W-7 and W-5 on TNS fluorescence in the presence of calmodulin and calcium. Experimental conditions were as described elsewhere (30).

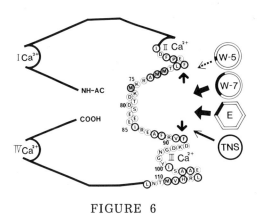

FIGURE 6

Possible binding sites of calmodulin antagonists on calmodulin. The sequence of bovine brain calmodulin (34) is shown utilizing the one-letter code for amino acid residues.

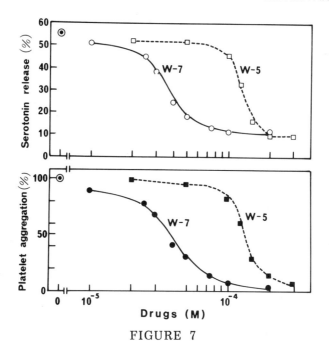

FIGURE 7

Effects of W-7 (○—●) and its analog, W-5 (□—■) on platelet aggregation (closed symbols) and [^{14}C]-serotonin release (open symbols) induced by collagen (2 μg/ml). Each point is the mean of three experiments. The control values (⊙) on aggregation and release were obtained without addition of the compounds.

autoradiographically confirmed to enter the cytoplasm. These findings are shown elsewhere (10).

Platelet aggregation inhibition: We found that W-7 inhibits human platelet aggregation and ATP or [^{14}C]-serotonin release induced by various aggregating agents such as collagen (2 μg/ml), ADP (5 μM), thrombin (0.1 U/ml), epinephrine (0.1 μg/ml) and sodium arachidonate (0.83 mM). The IC$_{50}$ values of W-7 for ATP and [^{14}C]-serotonin release induced by various agonists were indistinguishable. On the other hand, W-5, chlorine-deficient naphthalenesulfonamide was less potent in inhibiting platelet aggregation and [^{14}C-serotonin secretion, as shown in Fig. 7. This is probably due to the difference in their affinities for calmodulin, and resultant differences in inhibition of myosin light chain kinase. Moreover, W-7 produced a concentration dependent inhibition of Ca^{++}-dependent ATPase of platelet actomyosin. The function of the actomyosin system in platelets is related to its capacity for contraction and presumably mediates the release reaction (3,6). These results suggest that a calmodulin-mediated system plays an important role in platelet release reaction.

Vascular relaxation: Calmodulin antagonists such as W-7 produced relaxation of isolated aortic strips contracted by various agonists such as KCl, $CaCl_2$, norepinephrine, histamine, and prostaglandin $F_{2\alpha}$ (8). The relaxation induced by calmodulin antagonist was not affected by treatment with β-adrenergic or cholinergic blocking agents (8). W-5, which has a low affinity for calmodulin and which is a weak inhibitor of myosin light chain kinase produced no effect on contracted vascular strips, up to concentrations of 1×10^{-4} M. On the contrary, W-7, at a concentration of 1×10^{-4} M antagonized significantly the contractile responses of aortic strips, to the same extent seen with norepinephrine, prostaglandin $F_{2\alpha}$ and KCl. On the other hand, another type of calmodulin antagonist, phenothiazine (chlorpromazine), at a concentration of 1×10^{-6} M, specifically antagonized the contractions induced by norepinephrine, serotonin and histamine, but did not antagonize contractions induced by prostaglandin $F_{2\alpha}$ and angiotension II. If phenothiazine inhibits calmodulin related reactions at 10^{-6} M, this drug at this concentration probably inhibits all contractions produced by various agonists. Thus, W-7 and CPZ have a different spectrum of action as vascular relaxants, and W-7 may be a pertinent tool for application in calmodulin-related research of the vascular system.

Cell proliferation: Recently, Means and associates (24) reported that calmodulin may be a dynamic component of the mitotic apparatus, as determined by immunofluorescence, biochemical and ultrastructural observations. Calcium ion and calmodulin are considered to play important roles in regulating cell proliferation and are probably essential for the early DNA synthesis phase of the cell cycle (21-23). A calmodulin antagonist such as W-7 was found to inhibit proliferation of Chinese hamster ovary (CHO) K_1 cells, concentration dependently, and IC_{50} values of the cell proliferation of CHO-K_1 cells in culture for W-7 and W-5 were 32 and 200 μM, respectively. The concentrations of W-7 and W-5 producing 50% inhibition of cell proliferation were comparable with their affinity to calmodulin (Fig. 2). When the synchronous cells were treated with 25 μM W-7 or 2.5 mM thymidine for 12 hr with exclusion of W-7 or thymidine from the culture medium, the cell division of these synchronized cells was again observed at about 6 hr. A sharp increase in DNA synthesis, determined by [^3H]-thymidine incorporation, was observed immediately after exclusion of W-7, indicating that the effect of W-7 on cell proliferation might be through selective inhibition of the G_1/S boundary phase, such being similar to the effect of conditions involving excess thymidine.

CONCLUSION

Calmodulin is an ubiquitous intracellular calcium receptor protein that regulates a number of important enzymes (2,17,24) and microtubules disassembly (24). Biochemical research on calmodulin such as protein chemistry or enzymology is now established. The physiological function of calmodulin in living cells and tissues, nevertheless, remains obscure. Calmodulin antagonists such as naphthalenesulfonamide derivatives and dechlorinated naphthalenesulfonamides are useful tools for elucidating the functions of calmodulin in cells and tissues. However, there are disadvantages to using pharmacological tools and these can be overcome by synthesizing even more potent and more specific calmodulin antagonists.

ACKNOWLEDGEMENTS

TN-I was a generous gift from Dr. R. Kobayashi. This research was supported in part by grants from the Scientific Research Fund of the Ministry of Education, Science and Culture, Japan (1979-81). We are grateful to M. Ohara, Kyushu University, for assistance with the manuscript.

REFERENCES

1. Calissano, P., Alema, S. & Fasella, P. (1974): Biochemistry 13:4533-4560.
2. Cheung, W.Y. (1980): Science 207:19-27.
3. Dabrowska, R. & Hartshorne, D.J. (1978): Biochem. Biophys. Res. Commun. 85:1352-1359.
4. Dabrowska, R.J., Sherry, J.M.F., Aromatorio, D.K. & Hartshorne, D.J. (1978): Biochemistry 17:253-258.
5. Endo, T. & Hidaka, H. (1980): Biochem. Biophys. Res. Commun. 97:553-558.
6. Hathaway, D.R. & Adelstein, R.S. (1979): Proc. Nat. Acad. Sci. 76:1653-1657.
7. Hidaka, H. & Asano, T. (1976): J. Biol. Chem. 251:7508-7516.
8. Hidaka, H., Asano, M., Iwadare, S., Matsumoto, I., Totsuka, T. & Aoki, A. (1978): J. Pharmacol. Exp. Ther. 207:8-15.
9. Hidaka, H., Naka, M. & Yamaki, T. (1979): Biochem. Biophys. Res. Commun. 90:694-699.
10. Hidaka, H., Sasaki, Y., Tanaka, T., Endo, T., Ohno, S., Fujii, Y. & Nagata, T. (1981): Proc. Nat. Acad. Sci. 78:4354-4357.
11. Hidaka, H., Yamaki, T., Asano, M. & Totsuka, T. (1978): Blood Vessels 15:55-64.
12. Hidaka, H., Yamaki, T., Naka, M., Tanaka, T., Hayashi, H. & Kobayashi, R. (1980): Molec. Pharmacol. 17:66-72.
13. Hidaka, H., Yamaki, T., Totsuka, T. & Asano, M. (1979): Molec. Pharmacol. 15:49-59.

14. Hidaka, H., Yamaki, T. & Yamabe, H. (1978): Arch. Biochem. Biophys. 198:315-321.
15. Honda, F, Katsuki, S. & Sakai, N. (1973): Jap. J. Pharmacol. Suppl. 23:27.
16. Hummel, J.P. & Dreyer, W.J. (1962): Biochim. Biophys. Acta 63:530-532.
17. Klee, C.B., Crouch, T.H. & Richman, P.G. (1980): Ann. Rev. Biochem. 49:489-515.
18. Kobayashi, R., Tawata, M. & Hidaka, H. (1979): Biochem. Biophys. Res. Commun. 88:1037-1045.
19. Levin, R.M. & Weiss, B. (1977): Molec. Pharmacol. 13:690-697.
20. Leo, A., Hansch, C. & Elkins, D. (1971): Chem. Rev. 71: 525-554.
21. MacManus, J.P., Boynton, A.L. & Whitfield, J.F. (1978): Adv. Cyclic Nucleotide Res. 9:485-491.
22. MacManus, J.P., Whitfield, J.J., Boynton, A.L. & Rixon, R.H. (1975): Adv. Cyclic Nucleotide Res. 5:719-734.
23. Means, A.R., Chafouleas, J.G., Bolton, W.E., Hidaka, H. & Boyd, A.E. III (1981): Proc. West. Pharmacol. Soc. 24: 209-212.
24. Means, A.R. & Dedman, J.R. (1980): Nature 285:73-77.
25. Nishikawa, M., Tanaka, T. & Hidaka, H. (1980): Nature 287: 863-865.
26. Norman, J.A., Drummond, A.H. & Moser, P. (1979): Molec. Pharmacol. 16:1089-1094.
27. Perry, S.V. (1980): IN Muscle Contraction: Its Regulatory Mechanisms. (ed) S. Ebashi, Japan Science Society Press, Springer-Verlag, Berlin, pp. 207-220.
28. Salisbury, J.L., Condeelis, J.S. & Satir, P. (1980): J. Cell. Biol. 87:132-142.
29. Schubart, U.K., Erlichman, J. & Fleischer, J. (1980): J. Biol. Chem. 255:4120-4124.
30. Tanaka, T. & Hidaka, H. (1980): J. Biol. Chem. 255:11078-11080.
31. Tanaka, T. & Hidaka, H. (1981): Biochem. Int. 2:71-75.
32. Teo, T.S., Wang, T.H. & Wang, J.H. (1973): J. Biol. Chem. 248:588-595.
33. Walsh, M. & Stevens, F.C. (1978): Biochemistry 17:3924-3930.
34. Watterson, D.M., Sharief, F. & Vanaman, T.C. (1980): J. Biol. Chem. 255:962-975.
35. Weiss, B., Fertel, R., Figlin, R. & Uzunov, P. (1974): Molec. Pharmacol. 10:615-625.

INTRACELLULAR LOCALIZATION OF CALMODULIN ANTAGONIST (W-7)

S. Ohno, Y. Fujii, N. Usuda, T. Nagata, T. Endo*, T. Tanaka* and H. Hidaka*

Department of Anatomy, Shinshu University School of Medicine, Matsumoto 390, Japan, and
*Department of Pharmacology, Mie University School of Medicine, Tsu 514, Japan

INTRODUCTION

It is generally accepted that calmodulin (Ca^{++}-regulated modulator protein, modulator protein, Ca^{++}-dependent regulator) closely binds to calcium and specifically serves as an intracellular Ca^{++}-receptor (2,3,19). This protein, which is widely distributed in eukaryotic cells is usually localized in the cytoplasm of interphase cells, particularly around the mitotic spindle during division of certain cultured cells (3,17-19,35-37).

The synthetic compound, N-(6-aminohexyl)-5-chloro-1-naphthalenesulfonamide, referred to as W-7, reportedly causes a significant relaxation of vascular smooth muscle contracted by KCl, prostaglandin $F_{2\alpha}$, norepinephrine, histamine, $CaCl_2$, serotonin or angiotensin II (6,7). The relaxation was not affected by treatment with adrenergic or cholinergic blocking agents such as propranolol and atropine (7,8). Therefore, the pharmacological action of W-7 is not operative through surface membrane receptors. It was also reported that W-7 inhibits several Ca^{++}-calmodulin dependent enzymes selectively, including Ca^{++}-dependent cyclic nucleotide phosphodiesterase, myosin light chain kinase and Ca^{++}-Mg^{++}-ATPase activity of human red blood ghost by Ca^{++}-dependent binding to calmodulin with Kd value of 11 µM (6,8,9,15).

Although W-7 inhibited the in vitro reactions induced by enzymes requiring the Ca^{++}-calmodulin complex, the use of this agent is restricted until its binding to intracellular calmodulin has been clearly defined in living cells. Therefore, we are particularly interested in the cellular and subcellular localization of W-7. If the localization of W-7 can be clarified, W-7 will be a most useful tool to elucidate the physiological role of calmodulin. The localization within cells may however be difficult to determine as this compound has a small molecular weight (7).

In order to demonstrate the cellular and subcellular localization of some compounds related to cell activity, the technique of radioautography is available. However, compounds such as W-7 are not incorporated into macromolecular substances of cells and remain soluble after conventional fixative procedure. Therefore conventional radioautographical approaches are of little value (32). The conventional radioautographic methods which require fixation and dehydration of tissues appear to be inadequate for studying the cellular localization of radioisotope-labelled W-7, since artifacts due to their diffusion and extraction usually occur (4,33).

We designed a radioautographic technique, a "chemical fixation" method, using glutaraldehyde which allows for demonstration of the localization of a small molecular agent in cultured Chinese hamster ovary (CHO) cells. The CHO cells appear to offer an excellent model system for delineating the problems because these cells have been examined from viewpoints of morphology, biochemistry, transformation and cell cycles (11-13,25,26,29,30). Calmodulin was found to be densely distributed around the mitotic apparatus of CHO cells (36).

MATERIALS AND METHODS

Chinese hamster ovary (CHO) cells, clone kl (referred to as CHO-kl cells), were cultured in a medium consisting of Ham's F12 (Flow Labs, USA) supplemented with 10% fetal bovine serum (Gibco, USA), penicillin (100 IU/ml) and streptomycin (0.1 mg/ml) at 37°C in a fully humidified atmosphere of 5% CO_2 in air. The cells were seeded in petri dishes (Flat, Japan), 50 mm in diameter, for 2-3 days, after which there was an obvious formation of monolayers. At this point, 0.2 ml of tritiated W-7 solution, which had been dissolved in sterilized water at a concentration of 100 μCi/ml, was added to the 1.8 ml Ham's F12 conditioned medium (final concentration of 10 μCi/ml), in order to label the cells for 1 hr. The tritiated W-7 (specific activity 59.7 Ci/mol) was obtained from Banyu Pharmaceutical Company, (Japan).

The experiment was composed of the following three procedures: Procedure 1 - after labelling, CHO-kl cells, attached to the bottom of a petri dish, were thoroughly rinsed in 0.1 M phosphate buffer solution (pH 7.4) at 4°C three times, 3 min each, to remove unbound radioactive compounds around the cells and then additionally incubated in the phosphate buffer solution at 4°C for 1 hr. The cells were detached with a rubber policeman, collected as pellets

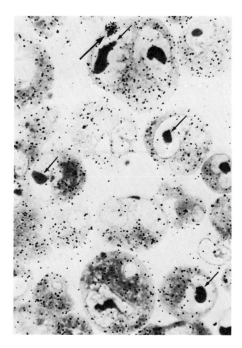

FIGURE 1

Photomicrograph of radioautogram of labelled CHO-kl cells doubly fixed in buffered glutaraldehyde and osmium tetroxide after the incubation of phosphate buffer solution for 1 hr, embedded in Epon, sectioned at 2 µM thickness, processed through wet mounting radioautography and stained with toluidine blue. Large amounts of silver grains are mainly localized in the round cytoplasms, while a few silver grains are in the nucleoli (small arrows). Silver grains are concentrated in one area of the cytoplasm of a cell (large arrow). The background level of grain density is low. X1,400.

by centrifugation, prefixed in 2.5% glutaraldehyde buffered with 0.1 M phosphate buffer solution (pH 7.4) at 4°C for 30 min and postfixed in 1% osmium tetroxide buffered with the same buffer solution for 1 hr.

Procedure 2 - After labelling, CHO-k1 cells, attached to the bottom of three petri dishes, were throughly rinsed in 0.1 M phosphate buffer solution (pH 7.4) at 4°C three times, 3 min each. Afterwards these cells were grouped into three. The first group was prefixed in 2.5% glutaraldehyde buffered with 0.1 M phosphate buffer solution (pH 7.4) at 4°C for 1 hr. The second group was prefixed in 4% paraformaldehyde buffered with 0.1 M phosphate buffer solution (pH 7.4) at 4°C for 1 hr. The third

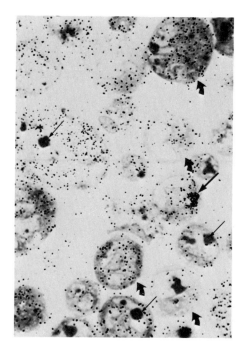

FIGURE 2

A photomicrograph of the radioautogram of labelled CHO-kl cells, processed as in Fig. 1. Large amounts of silver grains are localized in the cytoplasms, while a few silver grains are localized in the nucleoli (small arrows). Silver grains are concentrated in one area of the cytoplasm of a cell (large arrow). There are different grain densities between some cells (curved arrows). X 1,400.

group was prefixed in 1% osmium tetroxide buffered with the same buffer solution at 4°C for 1 hr. The cells belonging to each group were detached with a rubber policeman, respectively, collected as pellets by centrifugation and postfixed in buffered 1% osmium tetroxide solution for 1 hr.

Procedure 3 - After labelling of tritiated W-7 for 1 hr, CHO-k1 cells attached to the bottom of four petri dishes were chased in fresh conditioned medium without W-7 for 30 min, 1 hr, 3 hr, and 5 hr, respectively. These cells were then prefixed in 2.5% glutaraldehyde buffered with 0.1 M phosphate buffer solution (pH 7.4) at 4°C for 1 hr, detached with a rubber policeman, collected as pellets by centrifugation and postfixed in 1% osmium tetroxide solution for 1 hr.

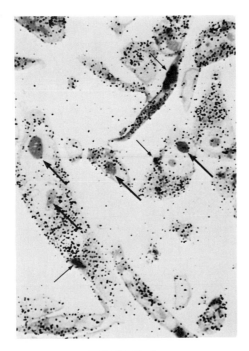

FIGURE 3

Photomicrograph of the radioautogram of labelled CHO-k1 cells, which were doubly fixed in buffered glutaraldehyde and osmium tetroxide, embedded in Epon, sectioned at 2 µm thickness, processed through wet mounting radioautography and stained with toluidine blue. Large amounts of silver grains are mainly localized in the spindle cytoplasms, while a few silver grains are localized in the nuclei, as shown in Figs. 1 and 2. Lipid droplets are not labelled (small arrows). Some cells show very dense labelling in their cytoplasm and definite labelling in their nucleoli (large arrows). X 1,400.

All the pellets obtained from each procedure were dehydrated in a series of ethanol and embedded in Epon. The 2 µm thick sections were cut with an ultramicrotome (Sorvall MT2-B) and mounted on glass slides. For light microscopic radioautography, Sakura RN-H2 liquid emulsion (Konishiroku Photo Industry Co., Tokyo) was warmed at 45°C in a thermobath and diluted with distilled water 1:1. The sections were coated with the same liquid emulsion by a dipping method, exposed in dark boxes containing silica gel at 4°C for 1 month, developed in SDX-1 developer (Sakura) at 20°C for 5 min, fixed in acid bath hypofixer and washed in running water and then stained with toluidine blue solution and observed under light microscopy (Olympus Vanox ABH-LB).

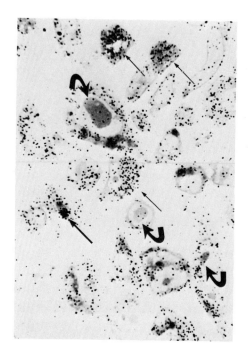

FIGURE 4

A photomicrograph of the radioautogram of labelled CHO-k1 cells, processed as in Fig. 3. Many silver grains are also localized in the cytoplasms. Some cells (small arrows) are labelled more densely than others. Silver grains are concentrated in one area of the cytoplasm (large arrow). Nucleoli of some cells are also labelled (curved arrows). X 1,400.

RESULTS

Procedure 1 - Many silver grains were retained in cytoplasms of CHO-k1 cells, even though the labelled cells had been incubated in the phosphate buffer solution for 1 hr. Almost all of the silver grains were localized in the cytoplasms of CHO-k1 cells, while a few grains were observed in the nuclei, particularly in the nucleoli (Figs. 1 and 2). The lipid droplets were not labelled.

Procedure 2 - Radioautograms of sectional CHO-k1 cells obtained from three groups showed different levels of grain density in their cytoplasms. When 2.5% glutaraldehyde was used as a fixative, large amounts of silver grains were distributed in their cytoplasms (Figs. 3 and 4). It is obvious that silver grains were distributed densely in the cytoplasm with a few located in the nucleoli. The radioautograms of the cells, which were first fixed in 4% paraformaldehyde or 1% osmium tetroxide after the incubation of the

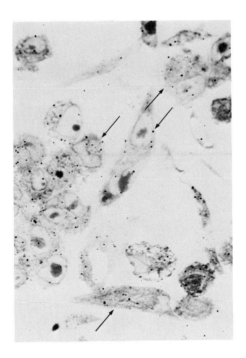

FIGURE 5

Photomicrograph of the radioautogram of labelled CHO-k1 cells, which were first fixed in buffered paraformaldehyde, embedded in Epon, sectioned at 2 µm thickness, processed through wet mounting radioautography and stained with toluidine blue. Silver grains are rarely localized in the cytoplasms of these cells, as compared with Figs. 1-4. Only small amounts of silver grains are observed in cytoplasms and nuclei of some cells (arrows). The background level of grain density is very low. X 1,400.

phosphate buffer solution, revealed that small amounts of silver grains were localized in cytoplasm (Figs. 5 and 6).

Procedure 3 - After a 30 min chase, numerous silver grains were still localized in the cytoplasms of CHO-k1 cells. In general, the grain density in the cells appeared to be lower than that in non-chased cells (Fig. 7). The amount of silver grains decreased, as the time of chase was lengthened (Figs. 8-10).

In summary, these radioautograms showed large amounts of tritiated W-7 in the cytoplasm and a few in the nucleoli of the CHO-k1 cells, when glutaraldehyde was used as a fixative.

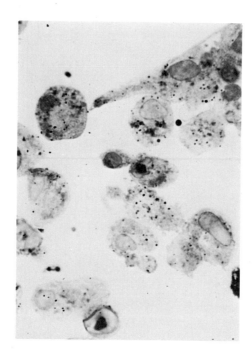

FIGURE 6

Photomicrograph of the radioautogram of labelled CHO-k1 cells, which were first fixed in buffered osmium tetroxide, embedded in Epon, sectioned at 2 μm thickness, processed through wet mounting radioautography and stained with toluidine blue. Silver grains are rarely localized in cytoplasms of these cells, as shown in Fig. 5. X 1,400.

DISCUSSION

This approach to radioautography of small molecular compounds does not involve the rapid freezing method of fresh tissues to prevent ice crystal formation (1,5,14,20,22-24,33,34). Wet mounting radioautograms showed definite cytoplasmic localization of W-7 in sectional CHO-k1 cells in the conventional Epon thick sections. With the massive labelling of the cytoplasm seen in the Epon sections, it is reasonable to assume a cytoplasmic localization of radioactive W-7. In the present study of tritiated W-7 incorporation into the cultured CHO-k1 cells, the abundant intracellular silver grains revealed by glutaraldehyde fixation might be due to binding of W-7 with insoluble macromolecules of the cells. The silver grains arose mostly from tritiated W-7 bound with cell structures by the action of glutaraldehyde.

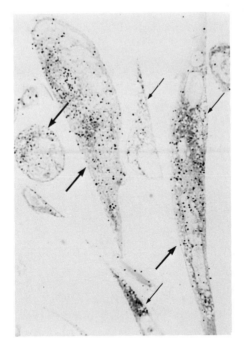

FIGURE 7

Photomicrograph of the radioautogram of CHO-k1 cells, which were chased in fresh medium for 30 min after labelling, doubly fixed in buffered glutaraldehyde and osmium tetroxide, embedded in Epon, sectioned at 2 μm thickness, processed through wet mounting radioautography and stained with toluidine blue. Large amounts of silver grains are also localized in the cytoplasms, as shown in Figs. 3 and 4. There are many small vacuoles (small arrows). The difference in silver grain density of each cell is obvious (large arrows). X 1,400.

It is essential to be aware of the effect of fixatives on the the reactive groups of W-7 and various tissue components. It is maintained that the aldehyde fixatives cause indissoluble linkage of proteins, lipids, carbohydrates and other cellular components in the form of macromolecular network. The proteins are the most important of these components (16). Glutaraldehyde, with two aldehyde groups, has been used extensively as a reagent for modification of protein-protein linking and hence for fixation (31). In reactive groups of proteins, amino acid residues readily react with aldehyde groups (27). As W-7 does have amino acid residue in its molecular structure (7), the retained end products might form protein-glutaraldehyde-(W-7) linkages. The covalent binding

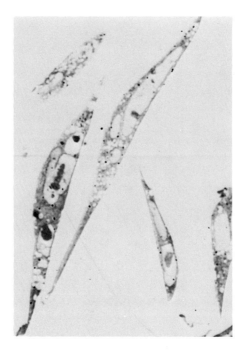

FIGURE 8

Photomicrograph of the radioautogram of CHO-kl cells, which were chased in fresh medium for 1 hr after labelling and processed as in Fig. 7. Small amounts of silver grains are localized in their cytoplasms, as compared with Figs. 3, 4 and 7. The grain density of labelled cells is dramatically decreased. X 1,400.

is firm and not broken by washing with buffer solution and organic solvents such as ethanol or acetone (28). Only a few silver grains were observed in CHO-k1 cells, when paraformaldehyde and osmium tetroxide were used as fixatives (Figs. 5 and 6). The mechanism of fixation, leading to the deposition of tritiated W-7, is unknown. Almost all of the tritiated W-7 was lost during procedures, because the fixatives altered the conformation of receptor proteins without linking up with W-7.

The radioautographic silver grains were found mostly in cytoplasms rather than nuclei, as is clear in Figs. 1-4. These results provide direct evidence for cytoplasmic localization of W-7, which is consistent with the notion of calmodulin being mainly localized in the cytoplasm. We speculate that these binding sites may be calmodulin proteins. Biochemically, calmodulin was found to include two classes of W-7 binding sites (10). The difference in

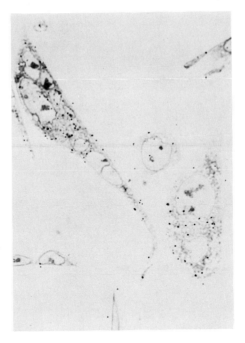

FIGURE 9

Photomicrograph of the radioautogram of CHO-kl cells, which were chased in fresh medium for 3 hr after labelling and processed as in Fig. 7. Small amounts of silver grains are also localized in their cytoplasms, as shown in Fig. 8. X 1,400.

uptake of radioactivity into each cell may be attributed to various amounts of calmodulin proteins. Therefore, it had to be determined whether W-7 within the cells is bound to specific receptors such as calmodulin proteins, or free in the cytoplasmic solution. The chasing results shown in Figs. 7-10 illustrate that W-7 incorporated into the cells binds to certain structures, as being a free small molecular compound, W-7, can be easily and rapidly elucidated with chasing. It is probable that calmodulin proteins play the role of receptors for W-7, while the cells are incubated in phosphate buffer solution, as shown in Figs. 1 and 2. We are now extending the study to analyze the binding tendency of W-7 to other proteins. In addition, the use of unlabelled W-7 as an inhibitor of incorporation of isotope-labelled W-7 serves the purpose of demonstrating the specificity of W-7 towards calmodulin (21). The use of inhibitors may be advantageous, not only to block the desired reaction in order to check for artifacts, but also to study diffusable substances to prevent their utilization or incorporation.

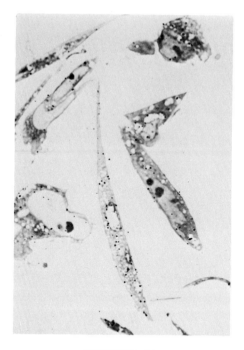

FIGURE 10

Photomicrograph of the radioautogram of CHO-k1 cells, which were chased in fresh medium for 5 hr after labelling and processed as in Fig. 7. Small amounts of silver grains are localized in their cytoplasms, as shown in Figs. 8 and 9. X 1,400.

The present report demonstrated a selective cellular accumulation of W-7 in CHO-k1 cells. In combination with earlier biochemical reports by one of the present authors, wet mounting radioautography yielded positive results for W-7 localized mainly in the cytoplasm (10). However, the resolution of light microscopic radioautography did not permit a precise ultrastructural localization. Electron microscopic studies on the same material should be highly informative (17). Studies are on-going to determine if W-7 is present around intracellular membranes in the interphase cells and central bodies at mitosis where calmodulin is densely localized (18, 19). This radioautographic technique shows promise for studying various aspects of the interactions between W-7 and the Ca^{++}-calmodulin complex.

ACKNOWLEDGEMENTS

Dr. Y. Sasaki (Research Institute, Asahi Chemical Co., Japan) provided the CHO-k1 cells, and helpful advice on tissue cultures. We thank Ms. E. Furihata and Mrs. K. Momose for skilful technical assistance and M. Ohara (Kyushu University) for critical reading of the manuscript.

REFERENCES

1. Andros, G. & Wolman, S.H. (1965): J. Histochem. Cytochem. 13:390-395.
2. Cheung, W.Y. (1980): Science 207:19-27.
3. Dedman, J.R., Welsh, M.J. & Means, A.R. (1978): J. Biol. Chem. 253:7515-7521.
4. Fischman, D.A. & Gershon, M.D. (1964): J. Cell Biol. 21: 139-143.
5. Hammarstrom, L., Appelgren, L.E. & Ullberg, S. (1965): Exp. Cell Res. 37:608-613.
6. Hidaka, H., Yamaki, T., Asano, M. & Totsuka, T. (1978): Blood Vessels 15:55-64.
7. Hidaka, H., Asano, M., Iwadare, S., Matsumoto, I., Totsuka, T. and Aoki, N. (1978): J. Pharmacol. Exp. Ther. 207:8-15.
8. Hidaka, H., Yamaki, T., Totsuka, T. and Asano, M. (1979): Molec. Pharmacol. 15:49-59.
9. Hidaka, H., Naka, M. and Yamaki, T. (1979): Biochem. Biophys. Res. Commun. 90:694-699.
10. Hidaka, H., Yamaki, T., Naka, M., Tanaka, T., Hayashi, H. and Kobayashi, R. (1980): Molec. Pharmacol. 17:66-72.
11. Hsie, A.W., Jones, C. & Puck, T.T. (1971): Proc. Nat. Acad. Sci. 68:1648-1652.
12. Hsie, A.W. and Puck, T.T. (1971): Proc. Nat. Acad. Sci. 68:358-361.
13. Hsie, A.W., Kawashima, K., O'Neill, J.P. & Schroder, C.H. (1975): J. Biol. Chem. 250:984-989.
14. Keefer, D.A., Stumpf, W.E. & Petrusz, P. (1976): Cell Tissue Res. 166:25-35.
15. Kobayashi, R., Tawata, M. & Hidaka, H. (1979): Biochem. Biophys. Res. Commun. 88:1037-1045.
16. Lenard, J. & Singer, S.J. (1968): J. Cell Biol. 37:117-121.
17. Lin, C.T., Dedman, J.R., Brinkley, B.R. & Means, A.R. (1980): J. Cell Biol. 85:473-480.
18. Marcum, J.M., Dedman, J.R., Brinkley, B.R. & Means, A.R. (1978): Proc. Nat. Acad. Sci. 75:3771-3775.
19. Means, A.R. & Dedman, J.R. (1980): Nature 285:73-77.
20. Miller, O.L., Stone, G.E. Jr. & Prescott, D.M. (1964): J. Cell Biol. 23:654-658.
21. Murrin, L.C., Enna, S.J. & Kuhar, M.J. (1977): J. Pharmacol. Exp. Ther. 203:564-574.
22. Nagata, T. & Nawa, T. (1966): Histochemie 7:370-371.

23. Nagata, T., Nawa, T. & Yokota, S. (1969): Histochemie 18: 241-249.
24. Nagata, T. & Murata, F. (1977): Histochemistry 54:75-82.
25. Niclson, S.E. & Puck, T.T. (1980): Proc. Nat. Acad. Sci. 77:985-989.
26. O'Neill, J.P., Schroder, C.H., Riddle, J.D. & Hsie, A.W. (1976): Exp. Cell Res. 97:213-217.
27. Pearse, A.G.E. (1980): IN Histochemistry, Theoretical and Applied, Vol.1 (ed) A.G.E. Pearse, Churchill-Livingstone, Edinburgh, London, pp. 97-158.
28. Peters, T. Jr. & Ashley, C.A. (1967): J. Cell Biol. 33:53-60.
29. Porter, K., Prescott, D. & Frye, J. (1973): J. Cell Biol. 57: 815-836.
30. Porter, K.R., Puck, T.T., Hsie, A.W. & Kelley, D. (1974): Cell 2:145-162.
31. Sabatini, D.D., Bensch, K. & Barrnett, R.J. (1963): J. Cell Biol. 17:19-58.
32. Sar, M. & Stumpf, W.E. (1979): Cell Tissue Res. 203:1-7.
33. Stumpf, W.E. & Roth, L.J. (1966): J. Histochem. Cytochem. 14:274-287.
34. Weiller, S., Le Goascogne, C. & Baulieu, E.E. (1976): Exp. Cell Res. 102:43-50.
35. Welsh, M.J., Dedman, J.R., Brinkley, B.R. & Means, A.R. (1978): Proc. Nat. Acad. Sci. 75:1867-1871.
36. Welsh, M.J., Dedman, J.R., Brinkley, B.R. & Means, A.R. (1979): J. Cell Biol. 81:624-634.
37. Wood, J.G., Wallace, R.W., Whitaker, J.N. & Cheung, W.Y. (1980): J. Cell Biol. 84:66-76.

POSSIBLE DUAL REGULATORY MECHANISM IN SMOOTH MUSCLE ACTOMYOSIN

M. Maruyama and S. Shibata

Department of Pharmacology, Mitsubishi-Kasei Institute of Life Sciences, Machida, Tokyo 194, Japan, and Department of Pharmacology, School of Medicine, University of Hawaii, Honolulu, Hawaii 96822, USA

In order to elucidate the regulatory mechanism in smooth muscles, we studied the inhibitory effect of calmodulin inhibitors (N^2-dansyl-L-arginine-4-t-butylpiperidine amide (TI 233) and chlorpromazine) on the divalent cation induced myosin ATPase activity of the chicken gizzard myosin B. Divalent cations (Mg^{++}, Co^{++}, Mn^{++}) activated the actin activated type myosin ATPase in the absence of Ca^{++}. Calmodulin inhibitors inhibited the divalent cation induced myosin ATPase only in the presence of Ca^{++}. Divalent cations caused the phosphorylation of the 20,000 dalton myosin light chain molecule, whereas in the absence of Ca^{++}, the phosphorylation of myosin molecule did not occur. This suggested that the phosphorylation of the myosin light chain does not play an essential role for the activation of myosin ATPase. The present experiment indicated that there are at least two possible regulatory mechanisms for the tension development in the smooth muscle, which are the calmodulin inhibitor-sensitive regulation and the inhibitor-insensitive one.

INTRODUCTION

It is well known that the contractile mechanism in the skeletal muscle are explained by the tropomyosin-troponin theory (3) for the Ca^{++} regulation and the sliding theory (6,7) for the force generation. In the smooth muscle, force generation may be explained by the sliding theory, whereas the possible mechanism of the Ca^{++} dependent regulation are not consistent among investigators (1.4,11). It is reported that certain anti-psychotic drugs prevent the superprecipitation and the myosin ATPase activity of the smooth muscle actomyosin through the inhibition of

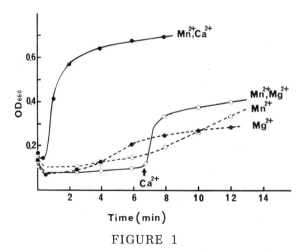

FIGURE 1

Superprecipitation of chicken gizzard myosin B induced by divalent cations. Reaction mixture contained following composition: 0.3 mg/ml myosin B, 60 mM KCl, 30 mM Tris-maleate (pH 6.8). The concentration of ions was 1 mM and the reaction started by adding 1 mM ATP.

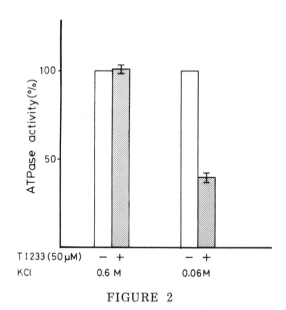

FIGURE 2

Effect of TI 233 on the two types of myosin ATPase activity. Reaction mixture contained the following composition: 0.1 mg/ml myosin B, 30 mM Tris-maleate (pH 6.8), 1 mM $CaCl_2$, 5 mM $MgCl_2$, KCl and TI 233. 100% of the activities were 165 nmol/min/mg protein (in 0.6 M KCl) and 35 nmol/min/mg protein (60 mM KCl).

calmodulin mediated myosin light chain kinase activity (5). Maruyama and Shibata also demonstrated the relaxing effect of TI 233 (N^2-dansyl-L-arginine-4-t-butylpiperidine amide) and chlorpromazine on the Ca^{++} induced contraction in saponin-treated skinned guinea pig taenia caecum and rabbit aorta (12). The present experiment was undertaken to study the influence of the calmodulin inhibitors on the divalent cation activated myosin ATPase activity.

MATERIALS AND METHODS

The preparation of chicken gizzard myosin B, the measurement of the superprecipitation and the myosin ATPase activity were carried out according to the methods of Nonomura and Ebashi (14). A Polytron (PT-20) was used for the homogenization of minced tissue, and the centrifugation in high ionic strength (0.6 M KCl) was carried out at 100,000 x g for 30 min. Using this procedure, myosin B contained much greater ATPase activity than that using the conventional method. Urea gel electrophoresis was carried out according to the method of Perrie and Perry (15) for the examination of the phosphorylation of the myosin light chain molecule. Protein concentration was determined using a protein kit from Bio-Rad Laboratories (Richmond, USA) as described by Bradford (2). TI 233 (8) was supplied by the Central Research Laboratories, Mitsubishi Chemical Co.Ltd.Japan. Chlorpromazine and other reagents (analytical grade) were obtained from Sigma (USA).

RESULTS AND DISCUSSION

Divalent cations (Mg^{++}, Co^{++}, Mn^{++}) caused the contraction of the saponin treated skinned guinea pig taenia caecum as previously reported (13,16). Calmodulin inhibitors, TI 233 and chlorpromazine, caused relaxation in the skinned fibers previously contracted by divalent cations only in the presence of Ca^{++} (13). Furthermore, the single treatment with divalent cations induced the superprecipitation and the myosin ATPase activity of chicken gizzard myosin B. In the presence of Ca^{++} (1 mM), the divalent cations (1 mM) caused much greater superprecipitation than that in the absence of Ca^{++} (Fig. 1). It is indicated that Ca^{++} enhanced the action of divalent cations on the superprecipitation of myosin B.

Calmodulin inhibitors decreased the actin activated type myosin ATPase activity in low ionic strength (0.06 M KCl), but they had no effect on the myosin type ATPase in high ionic strength (0.6 M KCl)(Fig. 2). The divalent cations (1 mM) activated the actin-activated type myosin ATPase, and only in the presence of Ca^{++}, calmodulin inhibitors inhibited these ATPase (Table 1). In the presence of Mg^{++} (5 mM) and Ca^{++} (10^{-4} M), the myosin ATPase exhibited the maximum activity and after the treatment of TI 233 reduced approximately 30% of the full value (Table 2). The result

TABLE 1: Effect of divalent cations and TI 233 on myosin ATPase.

1 mM	ATPase Activity (percent)
Co^{++}	58.0 ± 5.0
Co^{++} + 50 µM TI 233	57.0 ± 5.0
Mn^{++}	44.5 ± 3.5
Mn^{++} + 50 µM TI 233	42.0 ± 5.0
Ca^{++}	100.0
Ca^{++} + 50 µM TI 233	30.0 ± 5.5

Reaction mixture contained following composition: 0.1 mg/ml myosin B, 60 mM Tris-maleate (pH 6.8), 5 mM $MgCl_2$ and 1 mM ATP. 100% of ATPase activity was 35 nmol/min/mg protein.

TABLE 2: Effect of TI 233 on Ca^{++}-activated myosin ATPase.

	ATPase Activity (percent)
0 M Ca^{++}	18.0 ± 5.5
0 M Ca^{++} + 50 µM TI 233	17.5 ± 6.5
10^{-7} M Ca^{++}	15.0 ± 5.5
10^{-7} M Ca^{++} + 50 µM TI 233	14.5 ± 6.0
10^{-6} M Ca^{++}	48.5 ± 7.0
10^{-6} M Ca^{++} + 50 µM TI 233	22.2 ± 5.0
10^{-5} M Ca^{++}	95.0 ± 6.5
10^{-5} M Ca^{++} + 50 µM TI 233	27.5 ± 7.5
10^{-4} M Ca^{++}	100.0
10^{-4} M Ca^{++} + 50 µM TI 233	30.1 ± 5.0

Reaction mixture contained the following composition: 0.1 mg/ml myosin B, 60 mM KCl, 30 mM Tris-maleate (pH 6.8), 5 mM $MgCl_2$ and 1 mM ATP. 100 % of the myosin ATPase activity was 35 nmol/min/mg protein.

suggests that calmodulin inhibitors suppress the Ca^{++} dependent part of the myosin ATPase activity. Therefore Ca^{++} may play a major role on the activation of the myosin ATPase activity.

It is of interest to note that in the absence of Ca^{++} divalent cations did not induce the phosphorylation of the 20,000 dalton myosin light chain molecule. Thus it seems unlikely that the phosphorylation of the myosin light chain molecule play an essential role for the activation of myosin ATPase and the development of the smooth muscle contraction. The study of the interaction between calmodulin and divalent cations other than Ca^{++} has not as yet been reported, and only the binding constants between Mg^{++}, Mn^{++}, Ca^{++} and calmodulin have been studied (9,17).

The most accepted mechanism to account for the Ca^{++} dependent regulation of smooth muscle actomyosin is based on the phosphorylation and dephosphorylation of the 20,000 dalton myosin light chain molecule (1). However, Ebashi et al. (4) suggested that the regulation is due to a system termed leiotonin. On the other hand, Marston et al. (10,11) proposed a different hypothesis, that is, a dual regulatory mechanism by both thin and thick filaments in aorta actomyosin using the myosin competition test. The present study indicates that there are at least two types of regulatory mechanisms of smooth muscle contraction; one is TI 233 sensitive and the other is TI 233 insensitive. It therefore suggests that the phosphorylation of the 20,000 dalton myosin light chain molecule does not play an essential role in the contraction and the ATPase activity in smooth muscles. TI 233 sensitive regulation is probably related to the myosin linked regulation (1). However, it is not completely ruled out that there may be a possibility that TI 233 insensitive regulation may play some role on the actin linked regulation.

In conclusion, on the basis of the present experiments, it is strongly suggested that there are two different types of regulatory mechanisms in the smooth muscle contraction.

REFERENCES

1. Aksoy, M., Williams, D., Sharkey, E.M. & Hartshorne, D.J. (1976): Biochim. Biophys. Res. Commun. 69:35-41.
2. Bradford, M. (1976): Anal. Biochem. 72:248-254.
3. Ebashi, S. & Endo, M. (1968): Molec. Biol. 18:123-183.
4. Ebashi, S., Mikawa, T., Hirata, M. & Nonomura, Y. (1978): Ann. NY Acad. Sci. 307:307-321.
5. Hidaka, H., Yamaki, T., Naka, M., Tanaka, T., Hayashi, H. & Kobayashi, R. (1980): Molec. Pharmacol. 17:66-72.
6. Huxley, A.F. & Nidergerke, R. (1954): Nature 173:971-973.
7. Huxley, H.E. & Hanson, J. (1954): Nature 173:973-976.

8. Kikumoto, R., Tamao, Y., Ohkubo, K., Tonomura, S., Okamoto, S. & Hijikata, A. (1980): J. Med. Chem. 23:830-836.
9. Klee, C.B., Crouch, T.H. & Richman, R.C. (1980): Ann. Rev. Biochem. 49:489-515.
10. Lehman, W. & Szent-Gyorgyi, A.G. (1975): J. Gen. Physiol. 66:1-30.
11. Marston, S.B., Trevett, R.M. & Valters, M. (1980): Biochem. J. 185:355-365.
12. Maruyama, M. & Shibata, S. (1980): Blood Vessels 17:156.
13. Maruyama, M. & Shibata, S. (1981): Proc. 8th Int. Cong. Pharmacol., Tokyo, p. 680.
14. Nonomura, Y. & Ebashi, S. (1975): IN Method in Pharmacology, Vol. 3 (eds) E. Daniel & D.M. Paton, Pergamon Press, London, pp. 141-162.
15. Perrie, W.T. & Perry, S.V. (1970): Biochem. J. 119:31-38.
16. Saida, K. & Nonomura, Y. (1978): J. Gen. Physiol. 72:1-14.
17. Wolff, D.J., Poirier, P.J., Brostrom, C.A. & Brostrom, M.A. (1977): J. Biol. Chem. 252:4108-4117.

ION BINDING TO CALMODULIN

J. Haiech, M.C. Kilhoffer,* D. Gerard* and J.G. Lemaille

U-249 INSERM and Centre de Recherches de Biochimie macromoleculaire du CNRS, BP 5015, 34033 Montpellier, France, and
*ERA CNRS 551, Laboratoire de Biophysique, Faculte de Pharmacie, Universite Louis Pasteur, BP 10, 67048 Strasbourg, France

INTRODUCTION

Calmodulin (CaM) is the Ca^{++} dependent regulator of a number of intracellular enzymes including cyclic nucleotide phosphodiesterase (36,47,49), brain and pancreatic islet adenylate cyclase (4,50), plasma membrane (Ca^{++}-Mg^{++}) ATPases (19,28), myosin light chain kinases (12,52,57), phospholamban kinase (35) and glycogen phosphorylase b kinase (7,8). CaM is therefore a multifunctional protein that is believed to mediate the effects of the second messenger Ca^{++} in various physiological processes such as excitation-contraction or excitation-secretion coupling, or receptor-mediated endocytosis (for reviews see 6,34,38).

CaM is present in all tissues (20) and in all eukaryotic species, from Dictyostelium discoideum (2) or Tetrahymena pyriformis (29) to mammals. The ubiquitous protein was strongly conserved in evolution and there is no evidence for differences between calmodulin isolated from somatic cells and that obtained from gametes either oocyte (5) or sperm (16). When its primary structure became available (55), it was readily apparent that it belongs to the family of intracellular low molecular weight calcium-binding proteins, including troponin C, parvalbumin, myosin light chains, S 100 protein and the intestinal Ca^{++}-binding protein (14,17,18). By using the maximum parsimony method, CaM was found to exhibit a close cladistic relationship with troponin C, with which it shares extensive primary structure homology (55). The rate of evolution of CaM is the slowest among calciproteins, with only 0.3 nucleotide replacement per 100 codons per 10^8 years in the last 700 million years (14). CaM therefore appears as the closest to the ancestral four-domain calcium-binding protein from which the family evolved.

A two-step scheme accounts for the activation by Ca^{++} of CaM-dependent enzymes (E):

1) $\quad CaM + nCa^{++} \rightleftharpoons CaM \cdot Ca_n^{++} \rightleftharpoons CaM^* \cdot Ca_n^{++}$

2) $\quad CaM^* \cdot Ca_n^{++} + E \rightleftharpoons E \cdot CaM^* \cdot Ca_n^{++} \rightleftharpoons E^* \cdot CaM^* \cdot Ca_n^{++}$

where the asterisk indicates active species of CaM and enzyme.

For CaM to serve as a Ca^{++} trigger, it must fulfill the following requirements (18,23): first, CaM must bind Ca^{++} with diffusion-limited kinetics at one or more specific site(s), i.e. site(s) unoccupied by Mg^{++} in the resting cell, where Mg^{++} and Ca^{++} concentrations are 0.6 mM (21,22), and ca 0.1 µM, respectively (37). Second, Ca^{++} binding must induce a conformational change in CaM that enables it to bind and activate the target enzyme. Finally, CaM transconformation must propagate to the target enzyme which shifts to an active conformation, e.g. by unmasking of the catalytic site. This last step in the activation process is still largely unexplored, either because target enzymes, such as cyclic nucleotide phosphodiesterase or myosin light chain kinase, are difficult to purify in large quantities or because, as is the case for glycogen phosphorylase b kinase, they form large oligomeric entities (7) that are not readily accessible to physico-chemical analysis.

In contrast, extensive studies were devoted to either Ca^{++}-binding parameters or to the Ca^{++} induced conformational changes of CaM. We wish herein to describe the sequential and ordered binding of ions to CaM, and the conformational changes brought about by this sequential binding. Most of the information that will be detailed below was obtained from thermodynamic studies (24,30-32).

Binding of Ca^{++}, Mg^{++} and K^+ to Calmodulin

The Ca^{++} binding properties of CaM were first reported by Teo and Wang (48), and thereafter examined by others (28,33,36, 54,56). In agreement with the four Ca^{++} binding domains found in the primary structure of the protein (55), all studies point to the presence of four Ca^{++} binding sites. Domains I and III exhibit only three liganding carboxylate groups (X, Y and -Z) while domains II and IV show four liganding carboxylate groups (X, Y,

TABLE 1: The four Ca^{++} binding domains of calmodulin.

	12																											39
Domain I	F	K	E	A	F	S	L	F	D	K	D	G	N	G	T	I	T	T	K	E	L	G	T	V	M	R	S	L
	48																											75
Domain II	L	Q	D	M	I	N	E	V	D	A	D	G	N	G	T	I	D	F	P	E	F	L	T	M	M	A	R	K
	85																											112
Domain III	I	R	E	A	F	R	V	F	D	K	D	G	N	G	Y	I	S	A	A	E	L	R	H	V	M	T	N	L
	121																											148
Domain IV	V	D	E	M	I	R	E	A	N	I	D	G	D	G	E	V	N	Y	E	E	F	V	Q	M	M	T	A	K

α-helix						α-helix
	X	Y	Z	-Y	-X	-Z

Table from Ref. 55. Ca^{++} ligands X, Y, Z, -Y, -X and -Z are defined as described by Moews & Kretsinger (39).

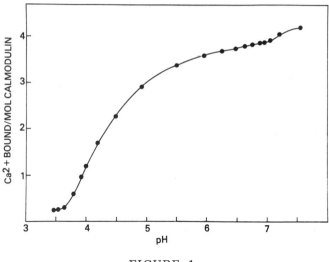

FIGURE 1

Effect of pH on the affinity of calmodulin for Ca^{++}. The number of mol calcium bound per mol of protein was monitored by flow dialysis. The pH was adjusted by successive additions of HCl. The total increase in volume due to addition of HCl was less than 5%, and therefore the negligible decrease in calmodulin concentration was not corrected. Calmodulin concentration was 3.9×10^{-5} M in 10 mM Hepes-KOH buffer, pH 7.55, 15 mM KCl. T = 25°C. Ca^{++} concentration was 2.15×10^{-4} M. Reprinted from Ref. 24 with permission of the American Chemical Society.

-X and -Z, and Y, Z, -Y and -Z, respectively (Table 1). Two major problems must be solved in the study of Ca^{++} binding to CaM. First, are all four sites equivalent and are they saturated at random, as proposed in one report (41)? Second, do other ions compete with Ca^{++} in the binding process, thereby explaining the discrepancies between previous reports in which experiments were performed at various ionic strengths and Mg^{++} concentrations (28, 33,36,41,54,56)?

CaM can be easily freed from Ca^{++} by acid treatment since its affinity for Ca^{++} becomes negligible below pH 3.5. As shown in Fig. 1 Ca^{++} binding is completely restored at neutral pH (24). Trichloroacetic acid precipitation could therefore be used to prepare calcium-free CaM for flow dialysis experiments. Table 2 shows that K^+ ions decrease the affinity of CaM for Ca^{++}, even at concentrations as low as 20 mM. Values of k_i were obtained from linear regression analysis of:

TABLE 2: Effect of K^+ on macroscopic constants describing Ca^{++} binding to calmodulin*.

KCl	K_1	K_2	K_3	K_4	K_5
(mM)	(μM)	(μM)	(μM)	(μM)	(μM)
0	.13	.14	.60	1.3	80
20	.37	.46	1.6	8.5	460
40	.73	.79	3.8	31	320
100	2	1.9	7.3	61	ND**
200	3.7	3.3	22***	580***	ND**
k'(mM)****	4	11	9	1.5	
r^2	1.0	1.0	0.99	0.98	

*	Experimental conditions as described in Fig. 2 legend.
**	At these K^+ concentrations, we were unable to detect more than 4 sites.
***	These numbers were not taken into account in the determination of k' and r^2.
****	k' is the intrinsic binding constant of K^+ for each Ca^{++} site.

Reprinted from Ref. 24, with permission of the American Chemical Society.

$\beta_{i-1}^{app}/\beta_i^{app}$, $[K^+]$ (24).

Interesting enough, K^+ addition abolished binding at a fifth unspecific binding site (Fig. 2). Mg^{++} ions also compete with Ca^{++} ions at low K^+ concentration (Table 3). When Mg^{++} ions were added at high K^+ concentration (200 mM), only three specific sites were titrated, as shown in Fig. 3, and the apparent dissociation constants for Mg^{++} are higher than those obtained at 20 mM K^+, suggesting a simple competition between K^+ and Mg^{++} for the same Ca^{++} binding sites (Table 4). The above results are in agreement with a model in which each of the four CaM sites is able to bind Ca^{++}, Mg^{++} and K^+ competitively. While Mg^{++} and K^+ binding is not sequential, one of the sites in the Ca^{++}-free state exhibits a higher affinity than the other three and Ca^{++} binding is sequential. This model allowed the calculation of the intrinsic binding constants of each site for each ion, as shown in Table 5. K^+ effects are of major importance not only in suppressing unspecific and probably physiologically irrelevant binding sites (28), but also in inducing at physiological ionic strength an apparent positive cooperativity for Ca^{++} ions which may result in a steeper activation of CaM-dependent enzymes (11).

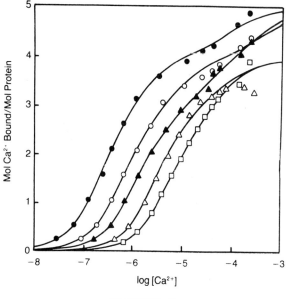

FIGURE 2

Effect of KCl on the affinity of calmodulin for Ca^{++}. The average number of Ca^{++}·(mol bound per mol of protein) is represented as a function of the negative logarithm of the free ligand concentration. Calmodulin concentration was 3.8×10^{-5} M. Ca^{++} content of the TCA-treated calmodulin was less than 10^{-6} M and was not taken into account. The buffer solution was 10 mM Tris-HCl, pH 7.55, T = 25°C. KCl concentrations were: 0 mM KCl (●—●), 20 mM KCl (○—○), 40 mM KCl (▲—▲), 100 mM KCl (△—△), 200 mM KCl (□—□). The solid lines were drawn using the general Adair-Klotz equation and the constants reported in Table 2.
Reprinted from Ref. 24, with permission of the American Chemical Society.

In contrast, competition by magnesium under physiological conditions (see Table 4) is likely to be minimal, since the apparent binding constants for Mg^{++} are then significantly higher than the free Mg^{++} concentration, that was estimated to be approximately 0.6 mM (21,22). In the resting cell (0.6 mM Mg^{++}, 0.1 µM Ca^{++}, 150 mM KCl), CaM would be under the form $CaM \cdot Ca^{++}_0 \cdot Mg^{++}_{0.7} \cdot K^{+}_{3.2}$, while the activated cell (0.6 mM Mg^{++}, 10 µM Ca^{++}, 150 mM KCl) would contain CaM mostly as $CaM \cdot Ca^{++}_{2.8} \cdot Mg^{++}_{0.1} \cdot K^{+}_{1.2}$, assuming thermodynamic equilibrium is reached.

TABLE 3: Effects of Mg^{++} on macroscopic constants describing Ca^{++} binding to calmodulin at 20 mM K^+.*

$MgCl_2$ (mM)	K_1 (μM)	K_2 (μM)	K_3 (μM)	K_4 (μM)
0	0.38	0.47	1.85	10.4
1	1.6	0.66	30	4.3
5	4.3	5.1	79	36
10	9.2	8.9	170	51
25	28	16	560**	70**
50	47	38	480**	230**
k'(mM)***	.6	.6	.4	1.5
r^2	0.99	0.99	0.99	0.93

* Calmodulin concentration was 5.7 x 10^{-5} M. Buffer was 10 mM Hepes-KOH, pH 7.55, 15 mM KCl, T = 25°C. Mg^{++} concentration was adjusted with $MgCl_2$ in the protein solution and in the flow dialysis buffer.
** These values were not used for the determination of k' and r^2.
*** k' is the apparent binding constant of Mg^{++} for each Ca^{++} site in the presence of 20 mM K^+.

Reprinted from Ref. 24, with permission of the American Chemical Society.

At any rate, CaM contains Ca^{++}-specific "triggering" sites (14,18,23) and not high affinity Ca^{++}-Mg^{++} sites involved in relaxation, such as those of parvalbumin (23) or troponin C domains III and IV (40). This may account for the fact that troponin C is unable to replace CaM in the activation of myosin light chain kinases (53). A crucial feature of the above model is that the binding of Ca^{++} to CaM is not only sequential but also ordered with a well defined and unique binding sequence. The next section deals with the determination of this unique binding sequence.

Sequence of Ion Binding to Calmodulin

A first approach to the determination of the ion binding sequence was based on the various affinities of the sites for K^+ ions, assuming that the affinity of a given site for K^+ increases with the number of carboxylates present in the ion binding loop (see Table 1). If this were the case, the highest affinity would be found in domain IV, containing five COO^- groups and the lowest

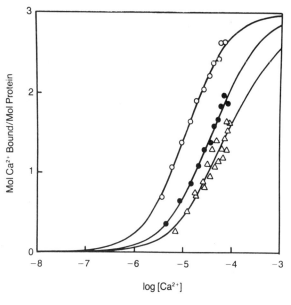

FIGURE 3

Effect of Mg^{++} on the affinity of calmodulin for Ca^{++} at 200 mM K^+. Same representation as in Fig. 2. Calmodulin concentration was 2.87×10^{-5} M. Buffer was 100 mM Hepes-KOH, pH 7.55, 150 mM KCl, T = 25°C. Mg^{++} concentration was adjusted with $MgCl_2$ in the protein solution and in the flow dialysis buffer as indicated for each curce. 0 mM Mg^{++} (○—○), 10 mM Mg^{++} (●—●), 25 mM Mg^{++} (△—△). Reprinted from Ref. 24, with permission of the American Chemical Society.

in domains I and III, containing three COO^- groups, while domain II (four COO^- groups) would exhibit intermediate affinity. The Ca^{++} binding sequence would then be either II→I→III→IV or II→III→I→IV from figures of Table 5 (24).

This ambiguity was resolved by the use of Tb^{+++} ions that compete with Ca^{++} ions for Ca^{++} binding sites (30,32). Upon excitation at 275 nm, Tb^{+++} ions become luminescent when bound to a protein close to a tyrosyl phenol group, as a result of a dipole-quadrupole energy transfer process. In contrast, free TB^{+++} is only poorly luminescent. When metal-free ram testis CaM was titrated by Tb^{+++}, the increase in luminescence at 545 nm remained weak up to 2 bound Tb^{+++} per mol CaM. Binding of

ION BINDING TO CALMODULIN 63

TABLE 4: Effects of Mg^{++} on macroscopic constants describing Ca^{++} binding to calmodulin at 200 mM K^+. *

$MgCl_2$	K_1	K_2	K_3
(mM)	(µM)	(µM)	(µM)
0	3.7	12	33
5	7.0	56	67
10	10	40	160
25	18	67	630**
50	42	190	170**
100	91**	110**	-
k'(mM)***	3.0	5.0	2.0
r^2	0.98	0.91	0.94

* Experimental conditions as in Fig. 3 legend.
** These values were not used for the determination of k' and r^2.
*** k' is the apparent binding constant of Mg^{++} for each Ca^{++} site in the presence of 200 mM K^+.
Reprinted from Ref. 24 with permission of the American Chemical Society.

additional Tb^{+++} resulted in a 400-fold enhancement of Tb^{+++} emission (Fig. 4). This already indicates that the two first Tb^{+++} ions were bound to sites which lack tyrosyl residues, a confirmation of the sequential and ordered binding. Vertebrate CaM contains only two tyrosyl residues. One is the -Y ligand of Ca^{++} in the domain III ion-binding loop, at position 99. The second is located at position 138, in the ion-binding loop of domain IV, between -X and -Z ligands of the ion (see Table 1). Domains I and II lack tyrosyl residues and therefore are the high affinity sites that bind Tb^{+++} first, establishing the ion binding sequence as II→I→III→IV. The strong quenching of the tyrosine fluorescence quantum yield observed upon binding more than two Tb^{+++} per mol should therefore be ascribed to tyrosine 99 of domain III.

This was confirmed by using a CaM isolated from octopus which appeared to contain only one tyrosyl residue (32). The tryptic digest of octopus CaM was fractionated by high performance liquid chromatography (1) and showed a single major tyrosine-containing peptide eluted at 20.5 min (Fig. 5). Its amino acid composition (Table 6) unambiguously identified this peptide as the C-terminal one, establishing that the single tyrosine was located at a position homologous to that of tyrosine 138 in domain IV of vertebrate CaMs (55).

TABLE 5: Intrinsic binding constants of calmodulin for Ca^{++}, K^+, Mg^{++}.*

Sites	1	2	3	4 **
Ca^{++}	$.67 \times 10^{-7}$	1.7×10^{-7}	6.0×10^{-7}	9.0×10^{-7}
K^+	3.7×10^{-3}	10.6×10^{-3}	8.7×10^{-3}	1.5×10^{-3}
Mg^{++}	$.70 \times 10^{-4}$	2.7×10^{-4}	1.0×10^{-4}	$.90 \times 10^{-4}$

* Intrinsic binding constants (M) derived from the theoretical model described in the appendix of reference 24.

The binding constants for Ca^{++} were extrapolated from data in Table 2. The binding constants for Mg^{++} were derived from the data obtained at 20 and 200 mM K^+ assuming a simple competition between K^+ and Mg^{++}.

** The numbers indicate the Ca^{++} binding sites according to the order in which they are saturated by Ca^{++}.

Reprinted from Ref. 24, with permission of the American Chemical Society.

When metal-free octopus CaM was titrated with Tb^{+++} ions (Fig. 6) no quenching of the tyrosine fluorescence quantum yield was observed and a plateau was reached at a stoichiometry of four Tb^{+++} per mol protein. This clearly shows that tyrosine 99 of domain III is involved in the quenching observed with vertebrate CaM, since this tyrosyl residue is missing in the octopus protein. The ion binding sequence is confirmed as II→I→III→IV. Another line of evidence was obtained, during Tb^{+++} titrations of octopus CaM, from the fact that Tb^{+++} luminescence increased, albeit weakly, only after three Tb^{+++} were already bound. The lower affinity domain IV is therefore the last to be saturated.

The major conclusions that emerge from the experiments described above (24,30,32) are the following. At each of the four ion-binding sites of CaM, Ca^{++}, Mg^{++} and K^+ ions bind competitively, and, at physiological K^+ concentration, competition by Mg^{++} is minimal and probably negligible. Binding of Ca^{++} to CaM is sequential and ordered with a unique binding sequence to successively domain II, I, III, then to lower affinity domain IV.

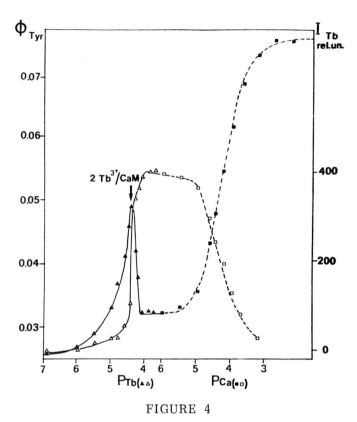

FIGURE 4

Terbium binding to ram testis calmodulin. Solid lines indicate the fluorescence quantum yield (Φ Tyr) of calmodulin (▲) and the concomitant variation of terbium luminescence at 545 nm: I_{Tb} (△) as a function of added Tb^{+++} (pTb = -log [Tb^{+++}]). Ca^{++}-induced Tb^{+++} removal is represented by dashed lines. (■) stands for tyrosine quantum yield and (□) for terbium luminescence. Calmodulin concentration was 2.2×10^{-5} M in 100 mM Tris buffer (pH 6.9). When Tb^{+++} was $>3 \times 10^{-4}$ M, a Tb^{+++}-induced precipitation of calmodulin occurred that prohibited the spectroscopic study.
Reprinted from Ref. 30, with permission of the Federation of European Biochemical Societies.

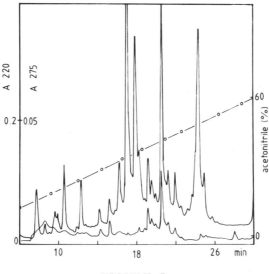

FIGURE 5

High performance liquid chromatography of an octopus calmodulin tryptic digest (10 nmol) on a μ Bondapak phenyl column. Upper trace: A_{220} nM of the eluate, 0.4 absorbance units full scale. Lower trace: A_{275} nM of the eluate, 0.1 absorbance units full scale. Lower and upper traces were obtained on different and consecutive runs. In the illustrated part of the run (6-30 min), the acetonitrile gradient (o—o) was from 12-60%.
Reprinted from Ref. 32, with permission of the Federation of European Biochemical Societies.

Ca^{++} Binding to Calmodulin Induces Sequential Conformational Changes

Calcium binding to CaM induces conformational changes which may be studied by a number of techniques including circular dichroism (13,33) and ultraviolet differential spectra (33), tyrosine fluorescence (13,15,28), proton nuclear magnetic resonance (46), susceptibility to proteolytic attack (25) or reactivity to various chemical modifications (42-45,51). In general, the α-helical content was found to increase upon Ca^{++} binding, while the environment of such residues as Tyr 138, Lys 77 and Met 71, 72 and 76 was found to be modified. More important perhaps is the fact that most spectral changes are complete after binding of 2 mol Ca^{++} per mol CaM (see, e.g. 11,46), which indicates that Ca^{++} binding occurs in at least two steps corresponding to the binding of a first pair of Ca^{++} ions to $CaM \cdot Ca_2^{++}$ and then to the binding of an additional pair of Ca^{++} ions. Since we have established that

TABLE 6: Amino acid composition of tyrosine-containing tryptic peptides from octopus and bovine brain calmodulin.*

Residue	Octopus Calmodulin peptide eluted at 20.5 min		Bovine Brain Calmodulin	
	Found	Integer	Peptide 127-148	Peptide 91-106
Asx	4.0	4	4	3
Thr	1.0	1	1	0
Ser	1.1	1	0	1
Glx	4.0	4	5	1
Gly	2.0	2	2	2
Ala	1.7	2	2	2
Val	1.9	2	2	1
Met	1.5	1-2	2	0
Ile	1.1	1	1	1
Leu	0.3	0	0	1
Tyr	0.8	1	1	1
Phe	0.8	1	1	1
Lys	0.9	1	1	1
Arg	0	0	0	1
Total		21-22	22	16

* Expressed as residues per mole, after 24 hr hydrolysis. Values for bovine brain calmodulin from Ref. 55.
Reprinted from Ref. 32, with permission of Federation of European Biochemical Societies.

Ca^{++} binds first to the high affinity domains I and II, the $CaM \cdot Ca_2^{++}$ form corresponds to the saturation of the first NH_2-terminal half of the molecule. The conformational change corresponding to the transition from $CaM \cdot Ca_0^{++}$ to $CaM \cdot Ca_2^{++}$ however, involves at least Tyr 138 in the COOH-terminal half of the CaM peptide chain, and is therefore a global transconformation. The detailed study of tyrosine fluorescence changes was recently reported (31). In the absence of divalent cations, the quantum yields are low for both ram testis ($\Phi = 0.03$) and octopus ($\Phi = 0.016$) CaM (Table 7). Addition of Ca^{++} results in a large increase of the tyrosine fluorescence quantum yield, up to a plateau of 0.08 and 0.048 for ram testis and octopus CaM, respectively. For both, the plateau is reached upon binding of two mol Ca^{++} per mol protein (Fig. 7). In contrast, Mg^{++} addition, though increasing the quantum yield, did not allow to reach the same plateau level observed with either Ca^{++} alone or Ca^{++} added after Mg^{++} (Fig. 8). Moreover, increasing K^+ concentrations decreased the efficiency of Mg^{++} ions in increasing the quantum yield, a confirmation of the competition between Mg^{++} and K^+ (see Table 7).

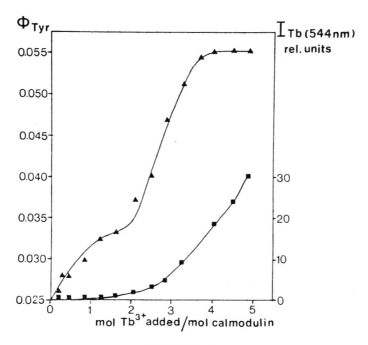

FIGURE 6

Titration of metal-free octopus calmodulin by Tb^{+++}. Tb^{+++} ($TbCl_3$) was added to 21.7 μM calmodulin in 20 mM MOPS buffer (pH 6.9) and was assumed to be protein bound. Φ_{Tyr} is the tyrosine fluorescence quantum yield (▲—▲). l_{Tb} is the variation of terbium luninescence (■—■) at 544 nm, expressed in relative units, upon excitation at the same wavelength (280 nm).
Reprinted from Ref. 32, with permission of the Federation of European Biochemical Societies.

This again suggests that, at physiological ionic strength, Mg^{++} ions have little effect on CaM conformation and explains why they are unable to activate CaM-dependent enzymes (56).

A detailed analysis of the fluorescence shows that the environment of both residues 99 and 138 is affected by calcium binding to sites I and II of CaM, while the binding of an additional pair of Ca^{++} ions to domains III and IV has little effect on the tyrosyl residues of these domains, in line with NMR data from Seamon (46).

TABLE 7: Fluorescence characteristics of ram testis and octopus calmodulins.

Conditions	−Ca^{++}				+Ca^{++}			
	Φ	τ(ns)	Φ/τ × 10^{-7} s^{-1}	ω	Φ	τ(ns)	Φ/τ × 10^{-7} s^{-1}	ω
Ram testis calmodulin								
10 mM Tris	0.030	1.7	1.75	0.56	0.078	2.3	3.4	0.15
100 mM Tris	0.027	1.7	1.60	0.60	−	−	−	−
10 mM Tris, 150 mM KCl	0.024	1.7	1.40	0.65	0.083	2.3	3.6	0.10
100 mM Tris, 150 mM KCl	0.025	1.7	1.45	0.63	0.076	2.3	3.3	0.17
100 mM Tris, 5 mM MgCl$_2$	0.046	1.9	2.40	0.40	0.082	2.3	3.6	0.10
100 mM Tris, 5 mM MgCl$_2$, 150 mM KCl	0.030	−	−	−	0.075	2.3	3.2	0.19
Denatured (6 M guanidine HCl)	0.047	1.8	2.60	0.35	−	−	−	−
Octopus calmodulin								
100 mM Tris	0.016	2.5	0.65	0.9	0.048	2.5	1.9	0.5
100 mM Tris, 5 mM MgCl$_2$	0.035	−	−	−	−	−	−	−
Free tyrosine, water, pH 7	0.14	3.5	4	0	−	−	−	−

Φ = fluorescence quantum yield (± 0.002); τ = fluorescence lifetime (± 0.2 ns); ω = fraction of static quenching. All protein solutions were at pH 7.5 ± 0.1. Reprinted from Ref. 31 with permission of American Chemical Society.

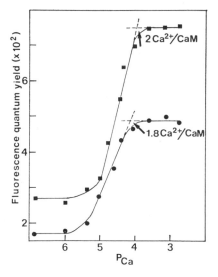

FIGURE 7

Effect of calcium on the fluorescence of ram testis calmodulin (45 μM in 100 mM Tris, pH 7.55)(■) and of octopus calmodulin (35 μM in 100 mM Tris, pH 6.9)(●). In both cases the plateau was reached upon binding of 1.8 to 2 mol Ca^{++} per mol of calmodulin. Reprinted from Ref. 31, with permission of American Chemical Society.

CONCLUSION

Recent reports show conclusively that $CaM \cdot Ca_2^{++}$ is unable to activate cyclic nucleotide phosphodiesterase (11), whereas the same enzyme was found to be activated by $CaM \cdot Ca_4^{++}$ (26) or by $CaM \cdot Ca_3^{++}$-$CaM \cdot Ca_4^{++}$ (10). Similarly, only $CaM \cdot Ca_4^{++}$ can activate myosin light chain kinase (3).

There is therefore evidence that the intermediary complex $CaM \cdot Ca_2^{++}$ in which sites I and II are saturated by Ca^{++} ions is unable either to bind to the enzyme or, if it binds, to alter the conformation of the target enzyme in such a way that it becomes active. This is in contrast with the fact that the Ca^{++} induced transconformation of CaM is almost complete at the stage of two bound Ca^{++}, at least when followed with the tyrosine probes.

Some of the questions raised in this paper will be solved only when 3-dimensional structures of Ca^{++}-free and Ca^{++}-loaded (9) CaMs will be available. Another approach to the problem, aimed

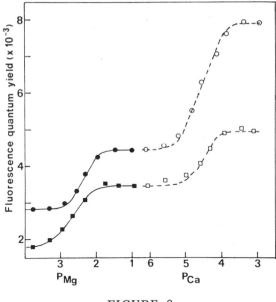

FIGURE 8

The effect of magnesium on the fluorescence of ram testis calmodulin (18 μM in 100 mM Tris, pH 7.6)(●) and octopus calmodulin (■) (10 μM in 100 mM Tris, pH 7.3) is represented by the solid line. The dashed line indicates the effect of further addition of Ca^{++} to the Mg^{++}-loaded ram testis (○) and octopus calmodulin (□). Reprinted from Ref. 31, with permission of the American Chemical Society.

at distinguishing between binding and activation per se, is the study of glycogen phosphorylase b kinase, to which CaM is attached as the δ-subunit irrespective of the Ca^{++} concentration (7,8).

ACKNOWLEDGEMENTS

This work was supported in part by grants to D.G.(ERA CNRS 551) and to J.G.D.(CNRS LP, INSERM ATP 63.78.95 and CRL 78. 4.086.1, 79.4.086.1, 79.4.151.3, 80.5.032, 81.3.023; DGRST ACC BFM and Parois arterieles and ACC Relations dynamiques entre macromolecules biologiques; NATO grant 1688; Muscular Dystrophy Association of America; and Fondation pour la Recherche Medicale Francaise). J.H. is Attache de Recherche CNRS and acknowledges receipt of DGRST and Fogarty Fellowships during his stay in the laboratory of Dr. C. Klee, NCI, NIH. M.C.K. is the recipient of a fellowship from Ministere des Affaires Etrangeres. The authors are grateful to Dr. C. Klee for stimulating discussion and advice. The expert editorial assistance of Ms. D. Waeckerle is also gratefully acknowledged.

REFERENCES

1. Autric, F., Ferraz, C., Kilhoffer, M.C., Cavadore, J.C. & Demaille, J.G. (1980): Biochim. Biophys. Acta 631:139-147.
2. Bazari, W.L. & Clarke, M. (1981): J. Biol. Chem. 256:3598-3603.
3. Blumenthal, D.K. & Stull, J.T. (1980): Biochemistry 19:5608-5614.
4. Brostrom, C.O., Huang, Y.C., Breckenridge, B. McL. & Wolff, D.J. (1975): Proc. Nat. Acad. Sci. 72:64-68.
5. Cartaud, A., Ozon, R., Walsh, M.P., Haiech, J. & Demaille, J.G. (1980): J. Biol. Chem. 255:9404-9408.
6. Cheung, W.Y. (1980): Science 207:19-27.
7. Cohen, P. (1980): Eur. J. Biochem. 111:563-574.
8. Cohen, P., Burchell, A., Foulkes, J.G., Cohen, P.T.W., Vanaman, T.C. & Nairn, A.C. (1978): FEBS Lett. 92:287-293.
9. Cook, W.J., Dedman, J.R., Means, A.R. & Bugg, C.E. (1980): J. Biol. Chem. 255:8152-8153.
10. Cox, J.A., Malnoe, A. & Stein, E.A. (1981): J. Biol. Chem. 256:3218-3222.
11. Crouch, T.H. & Klee, C.B. (1980): Biochemistry 19:3692-3698.
12. Dabrowska, R., Sherry, J.M.F., Aromatorio, D.K. & Hartshorne, D.J. (1978): Biochemistry 17:253-258.
13. Dedman, J.R., Potter, J.D., Jackson, R.L., Johnson, J.D. & Means, A.R. (1977): J. Biol. Chem. 252:8415-8422.
14. Demaille, J.G., Haiech, J. & Goodman, M. (1980): Protides Biol. Fluids 28:95-98.
15. Drabikowski, W., Kuznicki, J. & Grabarek, Z. (1977): Biochim. Biophys. Acta 485:124-133.
16. Feinberg, J., Weinman, J., Weinman, S., Walsh, M.P. Harricane, M.C. Gabrion, J. & Demaille, J.G. (1981): Biochim. Biophys. Acta 673:303-311.
17. Goodman, M. & Pechere, J.F. (1977): J. Molec. Evol. 9:131-158.
18. Goodman, M., Pechere, J.F., Haiech, J. & Demaille, J.G. (1979): J. Molec. Evol. 13:331-352.
19. Gopinath, R.M. & Vincenzi, F.F. (1977): Biochem. Biophys. Res. Commun. 77:1203-1209.
20. Grand, R.J.A., Perry, S.V. & Weeks, R.A. (1979): Biochem. J. 177:521-529.
21. Gupta, R.K. & Moore, R.D. (1980): J. Biol. Chem. 255:3987-3993.
22. Gupta, R.K. & Yushok, W.D. (1980): Proc. Nat. Acad. Sci. 77:2487-2491.
23. Haiech, J., Derancourt, J., Pechere, J.F. & Demaille, J.G. (1979): Biochemistry 18:2752-2758.
24. Haiech, J., Klee, C.B. & Demaille, J.G. (1981): Biochemistry 20:3890-3897.

25. Ho, H.C., Desai, R. & Wang, J.H. (1975): FEBS Lett. 50: 374-377.
26. Huang, C.Y., Chau, V., Chock, P.B., Wang, J.H. & Sharma, R.K. (1981): Proc. Nat. Acad. Sci. 78:871-874.
27. Jarrett, H.W. & Kyte, J. (1979): J. Biol. Chem. 254:8237-8244.
28. Jarrett, H.W. & Penniston, J.T. (1977): Biochem. Biophys. Res. Commun. 77:1210-1216.
29. Kakiuchi, S., Sobue, K., Yamazaki, R., Nagao, S., Umeki, S., Nozawa, Y., Yazawa, M. & Yagi, K. (1981): J. Biol. Chem. 256:19-22.
30. Kilhoffer, M.C., Demaille, J.G. & Gerard, D. (1980): FEBS Lett. 116:269-272.
31. Kilhoffer, M.C., Demaille, J.G. & Gerard, D. (1981): Biochemistry 20:4407-4414.
32. Kilhoffer, M.C., Gerard, D. & Demaille, J.G. (1980): FEBS Lett. 120:99-103.
33. Klee, C.B. (1977): Biochemistry 16:1017-1024.
34. Klee, C.B., Crouch, T.H. & Richman, P.G. (1980): Ann. Rev. Biochem. 49:489-515.
35. Le Peuch, C.J., Haiech, J. & Demaille, J.G. (1979): Biochemistry 18:5150-5157.
36. Lin, Y., Liu, Y.P. & Cheung, W.Y. (1974): J. Biol. Chem. 249:4943-4954.
37. Marban, E., Rink, T.J., Tsien, R.W. & Tsien, R.Y. (1980): Nature (Lond.) 286:845-850.
38. Means, A.R. & Dedman, J.R. (1980): Nature (Lond.) 285:73-77.
39. Moews, P.C. & Kretsinger, R.H. (1975): J. Molec. Biol. 91: 201-228.
40. Potter, J.D. & Gergely, J. (1975): J. Biol. Chem. 250:4628-4633.
41. Potter, J.D., Johnson, J.D., Dedman, J.R., Schreiber, W.E., Mandel, F., Jackson, R.L. & Means, A.R. (1977): IN Calcium Binding Proteins and Calcium Function. (eds) R.H. Wasserman, North-Holland, New York, pp. 239-250.
42. Richman, P.G. (1978): Biochemistry 17:3001-3005.
43. Richman, P.G. & Klee, C.B. (1978): J. Biol. Chem. 253:6323-6326.
44. Richman, P.G. & Klee, C.B. (1978): Biochemistry 17:928-935.
45. Richman, P.G. & Klee, C.B. (1979): J. Biol. Chem. 254:5372-5376.
46. Seamon, K.B. (1980): Biochemistry 19:207-215.
47. Stevens, F.C., Walsh, M., Ho, H.C., Teo, T.S. & Wang, J.H. (1976): J. Biol. Chem. 251:4495-4500.
48. Teo, T.S. & Wang, J.H. (1973): J. Biol. Chem. 248:5950-5955.
49. Teo, T.S. Wang, T.H. & Wang, J.H. (1973): J. Biol. Chem. 248:588-595.
50. Valverde, I., Vandermeers, A., Anjaneyulu, R. & Malaisse, W.J. (1979): Science 206:225-227.
51. Walsh, M. & Stevens, F.C. (1977): Biochemistry 16:2742-2749.
52. Walsh, M.P., Vallet, B. Autric, F. & Demaille, J.G. (1979): J. Biol. Chem. 254:12136-12144.
53. Walsh, M.P., Vallet, B., Cavadore, J.C. & Demaille, J.G. (1980): J. Biol. Chem. 255:335-337.

54. Watterson, D.M., Harrelson, W.G.Jr, Keller, P.M., Sharief, F. & Vanaman, T.C. (1976): J. Biol. Chem. 251:4501-4513.
55. Watterson, D.M., Sharief, F. & Vanaman, T.C. (1980): J. Biol. Chem. 255:962 975.
56. Wolff, D.J., Brostrom, M.A. & Brostrom, C.O. (1977): IN Calcium Binding Proteins and Calcium Function. (eds) R.H. Wasserman, North-Holland, New York, pp. 97-106.
57. Yagi, K., Yazawa, M., Kakiuchi, S., Ohshima, M. & Uenishi, K. (1978): J. Biol. Chem. 253:1338-1340.

STRUCTURE AND Ca^{++}-DEPENDENT CONFORMATIONAL CHANGE OF CALMODULIN

K. Yagi, S. Matsuda, H. Nagamoto, T. Mikuni and M. Yazawa

Department of Chemistry, Faculty of Science, Hokkaido University, Sapporo 060, Japan

INTRODUCTION

Primary structure of vertebrate calmodulin was first determined with one that was isolated from bovine brain by Watterson et al. (21) and primary structure of invertebrate calmodulin was determined with one isolated from sea anemone (19). The former contained two tyrosine residues at positions 99 and 138, while the latter contained one tyrosine residue at position 138. In spite of these differences in tyrosine content, an identical difference UV absorption spectrum, which was induced by the addition of Ca^{++}, was observed with the two calmodulins (26). It therefore indicates that the cause of difference spectrum at 280-290 nm is the change in microenvironment around Tyr 138.

The present study was performed with particular attention to Tyr 138. The distance between the nitroxide radical, combined with Tyr 138 (and Tyr 99) and the potential Ca^{++} binding site was measured after the Leigh theory (8). The estimated distance suggested that the high affinity Ca^{++} binding sites were located in the carboxyl terminal half (the third and fourth domains) of calmodulin molecule.

The abbreviations we will be using in this report are as follows: Tyr 99 and Tyr 138 are the tyrosine residues at the 99th and 138th positions from the N-terminal of calmodulin; Cys 98, Tyr 10 and Tyr 109 are the cysteine and tyrosine residues at the 98th, 10th and 109th positions from the N-terminal of troponin C; C-reagent is the spin-label reagent of cysteine residue, N-(2,2,6,6-tetramethyl-4-piperidine-1-oxyl)-maleimide; Y-reagent is a spin-label reagent of the tyrosine residue, N-(2,2,5,5-tetramethyl-3-carbonyl-pyrroline-1-oxyl)-imidazole; NN = 2-hydroxy-1-(2-hydroxy-4-sulfo-1-naphthylazo)-3-naphthoic acid.

METHODS AND MATERIALS

<u>Preparation of proteins</u>: Calmodulins prepared from pig brain, rabbit skeletal muscle, salmon testis, scallop adductor muscle, and scallop testis were used. Preparation procedure has been described elsewhere (26). Troponin C was prepared from rabbit skeletal muscle by the method of Perry and Cole (12). Parvalbumin was prepared from rabbit skeletal muscle as follows: Frozen muscle was homogenized in 2.5 vol of 50 mM sodium phosphate (pH 5.7) and 5 mM EDTA. After centrifugation (5,000 rpm, 30 min), proteins in the supernatant was precipitated in 3% TCA in a cold room. The precipitate was redissolved at pH 5.3 and the residual precipitate was discarded. Proteins in the supernatant was again precipitated in 3% TCA and the precipitate was dissolved in 0.2 M Tris-HCl (pH 8.0). It was dialyzed against 50 mM Tris-HCl (pH 8.0). The proteins were fractionated by DEAE-cellulose column chromatography and a fraction eluted at 0.03-0.04 M NaCl was collected. This protein fraction was subjected to gel filtration of sephadex G-75 equilibrated with 20 mM Tris-HCl (pH 7.5) and 70 mM NaCl. The second peak contains parvalbumin which showed a single band by SDS gel electrophoresis.

Ca^{++} measurements: $CaCO_3$ was used as a primary standard of Ca^{++} concentration. Ca^{++} concentration was determined by EDTA-titration using NN mixed with K_2SO_4 as a color indicator. NN is a product of the Dojindo Laboratory. The amount of contaminating Ca^{++} of the proteins was measured with an atomic absorption spectrophotometer (Hitachi 208). Parvalbumin preparation contained 0.15 moles Ca^{++} per mole. Calmodulin isolated from scallop testis using solvents treated with Chelex-100 contained 0.05 moles Ca^{++} per mole, but when Chelex 100-treated solvents were not used, calmodulin contained approximately 1.5-2 moles Ca^{++} per mole. TCA treatment appeared to be effective in decreasing the amount of contaminating Ca^{++}.

<u>Absorption spectrum</u>: UV absorption spectrum and difference UV absorption spectrum were measured using a recording spectrophotometer (Shimadzu UV350). For the latter measurement, the reference cell contained 1 mM EGTA and the sample cell contained appropriate amounts of EGTA and Ca^{++} to give a certain free Ca^{++} concentration. The binding constants of Ca^{++} to EGTA used in these experiments are described elsewhere (10).

<u>ESR</u>: ESR spectra were recorded at 9.44 GHz with 2.5 mW power using JEOL-FE3X spectrometer with a flat quartz liquid cell. The modulation width was 2.0 G. The binding of Mn^{++} to calmodulin and troponin C was investigated by ESR, using the procedure

of Yazawa and Morita (25). The bottom-to-peak height of the first line from low field of the hyperfine spectrum of Mn^{++} was employed to measure the signal intensity.

Leigh (8) has shown that the extent of quenching of the nitroxide spectrum provides an estimate of the distance between spin label and Mn^{++}. The Leigh theory is applied to estimate the distance between the spin label of tyrosine or cysteine residue and Mn^{++} bound to Ca^{++} binding sites of calmodulin or troponin C. The distance was estimated from the bottom-to-peak height of the center line of nitroxide spectrum using the figure calculated by Leigh (1) with the correlation time of 3 n sec. According to Dwek (4), the correlation time (τ) of dipolar interaction of Mn^{++} binding proteins is about 3 n sec. Relaxation time of nitroxide radical bound to the Ca^{++} binding proteins was estimated from the equation introduced by Stone et al. (17), and it lies in the range between 1.3 and 2.3 n sec (S. Matsude et al., in preparation).

The decrease in the center peak heights of nitroxide ESR spectrum was shown by the Ca^{++} (or other metal ion) binding to calmodulin and troponin C. The phenomenon resulted from conformational changes of proteins. In order to measure the distance from the decrease in center peak heights of ESR spectrum, it is necessary to minimize the conformational change which results from the binding of Mn^{++}. We used calmodulin and troponin C containing 1.6-2.0 moles Ca^{++} per mole of proteins in these experiments. It was then assumed that Ca^{++} bound to the high affinity sites is exchanged with the extraneously added Mn^{++}, and the metal ion-induced conformational change can be minimized.

The Y-reagent was synthesized (2) and was used for the spin labelling to tyrosine residues. The C-reagent is a product of Synvar and was used for the spin labelling to cysteine residue.

RESULTS

<u>Difference UV absorption spectrum observed at high and very low ionic strength</u>: Ca^{++}-induced difference UV absorption spectrum was first observed with porcine brain calmodulin by Klee (7) as a negative difference spectrum. We confirmed this result with several calmodulins isolated from vertebrate and invertebrate sources and it seemed to indicate that Tyr 138 was responsible for this difference spectrum (26). These measurements were carried out in a solvent containing 0.05 or 0.1 M NaCl and 10 mM Tris-HCl (pH 8.0). A negative difference spectrum obtained with scallop testis calmodulin is shown in Fig. 1, and it was observed upon the addition of Ca^{++} to the calmodulin dissolved in the solvent containing 0.1 M NaCl (or KCl). Difference spectrum around 260 nm corresponds to the perturbation arising from phenylalanine

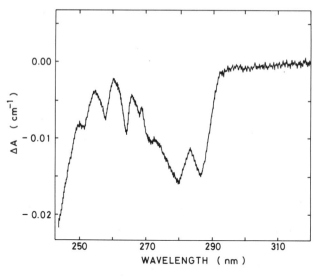

FIGURE 1

Ca^{++}-induced difference UV absorption spectrum of calmodulin in the presence of 0.1 M NaCl. Scallop testis calmodulin (1.26 mg/ml) was dissolved in 0.1 M NaCl and 20 mM MOPS-KOH (pH 7.2). Ca^{++} was added in a final concentration of 0.1 mM. Temperature was 20°C.

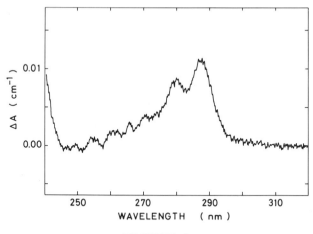

FIGURE 2

Na$^+$-induced difference UV absorption spectrum of calmodulin. Salmon testis calmodulin (0.78 mg/ml) was dissolved in 10 mM MOPS-KOH (pH 7.0). Na$^+$ was added in a final concentration of 0.1 M. Temperature was 8°C.

residues and it appears to be a positive difference spectrum because of the splitting of the peak at 259 nm (23).

On the other hand, Ca^{++}-induced positive difference spectrum had previously been observed with rabbit muscle and sea anemone calmodulins when measured in a solvent omitting 0.1 M NaCl (26). This result was confirmed with salmon testis calmodulin dissolved in 10 mM MOPS-KOH (pH 7.1), when 0.1 mM Ca^{++} or Mg^{++} was added. The difference spectrum around 260 nm was also distinctly observed and it also appeared positive. As shown in Fig. 2, the positive difference spectrum was also induced by the addition of a monovalent cation such as Na^+, K^+ and NH_4^+ in a final concentration of 0.1 M instead of 0.1 mM Ca^{++}. Since the primary structure of vertebrate calmodulins so far determined are identical to each other except for a few amide assignments, salmon testis calmodulin may not have a unique primary structure but may have a common structure of vertebrate calmodulins. The UV absorption spectrum of salmon testis calmodulin was nearly the same as that of other vertebrate calmodulins. These results seem to indicate that the positive difference spectrum is observed with calmodulins isolated from different sources by the addition of either monovalent or divalent cation when it is measured in a medium of very low ionic strength.

Even though the affinity of monovalent cation to calmodulin is low, as shown in the experiments of parvalbumin described below, Na^+ and K^+ at 0.1 M may be able to bind to the four Ca^{++} binding loops of calmodulin and the whole protein structure could then be stabilized. The effect of 0.1 mM Ca^{++} to the conformation of calmodulin at very low ionic strength may be similar to that of 0.1 M monovalent cation.

The difference spectrum induced by Ca^{++} at very low ionic strength (in 20 mM MOPS-KOH; pH 7.2) was measured with scallop testis calmodulin after removing the contaminating Ca^{++} of solvents by Chelex 100 treatment. The peak heights at 290 nm (ΔA_{290}) were measured at various free Ca^{++} concentrations. As shown in Fig. 3, ΔA_{290} vs pCa plot indicates that under these conditions Ca^{++} binds to calmodulin at free Ca^{++} concentrations as low as 0.1 μM. ΔA_{290} increased gradually by increasing the Ca^{++} concentration up to 0.5 mM after having once attained a plateau at a Ca^{++} concentration of about 1-10 μM.

The difference UV absorption spectrum induced by Ca^{++} in the presence of 0.1 M KCl and 50 mM MOPS-KOH (pH 7.2) was also measured with scallop testis calmodulin after removing the contaminating Ca^{++} of solvents.

The negative peak heights at 286 nm (ΔA_{286}) were measured at various free Ca^{++} concentrations and, as shown in Fig. 3, the

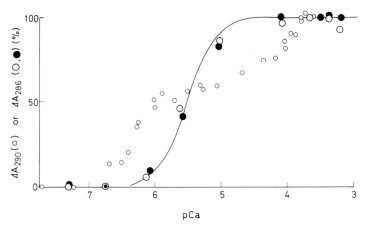

FIGURE 3

Spectrophotometric titration of Ca^{++} binding to calmodulin. Scallop testis calmodulin was used. Small circle (○) - the magnitude of positive difference spectrum at 290 nm is plotted on the ordinate. Calmodulin (1.17 mg) was dissolved in 20 mM MOPS-KOH (pH 7.2). Large circles (○ , ●) - the magnitude of negative difference spectrum at 286 nm is plotted on the ordinate. Calmodulin (0.76 mg/ml) was dissolved in 0.1 M KCl-50 mM MOPS-KOH (pH 7.2)(○) or in 0.1 M KCl-2 mM $MgCl_2$-50 mM MOPS-KOH (pH 7.2)(●). Free Ca^{++} concentration in the sample cell was determined using EGTA buffer. The reference cell contained 2 mM EGTA.

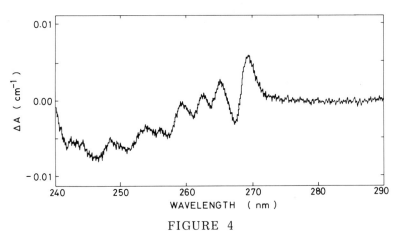

FIGURE 4

Ca^{++}-induced difference UV absorption spectrum of rabbit muscle parvalbumin. Parvalbumin (1.72×10^{-4} M) was dissolved in 10 mM Tris-HCl (pH 7.8). Ca^{++} was added in a final concentration of 0.1 mM.

relation between ΔA_{286} and pCa was a sigmoid curve having the Hill coefficient of 1.60. Two moles of Ca^{++} binding per mole of calmodulin have previously been demonstrated at free Ca^{++} concentration of 10 μM (24), where the maximum spectral change is attained (Fig. 3). Positive cooperativity between the two moles of Ca^{++} binding to calmodulin was thus demonstrated. Crouch and Klee (3) recently reported the positive cooperativity of Ca^{++} binding to bovine brain calmodulin.

The maximum spectral change, which was attained at free Ca^{++} concentration of about 10 μM, remained at the same level by the further addition of Ca^{++} up to 0.8 mM. The four Ca^{++} binding sites of calmodulin are expected to be saturated by Ca^{++} at free Ca^{++} concentration of 1 mM. Therefore, the change of microenvironment around Tyr 138, detected by difference spectrum, is maximum when two Ca^{++} binding sites among the four are occupied by Ca^{++}.

Difference spectrum of parvalbumin induced by monovalent and divalent cations: A positive difference spectrum was observed by the addition of Ca^{++} (0.1 mM) to parvalbumin dissolved in 10 mM Tris-HCl (pH 7.8), as shown in Fig. 4. The shape is very similar to the difference spectrum obtained with N-acetylphenylalanine-ethylester in 20% dimethylsulfoxide (23). Parvalbumin contains 9-10 phenylalanine residues, but it does not contain tyrosine and tryptophan. Therefore, the difference spectrum corresponds to the perturbation arising from phenylalanine residues.

A similar shape of difference spectrum was observed by the addition of Mg^{++} in place of Ca^{++}; 0.1 M Na^+ also induced a similar level of difference spectrum as that obtained with 0.1 mM divalent cations, but 0.1 M K^+ or NH_4^+ induced only a very small difference spectrum. Paravalbumin was dissolved in 30 mM His-KOH (pH 6.8) and 0.1 M KCl, and metal ion induced difference spectrum was measured. The peak height at 264 nm (ΔA_{264}) was plotted against the negative logarithm of free metal ion concentrations, and apparent binding constants of metal ions were estimated. The value of 33, 6.3×10^5 and 2.2×10^7 M^{-1} were obtained for the binding of Na^+, Mg^{++} and Ca^{++} to parvalbumin, respectively.

Mn^{++} binding to calmodulin and troponin C: The binding of Mn^{++} to troponin C and calmodulin was investigated by ESR. ESR spectrum of Mn^{++} was measured in the presence of 90 μM troponin C dissolved in 10 mM MOPS-KOH (pH 7.1) and 0.1 M KCl at 20°C. Mn^{++} was added in the range of 100 to 500 μM. The amount of Mn^{++} bound to troponin C was plotted against free Mn^{++} concentrations. As shown in Fig. 5, the binding was readily saturated at 2 moles of bound Mn^{++} and increased further by increase in the free Mn^{++} concentration. Free Mn^{++} concentration less than 1 μM could not

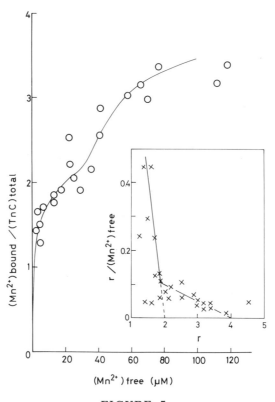

FIGURE 5

ESR measurements of Mn^{++} binding to troponin C. The amount of Mn^{++} bound to troponin C ($(Mn^{++})_{bound}$) was calculated from the following equation. Free Mn^{++} concentration ($(Mn^{++})_{free}$) was calculated using this value:

$$I_{obs} \cdot (Mn^{++})_{total} = I_{free} [(Mn^{++})_{total} - (Mn^{++})_{bound}] + I_{bound} \cdot (Mn^{++})_{bound},$$

where I is the signal intensity of Mn^{++}. I_{obs}, I_{free} and I_{bound} are the observed value, the intensity of free Mn^{++} (1.46 mm/μM) and the intensity of bound Mn^{++} (0.09 mm/μM), respectively. Troponin C (90 μM) was dissolved in 0.1 M KCl and 10 mM MOPS-KOH (pH 7.1). Mn^{++} was added in the concentration range from 100 to 500 μM. r = number of moles of Mn^{++} bound per mole of protein.

be measured by this method. The lowest value of r obtained was 1.28. The inset is a Scatchard plot calculated from the values shown in Fig. 5, which indicates the high and low affinity sites. Two different apparent binding constants (K'_{Mn}) were obtained. The high affinity sites have $K'_{Mn} = 9 \times 10^5$ M^{-1} for the first 2 moles Mn^{++} bound per mole of troponin C and the low affinity sites have $K'_{Mn} = 5.5 \times 10^4$ M^{-1} for another 2 moles Mn^{++}. This troponin C contained approximately 2 moles Ca^{++} per mole troponin C and the Ca^{++} reduced the apparent affinity for Mn^{++}. Assuming competition between Ca^{++} and Mn^{++}, the values of K_{Mn} of high and low affinity sites were obtained as 3.9×10^7 M^{-1} and 9×10^4 M^{-1}, respectively, for the next equation:

$$K_{Mn} = K'_{Mn} (1 + K_{Ca} [Ca^{++}])$$

where K_{Ca} of high and low affinity sites were those reported by Potter and Gergely (13).

The binding of Mn^{++} to calmodulin was also measured by ESR spectrum. Pig brain calmodulin (105 μM) was dissolved in 10 mM MOPS-NaOH (pH 7.1) and 0.1 M NaCl, at 20°C. Mn^{++} was added in the range from 80 to 700 μM. As shown in Fig. 6, the binding profile appeared to be the same as that of troponin C. The inset is a Scatchard plot of this result. Two different binding constants were obtained. The high affinity sites have $K'_{Mn} = 6.5 \times 10^5$ M^{-1} for the first 2 moles Mn^{++} bound per mole of calmodulin. The low affinity sites have $K'_{Mn} = 5.3 \times 10^4$ M^{-1} for another 2 moles Mn^{++}. According to the limit for experimental accuracy, the lowest value of r obtained was 1.13 and we could not judge the cooperative Mn^{++} binding to the high affinity sites from this result. Since this calmodulin contained approximately 1.5 moles Ca^{++} per mole of calmodulin, the same treatment as that applied to troponin C was performed for the estimation of K_{Mn} from K'_{Mn}. Using K_{Ca} reported by Klee (7), 9.5×10^5 M^{-1} and 6.2×10^4 M^{-1} was obtained for the K_{Mn} of high and low affinity sites, respectively.

<u>Selective spin labelling of tyrosine residues of pig brain calmodulin</u>: Spin labelling of tyrosine residues at positions 99 (Tyr 99) and 138 (Tyr 138) of calmodulin was attempted by incubation overnight with ten-fold molar excess Y-reagent over the calmodulin in 10 mM Tris-HCl (pH 8.2) and 1 mM EGTA at 0°C. This preparation was designated as Tyr 99·138-labelled calmodulin. ESR spectrum of Tyr 99·138-labelled calmodulin dissolved in 0.1 M NaCl-20 mM MOPS buffer (pH 7.1) is shown in Fig. 7. The line-shape of the spectrum indicated that the nitroxide radical was mobile with respect to the protein.

Spin labelling of Tyr 99 alone was attempted under the same experimental condition as above, except that 1 mM CaCl$_2$ was added in place of 1 mM EGTA. According to Richman (14) and

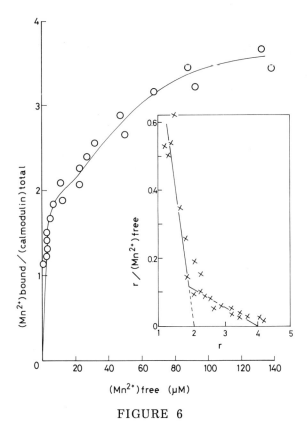

FIGURE 6

ESR measurements of Mn^{++} binding to calmodulin. The method of calculating (Mn^{++})bound and (Mn^{++})free as in legend of Fig. 5. I_{free} was 1.26 mm/μM. I_{bound} was 0.12 mm/μM. Pig brain calmodulin (105 μM) was dissolved in 0.1 M NaCl and 10 mM MOPS-NaOH (pH 7.1). Mn^{++} was added in the concentration range from 80 to 700 μM.

Richman and Klee (15), Tyr 99 is always exposed to solvent and the micro-environment is not substantially altered by the addition or removal of Ca^{++}. The amplitude of the center line of the ESR spectrum observed with Tyr 99·138-labelled calmodulin was twice as large as that of this preparation. Therefore, the tyrosine residue labelled in the latter preparation was regarded as Tyr 99 alone and this preparation was designated as Tyr 99-labelled calmodulin.

After the spin labelling of calmodulin in the presence of Ca^{++} at low ionic strength where Tyr 99 was selectively modified, the nitroxide radical was reduced by incubation with 1 mM ascorbic acid in 10 mM Tris-HCl (pH 7.6) for 1 hr at 20°C. ESR spectrum

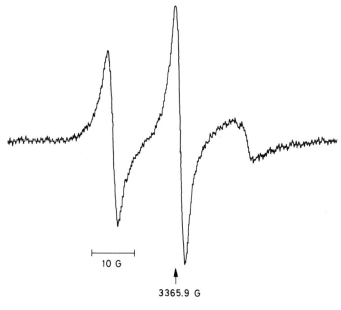

FIGURE 7

ESR spectrum of Tyr 99.138-labelled calmodulin. Solvent, 0.1 M NaCl and 20 mM MOPS-NaOH (pH 7.1).

of nitroxide radical was not observed after this treatment. After gel filtration using sephadex G-25 (fine) to remove unreacted reagent, the second spin labelling of this treated calmodulin was performed in the presence of 1 mM EGTA at 0°C, in order to modify Tyr 138.

Distance between the spin-label and Mn^{++} bound to calmodulin: Since calmodulin has high and low affinity sites against Mn^{++} and the binding constants are similar to those of Ca^{++} we assume that Mn^{++} can be used for the study of Ca^{++} binding to calmodulin.

ESR spectrum of nitroxide radical of three different spin labelled calmodulin preparations were measured; there were Tyr 99-labelled, Tyr 138-labelled, and Tyr 99·138-labelled calmodulins.

As shown in Fig. 8, quenching of nitroxide spectrum was observed by the addition of Mn^{++}. The amplitude of the center line was measured at various Mn^{++} concentrations. The relation between the amplitude and the molar ratio of added Mn^{++} to calmodulin is shown in Fig. 9a with two calmodulins; both Tyr 99 and Tyr 138 were spin labelled (Tyr 99·138-labelled calmodulin) and Tyr 99 alone was modified (Tyr 99-labelled calmodulin). The decrease

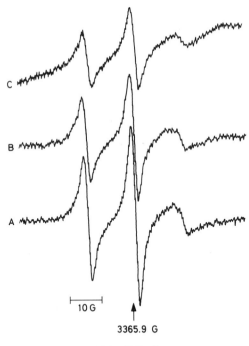

FIGURE 8

Quenching of nitroxide spectrum of calmodulin by the addition of Mn^{++}. Pig brain calmodulin (1 mg/ml) was dissolved in 0.1 M NaCl and 10 mM MOPS-NaOH (pH 7.1). Molar ratio of Mn^{++} to calmodulin: a) zero; b) 1.2; c) 4.7

in the amplitude with increase in the amount of Mn^{++} was large at the early stage and it became gentle above the molar ratio of Mn^{++} to calmodulin of 2.0 or 2.5 with Tyr 99·138-labelled calmodulin or Tyr 99-labelled calmodulin, respectively. The intersection of the former showed a decrease to 46% at 2 moles Mn^{++} and that of the latter showed 30% at 2.5 moles Mn^{++}. Since 2- or 2.5-fold molar excess Mn^{++} was added to calmodulin at the intersection, the Mn^{++} may be bound to the two high affinity sites among the four metal ion binding sites. Using the data calculated by Leigh (1), the distance between the label combined with Tyr 99 and Mn^{++} bound to the high affinity sites was estimated as 14 Å and the distance between the label combined with Tyr 138 and the Mn^{++} was longer than this value.

The result of calmodulin whose Tyr 138 alone was spin labelled (Tyr 138-labelled calmodulin) is shown in Fig. 10. The intersection was obtained at 3.0 moles Mn^{++} per mole of calmodulin and the

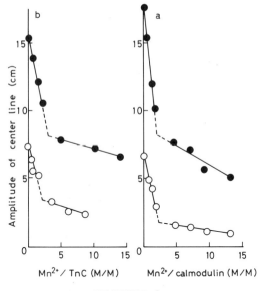

FIGURE 9

Titration of the amplitude of nitroxide ESR spectrum with Mn^{++} for pig brain calmodulin and rabbit muscle troponin C. a) Tyr 99-labelled calmodulin (○) or Tyr 99·138-labelled calmodulin (●) was dissolved in 0.1 M NaCl and 20 mM MOPS-NaOH (pH 7.1). Protein concentration = 1 mg/ml. Temperature = 23°C. b) Cys 90-labelled troponin C (1.0 mg/ml)(○) or Tyr 10·109-labelled troponin C (0.78 mg/ml) (●) was dissolved in 0.1 M KCl and 20 mM MOPS-KOH (pH 7.1). Temperature = 27°C.

decrease was up to 45%, indicating the distance between the label combined with Tyr 138 and Mn^{++} at the high affinity sites was about 17 Å.

Distance between the spin label and Mn^{++} bound to troponin C: Spin labelling of tyrosine residues at positions 10 (Tyr 10(and 109 (Tyr 109) of troponin C was attempted by the addition of tenfold molar excess Y-reagent over troponin C, which was dissolved in 10 mM Tris-HCl (pH 8.1). Incubation with Y-reagent was carried out overnight at 0°C. Spin labelling of cysteine residue at position 98 was performed with C-reagent equivalent to troponin C in molar ratio. The amplitude of the center line of the ESR spectrum of Tyr-labelled troponin C was nearly twice as large as that of the Cys-labelled one.

Relation between the amplitude and the molar ratio of added Mn^{++} to troponin C is shown in Fig. 9b. The intersection was at the

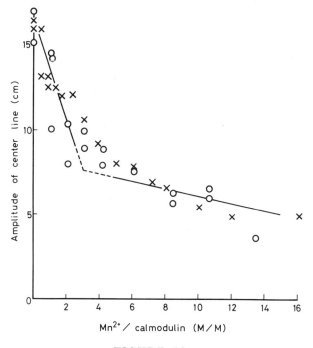

FIGURE 10

Titration of the amplitude of nitroxide ESR spectrum with Mn^{++} for Tyr 138-labelled calmodulin. Pig brain Tyr 138-labelled calmodulin (0.78 mg/ml) was dissolved in 0.1 M NaCl and 10 mM Tris-HCl (pH 7.1) (○) or in 10 mM Tris-HCl (pH 7.1) (●). Temperature = 20°C. JES-FE1X with a device of temperature control was used, and the signal intensity was higher than the results in Fig. 9.

molar ratio of Mn^{++} to troponin C of 2.0 or 3.0 with Cys 98-labelled or Tyr 10·109-labelled troponin C, respectively. At the intersection, a decrease in the amplitude to 45% was observed with Cys 98-labelled troponin C and that to 52% with Tyr 10·109-labelled troponin C. The decrease to 45% corresponds to a distance of 17 Å between the spin label of Cys 98 and Mn^{++} bound to the protein. In the case of Tyr 10·109-labelled troponin C, the distance can be estimated as follows. It was shown by Leavis et al (9) that fluorescence of Tyr 10 is not affected by the metal ion binding to the high affinity sites. Therefore, the location of Tyr 10 may be wide apart from the high affinity sites, and the label combined with Tyr 10 might cause a very slight change in the ESR spectrum after the addition of Mn^{++}. The observed decrease in amplitude to 52% may only be due to the label of Tyr 109. Therefore, the distance between the label of Tyr 109 and Mn^{++} bound to the troponin C can be estimated to be less than 10A.

DISCUSSION

A positive difference spectrum of tyrosine residue (Tyr 138) was observed with calmodulin by the addition of Na^+, K^+ or Ca^{++} at a very low ionic strength (Fig. 2). These cations can be bound to the carboxyl groups of proteins. It is then assumed that the monovalent and the divalent cations are preferentially bound to the Ca^{++} binding sites of calmodulin and the framework of metal ion binding domains are formed, resulting in the red shift of absorption spectrum of tyrosine residue.

In the presence of 0.1 M NaCl (or KCl), a negative difference spectrum of tyrosine residue (Tyr 138) of calmodulin was observed by the addition of Ca^{++} (Fig. 1). Since the framework of calmodulin molecule may already be stabilized in 0.1 M NaCl (or KCl), the Tyr 138 residue buried in the Ca^{++} binding domain would be exposed to the solvent by the exchange of Na^+ (or K^+) bound to the Ca^{++} binding domain with added Ca^{++}. According to Richman (15), N-acetylimidazole reacts with Tyr 138 of calmodulin in the presence of Ca^{++}, where Tyr 138 would be exposed to solvent. Furthermore, when a positive difference spectrum of tyrosine residue was observed, N-acetylimidazole was less reactive to Tyr 138. We could confirm these results with pig brain calmodulin when the modification reaction was carried out for 1 hr at 20°C or overnight at 0°C. When pig brain calmodulin was modified with Y-reagent by overnight incubation at 0°C, the same tendency as those obtained with acetylimidazole could be observed.

In the presence of 0.2 M KCl, a negative difference spectrum of tyrosine residue (Tyr 138) of octopus calmodulin was also observed by the addition of Ca^{++} by Seamon and Moore (18). In contrast to this result, however, Seamon (16) indicated from NMR measurements of bovine brain calmodulin, that Ca^{++} affected both Tyr 99 and Tyr 138 resonances and the substantial burying of both tyrosine residues was suggested. He also suggested a complex change in the environment around Tyr 138.

As shown in Fig. 3, the conformation change observed by the Ca^{++}-induced difference spectrum around 286 nm was saturated by two moles Ca^{++} binding, probably to the high affinity sites, and further addition of Ca^{++} gave no effect on the difference spectrum. The conformational transitions of calmodulin due to the binding of Ca^{++} at high and low affinity sites have already been shown by CD and difference spectrum (7,11,13,20,22).

Seamon (16) also concluded from NMR spectrum that the Ca^{++} induced conformational transitions occurred in two steps. The first transition accompanied the binding of two Ca^{++} and affected the resonances of both tyrosine residues (Tyr 99 and Tyr 138), phenylalanines, and ε-trimethyllysine-115. From the sensitivity of Tyr 99 resonance to the binding of the first two Ca^{++}, he suggested that domain 3 is a high affinity site. ε-Trimethyllysine occupies the position in a region of the sequence which links domains 3 and 4, and is only sensitive to the first conformational transition. He therefore concluded that both domains 3 and 4 are high affinity sites.

On the other hand, Kilhoffer et al. (5,6) used Tb^{+++}, a fluorescent probe, in place of Ca^{++} and they obtained results indicating that the high affinity sites of calmodulin are domains 1 and 2.

Our results seem to agree with the results of the NMR and CD experiments, strongly suggesting that the domains 3 and 4 are the high affinity sites. As shown in Figs. 9 and 10, the distance between Mn^{++} bound to the calmodulin and nitroxide radical of Tyr 99 or Tyr 138, was obtained as 14 Å or 17 Å, respectively, when the amount of bound Mn^{++} was about 2 moles per mole of calmodulin. Tyr 99 is involved in the Ca^{++} binding site of domain 3 and Tyr 138 is in the site of domain 4. The length of Y-reagent is about 11 Å. This evidence leads to the view that the distances measured as 14 Å and 17 Å is that between Mn^{++} bound to domain 3 and the nitroxide radical of Tyr 99 and that between Mn^{++} bound to domain 4 and the radical of Tyr 138, respectively.

Since it is already established that the domains 3 and 4 of troponin C are the high affinity sites (13), we applied the same procedure as above to troponin C in order to redetermine the high affinity sites of this protein. As shown in Fig. 9, we could obtain homologous results both with troponin C and calmodulin at about the distances between the bound Mn^{++} and the nitroxide radical of tyrosine residues which locate at the homologous position in both the proteins. The results appear to support our conclusion.

Two different conclusions were deduced from the different experiments, whether Tb^{+++} was used as the fluorescent probe (5) or Mn^{++} was used in our procedure. The different metal ions used in place of Ca^{++} might be the reason we came to different conclusions. Further investigation on the conformation of calmodulin using the procedures of ESR, NMR and difference UV absorption spectrum may be necessary.

ACKNOWLEDGEMENTS

This study was supported by grants from the Muscular Dystrophy Association of America, and from the Ministry of Education, Science and Culture of Japan.

REFERENCES

1. Azzi, A., Bragadin, M.A., Tamburro, A.M. & Santato, M. (1973): J. Biol. Chem. 248:5520-5526.
2. Barratt, M.D., Dodd, G.H. & Chapman, D. (1969): Biochim. Biophys. Acta 194:600-602.
3. Crouch, T.H. & Klee, C.B. (1980): Biochemistry 19:3692-3698.
4. Dwek, R.A. (1973): IN Nuclear Magnetic Resonances in Biochemistry. Application to Enzyme Systems. Oxford University Press, London, pp. 285-327.
5. Kilhoffer, M.-C., Demaille, J.G. & Gerard, D. (1980): FEBS Lett. 116:269-272.
6. Kilhoffer, M.-C., Gerard, D. & Demaille, J.G. (1980): FEBS Lett. 120:99-103.
7. Klee, C.B. (1977): Biochemistry 16:1017-1024.
8. Leigh, J.S. Jr. (1970): J. Chem. Phys. 52:2608-2612.
9. Leavis, P.C., Rosenfied, S.S., Gergely, J., Grabarek, Z. & Drabikowsky, W. (1978): J. Biol. Chem. 253:5425-5429.
10. Matsuda, S. & Yagi, K. (1980): J. Biochem. 88:1515-1520.
11. McCubbin, W.D., Hincke, M.T. & Kay, C.M. (1979): Canad. J. Biochem. 57:15-20.
12. Perry, S.V. & Cole, H.A. (1974): Biochem. J. 141:733-743.
13. Potter, J.D. & Gergely, J. (1975): J. Biol. Chem. 250:4628-4633.
14. Richman, P.G. (1978): Biochemistry 17:3001-3005.
15. Richman, P.G. & Klee, C.B. (1978): Biochemistry 17:928-935.
16. Seamon, K.B. (1980): Biochemistry 19:207-215.
17. Stone, T.T., Buckman, T., Nordio, P.L. & McConnell, H.M. (1965): Proc. Nat. Acad. Sci. 54:1010-1017.
18. Seamon, K.B. & Moore, B.W. (1980): J. Biol. Chem. 255:11644-11647.
19. Takagi, T., Nemoto, T., Konishi, K., Yazawa, M. & Yagi, K. (1980): Biochem. Biophys. Res. Commun. 96:377-381.
20. Walsh, M., Stevens, F.C., Oikawa, K. & Kay, C.M. (1978): Biochemistry 17:3924-3930.
21. Watterson, D.M., Sharief, F. & Vanaman, T.C. (1980): J. Biol. Chem. 255:962-975.
22. Wolff, D.J., Poirer, P.G., Brostrom, C.O. & Brostrom, M.A. (1977): J. Biol. Chem. 252:4108-4117.
23. Yanari, S. & Bovey, F.A. (1960): J. Biol. Chem. 235:2818-2826.
24. Yazawa, M., Kuwayama, H. & Yagi, K. (1978): J. Biochem. 84:1253-1258.
25. Yazawa, M. & Morita, F. (1974): J. Biochem. 76:217-219.
26. Yazawa, M., Sakuma, M. & Yagi, K. (1980): J. Biochem. 87:1313-1320.

GUANINE NUCLEOTIDE DEPENDENCE OF CALCIUM-CALMODULIN STIMULATION OF ADENYLATE CYCLASE IN RAT CEREBRAL CORTEX AND STRIATUM

K.B. Seamon and J.W. Daly

Laboratory of Bio-organic Chemistry, National Institute of Arthritis, Diabetes, Digestive and Kidney Diseases, National Institutes of Health, Bethesda, Maryland 20205, USA

INTRODUCTION

The physiological regulation of intracellular levels of cyclic AMP is complex and involves primarily the enzyme adenylate cyclase which converts the substrate ATP to cyclic AMP. Hormones regulate cyclic AMP levels by at least two apparently quite different mechanisms: 1) a direct coupling of a hormone-receptor complex with adenylate cyclase which can either increase or decrease enzyme activity (24), and 2) a mobilization or increase in intracellular calcium elicited by hormone binding to receptor followed by a calcium activation of adenylate cyclase (26).

Direct hormonal regulation of adenylate cyclase occurs both in intact cells and in broken cell preparations and requires the interaction of the hormone-receptor complex with a guanine nucleotide-binding protein (17,24). This guanine nucleotide binding subunit of adenylate cyclase binds the inhibitory nucleotide GDP very tightly in the absence of hormone resulting in an inactive form of the catalytic subunit. Hormone binding to receptor facilitates the exchange of guanine nucleotides at the binding site allowing GTP to displace GDP resulting in an active form of the catalytic subunit. Hydrolysis of the bound GTP by GTPase activity associated with the guanine nucleotide-binding subunit yields the GDP complex and returns the enzyme to the inactive form. Hormonal stimulations of adenylate cyclase thus require mediation by the guanine nucleotide-binding subunit and do not occur by direct interactions of the hormone-receptor complex with the catalytic subunit.

Calcium stimulation of adenylate cyclase occurs both in intact cells and in broken cell preparations and appears to be mediated by the action of the calcium binding protein calmodulin (cf. 35). Thus, membranes which are treated to remove endogenous calmodulin do not exhibit calcium stimulation (4,19,22). If calmodulin is added back to depleted membranes, enzyme activity can be stimulated by calcium. Calcium-dependent stimulation of mammalian adenylate cyclases has been reported for adenylate cyclase from brain (2,8), β-cells from pancreatic islets (33) and certain tumor cell lines (3).

The calcium-dependent stimulation of adenylate cyclase has now been compared in membranes from two brain regions, cerebral cortex and striatum. Although adenylate cyclase in both tissues exhibits calcium stimulation mediated by calmodulin, there are qualitative and quantitative differences. Furthermore, the characteristics of the calcium-dependent responses are quite distinct from the characteristics of the hormonal stimulations of striatal and cortical adenylate cyclase by dopamine and isoproterenol, respectively.

MATERIALS AND METHODS

$[\alpha\text{-}^{32}P]ATP$, $[\alpha\text{-}^{32}P]AppNHp$ (adenosine 5'[β,γ-imido]triphosphate), AppNH;, and GppNHp (guanosine 5'[β,γ-imido]triphosphate were purchased from International Chemical and Nuclear Corp. GDP-βS (guanosine 5'-0-(2-thiodiphosphate)) was purchased from Boehringer-Mannheim. All other reagents were of the highest quality available from standard commercial sources.

Membranes were prepared as described previously (30). Briefly rat cerebral cortex and striatum were removed from male Sprague-Dawley rats (150-175 g) and chilled briefly in ice-cold Krebs-Ringer bicarbonate glucose buffer. The tissues were homogenized in ice-cold Tris-HCl buffer, pH 7.5, 0.1 mM $CaCl_2$, in a Dounce homogenizer. The homogenate was centrifuged at 10,000 g for 10 min, the pellet was washed once in ice-cold buffer, centrifuged and resuspended in ice-cold 50 mM Tris-HCl, pH 7.5, 0.1 mM $CaCl_2$. Fresh membranes were used in all experiments.

Adenylate cyclase assays were carried out as described previously (30). Incubations were in a total volume of 250 μl containing 50 mM Tris-HCl buffer, pH 7.5 or 8.0, 1.0 mM 3-isobutyl-1-methylxanthine, 2 mM $MgCl_2$ and 0.2 mM EGTA. When ATP was used as a substrate, each assay contained 2 μCi $[\alpha\text{-}^{32}P]ATP$ and a nucleotide regenerating system of 5 units creatine-phosphokinase and 2 mM creatine phosphate. When AppNHp was used as substrate the assay contained 3 μCi $[\alpha\text{-}^{32}P]AppNHp$ and no regenerating system. Substrate concentrations were either 0.1 mM or 0.5 mM and are given in the figure legends. Assays with AppNHp were carried out at

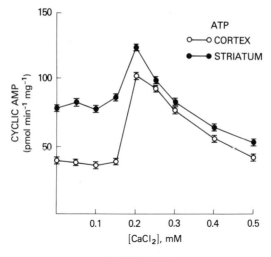

FIGURE 1

Effect of calcium on rat cortical and striatal adenylate cyclase with ATP as substrate. Adenylate cyclase activity in crude membranes was measured as described in Methods with 0.1 mM ATP as substrate at pH 8.0. Calcium chloride was added in addition to the 200 µM EGTA which was present in the assay system.

pH 8 as brain adenylate cyclase has low activity with AppNHp at pH 7.5 (20). Assays were carried out at 30°C for 10 min and were terminated by the addition of 0.5 ml trichloroacetic acid. Radioactive cyclic AMP was isolated and analyzed as described by Salomon et al. (27). Assays were carried out in triplicate and activity is expressed as pmol of cyclic AMP formed per minute per milligram of protein.

It has been demonstrated that the calcium activation of rat cortical and striatal adenylate cyclase is mediated by calmodulin (2,8,15, 19). Therefore in our experiments we refer to the calcium stimulated response as the calcium-calmodulin response, since undoubtedly the calcium stimulations are mediated by calmodulin.

RESULTS

The activity of adenylate cyclase in membranes from rat cerebral cortex and striatum is stimulated by low levels of calcium (Fig. 1), as has been previously reported by others (4,15,22, cf. 35). The amount of free calcium needed for maximal stimulation of the enzyme in both preparations is about 0.1 µM. The maximal increase in enzyme activity for each tissue is different with the cortical

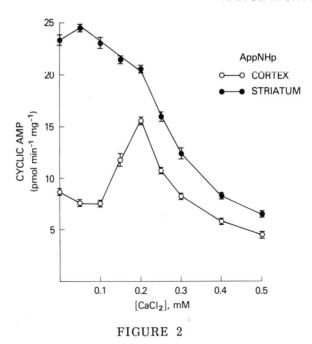

FIGURE 2

Effect of calcium on rat cortical and striatal adenylate cyclase with AppNHp as substrate. Adenylate cyclase in crude membranes was measured as described in Methods with 0.1 mM AppNHp as substrate at pH 8.0. In addition to the standard assay solution, 1 μg of purified bovine brain calmodulin was included in each assay. Calcium chloride was added in addition to the 200 μM EGTA present in the standard assay solution.

adenylate cyclase being stimulated 2.5-fold and the striatal adenylate cyclase being stimulated only about 1.5-fold over basal levels. When assays are carried out at pH 7.5 there is a greater calcium-stimulation: the rat cortical adenylate cyclase is stimulated 3.8-fold and the striatal adenylate cyclase is stimulated 1.8-fold (data not shown). The maximal levels of stimulation of adenylate cyclase in cortical or striatal membranes by calcium are unaffected by the addition of exogenous calmodulin (1 μg/assay) to the assays or by the addition of 10 μM GTP (data not shown). Thus, the extent of maximal stimulation by calcium is not limited by a lack of calmodulin or GTP in either cortical or striatal membranes. Adenylate cyclase in both cortical and striatal membranes is inhibited at higher levels of calcium (Fig. 1). Inhibition of adenylate cyclase by calcium ions has been observed in a number of different adenylate cyclase preparations (1) and has been proposed to be due to the interaction of calcium at a divalent cation site (31). Inhibition is certainly not due to a depletion of the substrate, MgATP, by complex formation with calcium: fully 85% of the ATP (0.1 mM) will be in the MgATP form even in the presence of 0.5 mM calcium.

When adenylate cyclase is assayed with ATP and nucleotide regenerating systems GTP can be formed by transphosphorylation from endogenous GDP. In order to minimize contributions of GTP the assays of brain adenylate cyclase were also conducted with the non-phosphorylating ATP analog, AppNHp. Adenylate cyclase in rat cerebral cortex membranes still exhibits calcium-stimulation when assayed with AppNHp as substrate (Fig. 2). Indeed, the fold stimulations of the adenylate cyclase of rat cortical membranes by calcium are slightly greater with AppNHp as substrate at both pH 8.0 (Fig. 2) and pH 7.5 (data not shown) than those observed with ATP as substrate. In marked contrast, adenylate cyclase of rat striatal membranes does not exhibit significant calcium stimulation when assayed with AppNHp as substrate either at pH 8.0 (Fig. 2) or at pH 7.5 (data not shown). The calcium dependent stimulation of adenylate cyclase in striatal membranes is not restored by the addition of exogenous calmodulin. The results depicted in Fig. 2 were from assays containing 1 µg purified bovine brain calmodulin. Both cortical and striatal adenylate cyclase are inhibited by higher concentrations of calcium when the enzyme is assayed with AppNHp, as was the case with ATP as substrate (see Figs. 1 and 2).

The calcium dependent stimulation of adenylate cyclase was now compared to hormone dependent stimulations under identical conditions: dopamine was used with striatal membranes and isoproterenol with cortical membranes.

Dopamine is well known to stimulate adenylate cyclase in membranes from striatum (cf. 28). The stimulation occurs through the interaction of the dopamine-receptor complex with the enzyme via a guanyl nucleotide dependent process (7). Dopamine stimulates striatal membrane adenylate cyclase 1.6-fold using the present assay conditions with ATP as substrate (Table 1). These assays were carried out at pH 8.0 which results in a smaller dopamine response than what is usually reported for experiments conducted at pH 7.5. When 100 µM GTP is included in the assays there is only a slight increase in the dopamine response (Table 1). When the assays are carried out using AppNHp as substrate in the absence of a regenerating system the dopamine response is completely lost (Table 1). This is not due to a shift of the dose-response curve since assays are with supramaximal concentrations of dopamine, 100 µM. The dopamine response of striatal adenylate cyclase when assayed with AppNHp is partially restored by the addition of exogeneous GTP (Table 1). The restoration of the dopamine response by GTP is dose-dependent with an EC_{50} of about 0.05 µM (Fig. 3). The addition of GTP to assays using ATP as substrate does not appreciably affect the dopamine response while slightly increasing the basal activity (Fig. 4). The results indicate that the levels

TABLE 1: Stimulation of striatal adenylate cyclase by dopamine.

	Activity (pmol min^{-1} mg^{-1})			
	Dopamine 0.5 mM ATP		Dopamine 0.5 mM AppNHp	
	−	+	−	+
No additions	142.0 ± 1.2	219.0 ± 2.2	27.9 ± 1.9	27.9 ± 1.8
GMP	125.8 ± 2.9	210.6 ± 4.6	24.7 ± 0.6	28.7 ± 0.4
GDP	117.0 ± 5.6	177.9 ± 2.3	28.3 ± 1.1	25.0 ± 2.4
GTP	118.9 ± 2.6	198.0 ± 4.2	52.0 ± 0.07	76.8 ± 2.1

The adenylate cyclase activity in striatal membranes was measured by the standard assay procedure with the indicated substrate at pH 8.0 in the presence or absence of 100 μM dopamine. The indicated guanine nucleotides were at a concentration of 100 μM.

TABLE 2: Stimulation of cortical adenylate cyclase by isoproterenol.

	Activity (pmol min^{-1} mg^{-1})			
	Isoproterenol 0.5 mM ATP		Isoproterenol 0.5 mM AppNHp	
	−	+	−	+
No additions	82.4 ± 1.4	117.1 ± 0.6	5.4 ± 0.6	6.3 ± 0.2
GMP	94.3 ± 2.1	138.0 ± 4.0	4.5 ± 0.4	4.7 ± 0.2
GDP	86.4 ± 3.6	129.8 ± 0.8	6.4 ± 0.7	6.9 ± 0.1
GTP	116.1 ± 2.5	175.6 ± 2.6	18.1 ± 0.8	23.6 ± 1.2

The adenylate cyclase activity in cortical membrane was measured by the standard assay procedure with the indicated substrate at pH 8.0 in the presence or absence of 100 μM isoproterenol. The indicated guanine nucleotides were at a concentration of 100 μM.

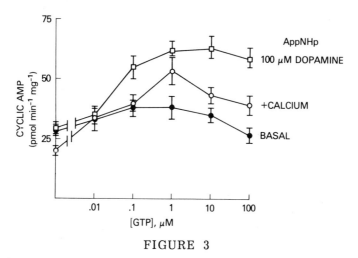

FIGURE 3

Effect of GTP on dopamine and calcium stimulation of striatal adenylate cyclase with AppNHp as substrate. Adenylate cyclase in rat striatal membranes was measured as described in Methods with 0.5 mM AppNHp as substrate at pH 8.0. The assays performed in the presence of calcium had 200 µM $CaCl_2$ added in addition to the 200 µM EGTA present in the assay solution. Dopamine (100 µM) was added to assays which contained the standard assay solution with no added calcium.

of endogenous GTP that are present when ATP is used as substrate for the enzyme are sufficient to fully support the coupling of a dopamine-receptor complex to the adenylate cyclase system. When AppNHp is used as substrate there is not enough endogenous GTP in the system to support the coupling and therefore no dopamine response is observed.

The stimulation of adenylate cyclase in cerebral cortical membranes by isoproterenol has been previously reported, but the response has not been consistent (10,16,21). The stimulation occurs via a β-adrenergic receptor. With the present assay system, isoproterenol consistently stimulates adenylate cyclase 1.4-fold in rat cortical membranes using 0.5 mM ATP as substrate (Table 2). The isoproterenol response is not evident when the enzyme is assayed with AppNHp as substrate (Table 2). However, the response can be partially restored by the addition of exogenous GTP when assayed with AppNHp (Table 2). It would thus appear that in cortical membrane preparations as in striatal preparations there is not enough endogenous GTP to support a hormonal response when AppNHp is used as substrate.

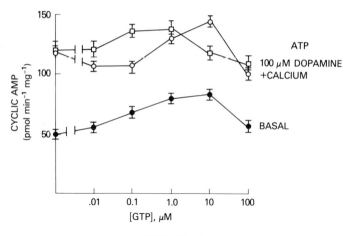

FIGURE 4

Effect of GTP on dopamine and calcium stimulation of striatal adenylate cyclase with ATP as substrate. Adenylate cyclase assays were carried out as described in Methods with 0.5 mM ATP as substrate at pH 8.0. Assays performed with calcium contained 200 µM $CaCl_2$ in addition to the 200 µM EGTA present in the assay solution. Dopamine (100 µM) was added to assays which contained the standard assay solution with no added calcium.

Since the calcium stimulation of rat cortical adenylate cyclase occurs with AppNHp as substrate, it would appear that GTP is not required for this response. Exogenous GTP, however, can lead to a slight augmentation of the calcium stimulation of rat cerebral cortical adenylate cyclase (Fig. 5). The loss of the calcium response of rat striatal membrane adenylate cyclase with AppNHp as substrate suggested that there is a requirement for GTP in this system. However, GTP is unable to restore the response to calcium while partially restoring responses to the hormone dopamine (Figs 3 and 5). Other guanine nucleotides (GMP, GDP, GppNHp) are also unable to restore the calcium dependent response of striatal adenylate cyclase (data not shown). Attempts to restore calcium responses with guanine nucleotides were carried out over the full calcium range even in the presence of added calmodulin (data not shown).

The effect of inhibitory guanine nucleotide GDP-βS on responses of adenylate cyclase to calcium, hormones, GppNHp and fluoride was now investigated in rat cortical and striatal membranes. GppNHp is a non-hydrolyzable analog of GTP which binds at the guanine nucleotide binding subunit resulting in a persistently activated state of the enzyme (18). Inhibition of hormonal and GppNHp stimulations of adenylate cyclase by GDP-βS occurs by virtue of its competitive binding at the guanine nucleotide binding

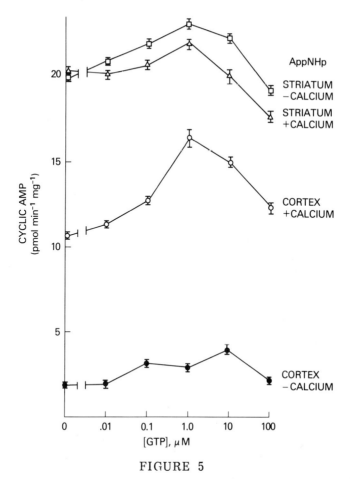

FIGURE 5

Effect of GTP on calcium stimulation of rat cortical and striatal adenylate cyclase with AppNHp as substrate. Adenylate cyclase was measured as described in Methods with 0.1 mM AppNHp at pH 7.5. Assays performed in the presence of calcium contained 200 µM calcium in addition to the standard assay solution which contained 200 µM EGTA.

site (6,12). Fluoride stimulation of adenylate cyclase requires the presence of the guanine nucleotide binding subunit, but can apparently stimulate the enzyme with GTP, GDP, or no guanine nucleotide at the binding site (12). The mechanism of fluoride stimulation of adenylate cyclase is poorly understood. However, it has been demonstrated with turkey erythrocytes that the fluoride response is inhibited when GDP-βS occupies the guanine nucleotide binding site (11,12). In the present study, stimulation of rat cortical adenylate cyclase by GppNHp and fluoride is inhibited markedly

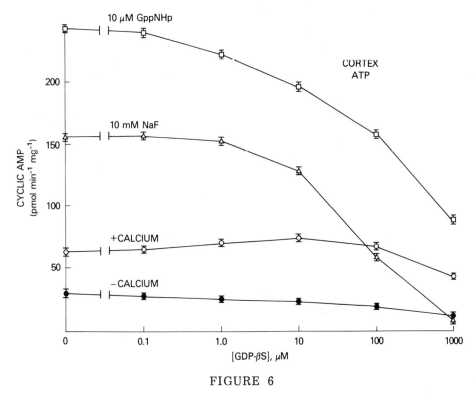

FIGURE 6

Effect of GDP-βS on GppNHp, NaF, and calcium stimulations of rat cortical adenylate cyclase with ATP as substrate. Rat cortical adenylate cyclase was assayed as described in Methods with 0.1 mM ATP as substrate at pH 7.5. Assays contained either 10 μM GppNHp, 10 mM NaF, or 200 μM $CaCl_2$ in addition to the standard assay solution.

by GDP-βS with half maximal inhibitions occurring at about 100 μM GDP-βS (Fig. 6). In contrast the calcium dependent stimulation of cortical adenylate cyclase is not inhibited except at very high concentrations of GDP-βS (Fig. 6). At such high concentrations basal activity of the enzyme is also inhibited. The lack of inhibition of the calcium response by GDP-βS pertains even when the enzyme is assayed with AppNHp in order to ensure that there is minimal contributions from endogenous levels of GTP (data not shown). The results indicate that binding of GDP-βS at the guanine nucleotide binding sites does not significantly affect the calcium dependent response. GDP-βS has similar effects on adenylate cyclase activity in rat striatal membranes (Fig. 7). The stimulations of the striatal enzyme by GppNHp and dopamine are inhibited by GDP-βS with half maximal inhibitions occurring at about 50 μM. In contrast to

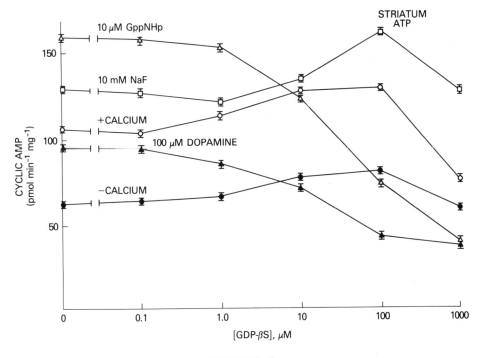

FIGURE 7

Effect of GDP-βS on GppNHp, NaF, calcium and dopamine stimulations of rat striatal adenylate cyclase with ATP as substrate. Rat striatal adenylate cyclase was assayed as described in Methods with 0.1 mM ATP as substrate at pH 7.5. Assays contained either 10 μM GppNHp, 10 mM NaF, 200 μM CaCl$_2$ or 100 μM dopamine in addition to the standard assay solution.

the results with the cortical membranes, the fluoride response in the striatal membranes does not appear to be inhibited by GDP-βS and is actually slightly augmented. Calcium responses in striatal membranes as in cortical membranes are inhibited only by very high concentrations of GDP-βS. It appears likely that the inhibition of calcium dependent activations which occur at high concentrations (1 mM) of GDP-βS is due to a non-specific effect of the nucleotide. Inhibition of enzyme activity by very high concentrations of GDP-βS also pertains for adenylate cyclase stimulated by the diterpene, forskolin (data not shown). The diterpene forskolin has been demonstrated to stimulate adenylate cyclase via interactions with the catalytic subunit (30).

DISCUSSION

The regulation of adenylate cyclase in vivo and the mechanisms of activations in vitro are complex and in many cases not well understood. Studies with cells which have a relatively homogeneous population of adenylate cyclase and associated receptor systems have provided valuable insights (cf. 25). The study of adenylate cyclase and cyclic AMP in brain tissues is, of course, complicated by the heterogeneous nature of the tissue which contains many different cell types. Each of these may contain adenylate cyclase which responds to different stimulatory as well as inhibitory agents dependent on the presence or absence of the relevant receptors in each class of cells. In addition, there may be inherent differences among cell types in the catalytic subunit of the enzyme or in the presence of other protein components which mediate the various modulatory effects. However, in spite of these difficulties, brain represents a tissue in which adenylate cyclase undoubtedly has important roles (cf. 9) and the mechanisms involved in the regulation of the enzyme in brain must be delineated. Certain of these mechanisms appear unique to brain. For example, a calcium dependent α-adrenergic receptor-mediated activation of cyclic AMP generating systems has been demonstrated only in brain tissue (29). Furthermore, brain adenylate cyclase is one of the few cyclases which has been shown to be activated by calcium-calmodulin (35). Activation of brain adenylate cyclase by calcium-calmodulin occurs at very low concentrations (1 µM or less) of calcium and is not well understood. The mechanism is apparently unrelated to the inhibition which occurs at high concentrations of calcium. Such enzyme inhibition by high concentrations of calcium occurs with adenylate cyclases from most sources including brain and presumably involves interaction at a divalent cation site (31). The relationship of the calcium-calmodulin stimulation of adenylate cyclase to subunits of the enzyme is under active investigation. Recent data with an enzyme preparation from brain suggested a dependence of the calcium response on the guanine nucleotide binding subunit (34). Other studies on stability and stimulation of adenylate cyclase in brain membranes also suggest an inter-relationship of calcium-calmodulin and guanyl nucleotide mechanisms (5). Thus GppNHp stablizes the calmodulin dependent activity of rat cortical membranes and GTP or GppNHp affect the curve for activation of rat cortical adenylate cyclase by calmodulin (5). Data with rat brain preparations have been interpreted in terms of a phosphorylation of a membrane site leading to both a reduction in calcium-calmodulin interaction with adenylate cyclase and to alterations in hormone activation of the enzyme (13-15). In order to provide further insights into these inter-relationships, the guanine nucleotide dependence of calcium-calmodulin stimulation of brain adenylate cyclase has been investigated in detail with membranes from rat cerebral cortex and striatum.

Adenylate Cyclase in Rat Cerebral Cortical Membranes

Calcium-calmodulin stimulation of adenylate cyclase in rat cortical membranes is clearly distinct from hormonal stimulation of adenylate cyclase, with regard to dependence on guanine nucleotides. Hormonal activation of adenylate cyclase in most tissues, if not all, depends on the interaction of the hormone-receptor complex with the guanine nucleotide regulatory subunit. The interaction of the hormone-receptor complex with the guanine nucleotide regulatory subunit leads to the facilitated exchange of GTP for GDP bound at the nucleotide binding site and consequent activation of the catalytic subunit. Thus hormonal stimulation of adenylate cyclase should not occur in the absence of GTP or in the presence of inhibitory guanine nucleotides such as GDP or GDP-βS. This was demonstrated for the β-adrenergic receptor stimulation of rat cerebral cortex adenylate cyclase by isoproterenol. The response to isoproterenol is not observed when assayed with AppNHp in the absence of exogenous GTP but does occur with ATP and a regenerating system (Table 2). With AppNHp as substrate for adenylate cyclase, transphorylation will not occur and endogenous guanine nucleotides would be present as GDP. Exogenous GTP does partially restore an isoproterenol response with AppNHp as substrate. Such results are consistent with a requirement for GTP for the β-receptor stimulation of adenylate cyclase in rat cerebral cortical membranes and indicate that β-receptor stimulation is mediated by a guanine nucleotide regulatory subunit which with AppNHp as substrate is occupied by inhibitory GDP. In contrast, the stimulation of adenylate cyclase in rat cerebral cortex by calcium-calmodulin occurs even with AppNHp as substrate. Thus, the calcium-calmodulin stimulation of adenylate cyclase in rat cerebral cortex does not appear to require the presence of GTP. It is tempting but premature to conclude that calmodulin stimulation of adenylate cyclase in rat cortical membranes might not require the interaction of the guanine nucleotide regulatory subunit. However, it remains possible that the stimulation of the cortical enzyme by calcium-calmodulin does require the guanine nucleotide binding subunit but that activation occurs with either GTP or GDP at the binding site. This would be analogous to stimulation by fluoride which does require the guanine nucleotide binding subunit but occurs with either GTP or GDP and even no nucleotide at the binding site (12). However, the potent inhibitory GDP analog GDP-βS inhibits even fluoride stimulation of adenylate cyclase (11,12). The effects of GDP-βS were, therefore, investigated in rat cerebral cortical membranes. GDP-βS inhibited the stimulation of adenylate cyclase by GppNHp and fluoride but had little effect on the calcium-calmodulin stimulation (Fig. 6). It must be assumed that exchangeable guanine nucleotide binding sites were being occupied by GDP-βS, since the GppNHp stimulation was completely inhibited under the same assay conditions. Therefore either calcium-calmodulin stimulation is not affected by GDP-βS at the guanine nucleotide binding site or calcium-calmodulin is acting at a population of adenylate cyclase which is not associated with the GppNHp-sensitive population of the adenylate cyclase. The latter seems unlikely in view of reported effects of guanine nucleotides on the stability of calmodulin-sensitive cyclases in rat cerebral cortical membranes (5).

The present results indicate that calcium-calmodulin stimulation of adenylate cyclase in a crude membrane fraction from rat cerebral cortex neither requires GTP nor is inhibited by GDP-βS. Calmodulin cannot, therefore, be stimulating adenylate cyclase sole by affecting the exchange of guanine nucleotides at the guanine nucleotide regulatory subunit. Recently a diterpene, forskolin, has been described which stimulates adenylate cyclase in all mammalian tissues through an interaction with the catalytic subunit (30). This stimulation by forskolin is similar to that of calmodulin in that neither is inhibited by GDP-βS and both stimulations can be observed in rat cerebral cortical membranes using AppNHp as substrate. It is tempting to speculate that calcium-calmodulin is acting in a similar manner as forskolin, that is, by an interaction with the catalytic subunit. However, one major difference between the forskolin and calcium-calmodulin responses is that forskolin stimulates adenylate cyclase in all mammalian tissues yet tested (30), while calcium-calmodulin stimulation of adenylate cyclase has been demonstrated for only a few tissues. The lack of general stimulation of adenylate cyclase by calmodulin suggests that either catalytic subunits are not identical with respect to calcium-calmodulin stimulation or that other protein factors may be involved in mediating the effects of calmodulin on adenylate cyclase. In this regard Westcott et al. (34) have isolated a solubilized fraction of rat cerebral cortical adenylate cyclase which is insensitive to stimulation by calmodulin, GppNHp, or NaF. These stimulations could be reconstituted when a cytosolic preparation is added (32). The active fraction appears to be protein and might indeed by the guanine nucleotide binding subunit. However, the guanine nucleotide binding subunit cannot be the only membrane component essential for stimulation of adenylate cyclase by calcium-calmodulin since all hormone responsive adenylate cyclase contain the guanine nucleotide binding subunit and not all respond to calcium-calmodulin. The present results with cerebral cortical membranes do indicate that if the guanine nucleotide binding subunit is essential for calcium-calmodulin stimulation as suggested by the work of Storm and coworkers (32,34), then the nature of the guanine nucleotide ligand is inconsequential.

Adenylate Cyclase in Rat Striatal Membranes

The stimulation of striatal adenylate cyclase by calcium-calmodulin proved to be both quantitatively and qualitatively different from the stimulation of the cortical enzyme. Thus, the striatal adenylate cyclase is stimulated by calmodulin to a much smaller extent than was the cortical enzyme. Such a quantitative difference might mean that striatal adenylate cyclase is inherently less sensitive to calmodulin activation than cortical adenylate cyclase. Alternatively the striatum may contain different forms of adenylate cyclase with only a small percentage being responsive to calmodulin. Further studies with more purified preparations of the enzyme will be required to resolve this question.

Qualitatively, the most important difference between the calcium-calmodulin stimulation of striatal adenylate cyclase and stimulation of the cortical enzyme was in the apparent dependence on GTP. Hormonal responses are clearly dependent on GTP in both striatal and cortical membrane. Indeed, it has been previously shown that dopamine requires GTP for effective stimulation of striatal adenylate cyclase (7). Undoubtedly, this dependence reflects the requisite involvement of the guanine nucleotide regulatory subunit. In the present study, dopamine stimulation of striatal adenylate cyclase was not manifest when assays are carried out with AppNHp as substrate (Table 1). Surprisingly, in view of the results with cortical membranes, calcium-calmodulin activation of striatal adenylate cyclase like the hormonal stimulation was not manifest when assays were carried out with AppNHp as substrate. With cortical adenylate cyclase the calcium-calmodulin stimulation was fully manifest with AppNHp as substrate. GTP was unable to restore the calcium-calmodulin response of striatal adenylate cyclase with AppNHp as substrate. The dopamine response of striatal adenylate cyclase was, however, fully restored by GTP with AppNHp as substrate. It is therefore difficult to rationalize the loss of calcium-calmodulin stimulation of striatal adenylate cyclase with the absence of GTP, unless the dopamine responsive enzyme utilizes exogenous GTP much more effectively than the calcium-calmodulin responsive enzyme. This appears unlikely, but must be considered as a possibility. The dopamine and calcium responses of adenylate cyclases in striatal membranes are additive (23) and hence involve either different mechanisms or different populations of adenylate cyclase. The present results support the idea that calmodulin stimulation of striatal adenylate cyclase, like activation of cortical adenylate cyclase, is unaffected by the nature of the guanine nucleotide present at the binding site. The loss of calcium responsiveness of striatal adenylate cyclase with AppNHp as substrate must then be concluded to be unrelated to lack of GTP generation under such conditions. Since AppNHp is a non-phosphorylating analog of ATP, it is tempting to speculate that phosphorylation by exogenous ATP of membrane proteins is requisite to calcium-calmodulin stimulation of striatal adenylate cyclase. A corollary to this hypothesis is that phosphorylation by exogenous ATP of membrane proteins is not requisite to calcium-calmodulin stimulation of cortical adenylate cyclase. The role of membrane phosphorylation to calcium-calmodulin stimulation of adenylate cyclase deserves further investigation.

SUMMARY

Hormonal and calcium-calmodulin stimulation of adenylate cyclase have been compared in rat cortical and striatal membranes. Isoproterenol was used in the cortical system as the stimulatory "hormone" and dopamine was used in the striatal system. The resultant hormonal stimulations were shown to require a GTP interaction with the guanine nucleotide regulatory subunit of adenylate cyclase. In contrast, calcium-calmodulin stimulation was apparently independent of the nature of the guanine nucleotide present at the guanine nucleotide regulatory subunit. Further studies will be required to define whether or not a guanine nucleotide binding subunit is requisite for calcium-calmodulin responses and whether or not phosphorylation and dephosphorylation of membrane components plays a significant role in calcium-calmodulin respnses. The present data with a crude membrane system cannot provide such answers which can only be obtained with isolated components of the calcium-calmodulin sensitive adenylate cyclase system.

ACKNOWLEDGEMENTS

The authors wish to thank Michelle Kibble and William Padgett for their excellent technical advice.

REFERENCES

1. Birnbaumer, L. (1973): Biochim. Biophys. Acta 300:129-158.
2. Brostrom, C.O., Huang, Y.C., Breckenridge, B.McL. & Wolff, D.J. (1975): Proc. Nat. Acad. Sci. 72:64-68.
3. Brostrom, M.A., Brostrom, C.O., Breckenridge, B.McL. & Wolff, D.J. (1976): J. Biol. Chem. 251:4744-4750.
4. Brostrom, C.O., Brostrom, M.A. & Wolff, D.J. (1977): J. Biol. Chem. 252:5677-5685.
5. Brostrom, M.A., Brostrom, C.O. & Wolff, D.J. (1978): Arch. Biochem. Biophys. 191:341-350.
6. Cassel, D., Eckstein, F., Lowe, M. & Selinger, Z. (1979): J. Biol. Chem. 254:9835-9838.
7. Chen, T.C., Cote, T.E. & Kebabian, J.W. (1980): Brain Res. 181:138-149.
8. Cheung, W.Y., Bradham, L.S., Lynch, T.J., Lin, Y.M. & Tallant, E.A. (1975): Biochem. Biophys. Res. Commun. 66: 1055-1062.
9. Daly, J.W. (1977): <u>Cyclic Nucleotides in the Nervous System</u>, Plenum Press, New York.
10. Dolphin, A., Adrian, J., Hamon, M. & Bockaert, J. (1979): Molec. Pharmacol. 15:1-15.
11. Downs, R.W. Jr., Spiegel, A.M., Singer, M., Reen, S. & Aurbach, G.D. (1980): J. Biol. Chem. 255:949-954.
12. Eckstein, F., Cassel, D., Levkovitz, H., Lowe, M. & Selinger, Z. (1979): J. Biol. Chem. 254:9829-9834.

13. Gnegy, M.E., Uzunov, P. & Costa, E. (1976): Proc. Nat. Acad. Sci. 73:3887-3890.
14. Gnegy, M.E., Nathanson, J.A. & Uzunov, P. (1977): Biochem. Biophys. Acta 497:75-85.
15. Gnegy, M. & Treisman, G. (1981): Molec. Pharmacol. 19:256-263.
16. Hegstrard, L.R., Minneman, K.P. & Molinoff, P.B. (1979): J. Pharmacol. Exp. Ther. 210:215-221.
17. Limbird, L.E. (1981): Biochem. J. 195:1-13.
18. Londos, C., Salomon, Y., Lin, M.C., Harwood, J.B., Schramm, M., Wolff, J. & Rodbell, M. (1974): Proc. Nat. Acad. Sci. 71:3087-3090.
19. Lynch, T.J., Tallant, E.A. & Cheung, W.Y. (1977): Arch. Biochem. Biophys. 182:124-133.
20. Maguire, M. & Gilman, A.G. (1974): Biochim. Biophys. Acta 358:154-163.
21. Partington, C.R. & Daly, J.W. (1979): Molec. Pharmacol. 15:484-491.
22. Piascik, M.T., Wisler, P.L., Johnson, C.L. & Potter, J.D. (1980): J. Biol. Chem. 255:4176-4182.
23. Premont, J., Perez, M. & Bockaert, J. (1977): Molec. Pharmacol. 13:662-670.
24. Rodbell, M. (1980): Nature 284:17-22.
25. Ross, E.M. & Gilman, A.G. (1980): Ann. Rev. Biochem. 49:533-564.
26. Rasmussen, H. & Goodman, D.B.P. (1977): Physiol. Rev. 57:421-509.
27. Salomon, Y., Londos, C. & Rodbell, M. (1974): Anal. Biochem. 58:541-548.
28. Schmidt, M.J. (1979): IN Neuropharmacology of Cyclic Nucleotides. (ed) G. Palmer, Urban & Schwarzenberg, Baltimore, pp. 1-52.
29. Schwabe, U. & Daly, J.W. (1977): J. Pharmacol. Exp. Ther. 202:134-143.
30. Seamon, K.B., Padgett, W. & Daly, J.W. (1981): Proc. Nat. Acad. Sci. 78:3363-3367.
31. Steer, M.L. & Levitski, A. (1975): J. Biol. Chem. 250:2080-2084.
32. Toscano, W.A. Jr., Westcott, K.R., LaPorte, D.C. & Storm, D.R. (1979): Proc. Nat. Acad. Sci. 76:5582-5586.
33. Valverde, I., Vandermeers, A., Anjaneyolu, R. & Malaisse, W.J. (1979): Science 206:225-227.
34. Westcott, K.R., LaPorte, D.C. & Storm, D.R. (1979): Proc. Nat. Acad. Sci. 76:204-208.
35. Wolff, D.J. & Brostrom, C.O. (1979): IN Advances in Cyclic Nucleotide Research, Vol. 11. (eds) P. Greengard & G.A. Robison, Raven Press, New York, pp. 27-88.

PATTERNS OF INTRACELLULAR Ca^{++} DISTRIBUTION AND MOBILIZATION IN SMOOTH MUSCLE

G.B. Weiss

Department of Pharmacology, University of Texas
Health Science Center at Dallas
Dallas, Texas 75235, USA

Mobilization of Ca^{++} into the intracellular compartment occurs by a number of different mechanisms involving multiple binding sites and entry channels. The various Ca^{++} sites and channels can be dissociated with combinations of several techniques. These include: a) use of lanthanum and low temperature to remove superficial Ca^{++} and retain cellular Ca^{++}, b) use of Scatchard-coordinate plots to identify incubation conditions appropriate for characterizing predominantly high or low affinity Ca^{++} components, c) measurement of ^{45}Ca efflux to ascertain rates for Ca^{++} loss or exchange, d) use of high concentrations of Sr^{++} to remove high but not low affinity Ca^{++}, and e) use of smooth muscle microsomal preparations to distinguish between specific binding and less specific accumulation of Ca^{++}. The effects of stimulatory agents (e.g. high K^+ and norepinephrine) on specific Ca^{++} fractions can be qualitatively and quantitatively measured. The Ca^{++}-related loci of action of a number of different inhibitory agents (e.g. aminoglycoside antibiotics, D-600 and other voltage sensitive channel inhibitors, nitroprusside, hydralazine, and antimycin A) can be ascertained. Thus a number of physiologically significant membrane Ca^{++} sites and entry channels in vascular smooth muscle can be delineated with currently employed specific agents and procedures. Use of other equally specific Ca^{++} and calmodulin inhibitors to help identify intracellular Ca^{++} interactions and utilization mechanisms should be of similar value in development of models for the equally complex events linking mobilization of membrane or extracellular Ca^{++} to biochemical mechanisms directly related to generation and regulation of contractile tension.

INTRODUCTION

Calcium ion has numerous important roles in biological processes. Among the most obvious of these is the long recognized need for Ca^{++} as an initiator and regulator of muscle contractility. Traditionally, these functions of Ca^{++} have been divided into two general areas. First, there is the requirement for the presence of Ca^{++} in excitation-contraction coupling. Secondly, there is the essential role of Ca^{++} in the function of intracellular contractile systems in all types of muscle. Though these two events are experimentally dissociable, it has always been evident that they are integral parts of a sequence of physiologically significant steps beginning with a stimulus external to the muscle cell and terminating with a measurable alteration in contractile tone or tension. Changes in contractile responsiveness have been assumed to be related in some manner to variations in the level of intracellular Ca^{++}.

The points within this sequence that have been of particular interest are those concerned with how Ca^{++} is mobilized from extracellular, membrane and cellular sites or stores in response to various stimuli. As expressed in an earlier review (31), the working hypothesis is that the qualitative and quantitative nature of the muscle response to a given agent is a direct consequence of the amounts of Ca^{++} mobilized into the intracellular compartment. Characterization of the differing ways Ca^{++} is mobilized by a number of membrane-active agents has been a productive approach to increased understanding of excitation-contraction coupling. Furthermore, it provides a framework within which the increasingly complex intracellular Ca^{++} regulatory mechanisms can be integrated.

In this discussion, I will attempt a brief review of what is known about the distribution and mobilization of Ca^{++} in smooth muscle, particularly vascular smooth muscle. First, I will outline some of the techniques and approaches employed to measure Ca^{++} binding and movements in isolated vascular smooth muscle. Subsequently, I will attempt to correlate anatomical Ca^{++} stores with patterns of Ca^{++} mobilization. The results are consistent with a cellular model including multiple Ca^{++} sites and channels (35,36). Whether the intracellular mobilization of Ca^{++} is a convergent event in which the intracellular concentration of free Ca^{++} is altered or whether multiple intracellular pathways exist for utilization of Ca^{++} in induction and maintenance of contractility is a subsequent consideration. It is at this point that an increased understanding of the roles of calmodulin in vascular smooth muscle is of critical importance.

Techniques for Dissociating Different Calcium Sites and Fractions

The presence of Ca^{++} in isolated vascular smooth muscle tissue is so widespread that dissociation of various Ca^{++}-related actions has been difficult. However, several recent technical approaches have separated different Ca^{++} fractions and facilitated the identification of specific functions for some of the component Ca^{++} sites or stores. The experimental techniques of particular interest are outlined in Table 1. Each of these approaches permits isolation of a Ca^{++} component that can be specifically altered by one or more pharmacological agents. In this manner the function of Ca^{++} from a particular store can be associated or correlated with an induced response.

Measurement of Ca^{++} uptake into the intracellular compartment with radioisotopic Ca^{++} (^{45}Ca) provides an excellent example of increasing technical precision. The major portion of ^{45}Ca taken up by isolated smooth muscle preparations was at extracellular sites and this prevented detection of quantitatively smaller changes in cellular ^{45}Ca uptake (see 30,31). Use of the trivalent lanthanides (especially La^{+++}) as Ca^{++} antagonists was an improvement because these ions replaced Ca^{++} at superficial (La^{+++}-accessible) sites, blocked Ca^{++} uptake, and decreased Ca^{++} efflux (7,27,28,37). Increased La^{+++} concentrations (5) and decreased temperatures (3) improved measurement of Ca^{++} uptake by enhancement of decreased Ca^{++} efflux. We have found that washout in an isosmotic (80.8 mM) La^{+++}-substituted solution at 0.5°C provides a very satisfactory method for retention of the major portion of cellular Ca^{++} during removal of La^{+++}-sensitive (superficial) Ca^{++} and inhibition of Ca^{++} uptake and efflux (13). An alternative approach employs buffered EGTA solutions and yields similar values for Ca^{++} retention (25,26).

Efflux of ^{45}Ca from vascular smooth muscle in the absence of significant exchange with extracellular Ca^{++} yields a two-component ^{45}Ca washout (desaturation) curve (31). Conventionally, the more rapid washout component was believed to include mainly extracellular and relatively superficial Ca^{++}, whereas the slower washout component was thought to be primarily cellular Ca^{++} (31). More recent evidence obtained from washouts of muscles previously incubated in solutions favoring accumulation of high or low affinity Ca^{++} indicates that the more rapid washout component includes most of the low affinity Ca^{++} and the slower component consists of the major portion of high affinity Ca^{++} (33,34).

Use of Scatchard-coordinate plots of equilibrated Ca^{++} uptake at different extracellular Ca^{++} concentrations results in identification of two distinct populations of accumulated Ca^{++} (32,33). By examining alterations in binding and mobilization of ^{45}Ca under conditions favoring high affinity (low extracellular Ca^{++}) or low

TABLE 1: Some techniques useful for dissociation of cellular Ca^{++} fractions.

Measurement	Modification	Components
^{45}Ca uptake	Removal of extracellular Ca^{++}	a. Total Ca^{++} uptake b. Cellular Ca^{++} content
^{45}Ca efflux	Varied ^{45}Ca specific activity	a. Fast washout component b. Slow washout component
Net Ca^{++} uptake	Scatchard-coordinate plot of equilibrated Ca^{++}	a. High affinity Ca^{++} uptake b. Low affinity Ca^{++} uptake
Inhibition of Ca^{++} uptake	Use of high concentrations of Sr^{++}	a. Displacement of high affinity Ca^{++} b. Little effect on low affinity Ca^{++}
Ca^{++} uptake in microsomal preparation	Presence and absence of ATP	a. Binding of Ca^{++} b. Total uptake of Ca^{++}

affinity (high extracellular Ca^{++}) accumulation of Ca^{++}, qualitatively different effects of high K$^+$ and norepinephrine can be observed (13). Further separation of high and low affinity Ca^{++} can be attained by displacement of Ca^{++} from high affinity sites with Sr^{++} (15). Apparently high (e.g. 5.0 mM) concentrations of Sr^{++} displace the much lower concentrations of high affinity Ca^{++} present but have little effect on higher concentrations of low affinity Ca^{++} (15,33).

Microsomal fractions obtained from canine aortic smooth muscle demonstrate an ATP-dependent accumulation of Ca^{++} that is increased by oxalate and decreased by omission of ATP (18,20). In the absence of ATP, uptake of Ca^{++} can be attributed to binding whereas, in the presence of ATP, accumulation of Ca^{++} results from energy-dependent processes as well as binding. Thus, whether a particular agent directly alters binding of Ca^{++} or either directly or indirectly affects another site or reaction important for Ca^{++}-related smooth muscle function can be ascertained with this technique.

Combinations of these various approaches can yield very specific mechanism-related information. For example, ^{45}Ca retention

(Ca^{++} uptake) after washout in La^{+++}-substituted solution at 0.5°C can be measured for muscles previously incubated with ^{45}Ca in solutions favoring either high or low affinity Ca^{++} accumulation (13). Under these conditions high K^+-induced depolarization approximately doubles the low affinity Ca^{++} uptake but has only a slight effect on high affinity Ca^{++} accumulation. Conversely norepinephrine does not alter resting uptake of low affinity Ca^{++} but decreases high affinity Ca^{++} retention by approximately one-third. Thus, clear qualitative differences between the effects of high K^+ and norepinephrine on Ca^{++} retention can be obtained by analysis of cellular Ca^{++} content under conditions favoring either high or low affinity Ca^{++}. The increased Ca^{++} uptake elicited with high K^+ indicates a mobilization of superficial (and/or extracellular) low affinity Ca^{++}, whereas the decrease in high affinity Ca^{++} retention obtained with norepinephrine may result from removal from the cell of a fraction of that cellular high affinity Ca^{++} released by norepinephrine.

Combination of ^{45}Ca efflux measurement with a preceding uptake of ^{45}Ca from solutions favoring either high (0.03 mM Ca^{++} solution) or low (1.5 or 5.0 mM Ca^{++} solution) also yields qualitatively specific information on Ca^{++} mobilization patterns. Depending upon which Ca^{++} fraction is present, norepinephrine increases ^{45}Ca efflux, decreases $^{45}Ca^{++}$ efflux or has no effect on ^{45}Ca efflux (15). This reflects varied effects of norepinephrine on different cellular and membrane Ca^{++} components (15,16,36).

Use of high (5.0 mM) concentrations of Sr^{++} in combination with each of the first three measurements outlined in Table 1 results in qualitatively distinct dissociations of Ca^{++} fractions in rabbit aortic smooth muscle. The uptake of ^{45}Ca from solutions containing low concentrations of Ca^{++} is specifically inhibited by Sr^{++} (15,33), the corresponding slow component of ^{45}Ca efflux is specifically reduced by Sr^{++} (33), and the high affinity component of the Scatchard-coordinate plot is specifically blocked by Sr^{++} (15,33). Conversely, Sr^{++} has no effect on: a) low affinity Ca^{++} uptake (15), b) the size of the fast component of ^{45}Ca efflux (33), c) the low affinity component of the Scatchard-coordinate plot (15), or d) the magnitude of the K^+-induced increase in low affinity Ca^{++} uptake (15).

Measurement of effects of K^+ and norepinephrine on Ca^{++} uptake in canine aortic smooth microsomal preparations does not yield qualitatively significant dissociations of Ca^{++} components when combined with the other techniques outlined in Table 1, even though similar two-compartment ^{45}Ca efflux curves and Scatchard-coordinate plots are obtained. Possibly this is because the voltage-sensitive low affinity Ca^{++} uptake channels do not function in this preparation, and the coupling between the norepinephrine receptor and the high affinity Ca^{++} sites is disrupted.

Effects of Potassium and Norepinephrine on Calcium Mobilization

The effects of high K^+ and norepinephrine clearly differ in mechanism in vascular smooth muscle (for review see 31). Because rabbit aortic smooth muscle binds relatively large quantities of high affinity Ca^{++} per unit tissue weight, this smooth muscle is particularly useful for experimentally dissociating this relatively small Ca^{++} uptake fraction (80-120 nmol/g) from the much larger low affinity Ca^{++} uptake fraction (300-600 nmol/g) present under appropriate conditions. In the initial study reporting differential effects of K^+ and of norepinephrine on rabbit aortic responsiveness, the major portion of divergent effects observed were attributed to a striking lack of sensitivity of a portion of the norepinephrine-induced response to Ca^{++} depletion. This difference, attributed to an ability of norepinephrine to release a relatively inaccessible (only slowly depleted) Ca^{++} store has since been more precisely defined by: a) electrophysiological measurements indicating that norepinephrine (in concentrations lower than 10^{-6} M) does not significantly reduce membrane polarization (22) even though approximately 80% of the maximum tension response is attained at the 10^{-6} M concentration, b) an uptake of low affinity Ca^{++} accompanying norepinephrine-induced changes in cell polarization at norepinephrine concentrations above 10^{-6} M (9,13), and c) an additional uptake of ^{45}Ca (and increased muscle tension) elicited with 10^{-6} M norepinephrine in aortic strips previously depolarized with high K^+.

The observed differences between effects of norepinephrine and those of high K^+ can be attributed to divergent actions on Ca^{++} sites and membrane Ca^{++} channels. Two principal sites and two entry channels of primary importance are delineated in Table 2. Low affinity Ca^{++} is mobilized from superficial (cell surface?) sites by high K^+ or depolarizing concentrations of norepinephrine; this Ca^{++} enters the cell through a voltage-sensitive Ca^{++} channel and initiates a series of reactions leading to a sustained tonic contraction. In a K^+-depolarized muscle, additional quantities of low affinity Ca^{++} enter the depolarized cell through a different (receptor linked) channel and lead to an augmented K^+-induced tension response. High affinity Ca^{++} is released from cellular sites by norepinephrine and the resulting tension response is rapid but not well maintained. This norepinephrine-induced Ca^{++} release and tension response can be prevented by the α-adrenergic receptor blocker, phentolamine (13). Thus the high affinity Ca^{++} store must be located at cellular sites coupled in some unspecified manner to the external cell surface receptor. Inner cell membrane surface sites would be more readily coupled to a cell surface reaction than would binding sites for Ca^{++} in or on other intracellular organelles. The displacement of high affinity Ca^{++} by Sr^{++} supports the idea that true binding sites exist.

TABLE 2: Functionally important Ca^{++} sites and entry channels identified in vascular smooth muscle.

Ca^{++} Site or Channel	Method of Identification	Type of Function
Superficial low affinity Ca^{++} site	Removed by La^{+++} Mobilized by high K^+ and by depolarizing norepinephrine concentrations	Stabilizing of membrane Ca^{++} entry with depolarization and tonic contraction
Cellular high affinity Ca^{++} site	La^{+++} resistant Released by norepinephrine	Phasic contraction without large polarization changes No Ca^{++} entry; not readily blocked by Ca^{++} depletion
Voltage sensitive Ca^{++} channel	Increased low affinity Ca^{++} uptake with high K^+	Increased tension with depolarization
Receptor linked Ca^{++} channel	Increased Ca^{++} uptake with norepinephrine in previously depolarized muscles	Additional tension response in depolarized muscles

The effects of high K^+ and norepinephrine on tension responses are relatively rapid and reach maximum levels within a few minutes at most. The rate at which low affinity Ca^{++} retention was increased by high K^+ and (additionally) by norepinephrine and the rate at which high affinity Ca^{++} retention was decreased by norepinephrine was estimated by adding the stimulatory agents to the muscle bathing solution for short time periods subsequent to equilibration of the muscles with ^{45}Ca (14). The increases or decreases in tension observed after 1-2 min incubation with norepinephrine or high K^+ were not significantly different from those seen in equilibrated muscles (14). Thus, mobilization of either high or low affinity Ca^{++} is sufficiently rapid to correlate with the time course for development of tension.

Some Effects of Specific Inhibitors on Calcium Mobilization

Use of agents that inhibit responsiveness to high K^+ or norepinephrine by acting at one or more Ca^{++} site or store have contributed to the delineation of mechanisms of Ca^{++} mobilization in vascular smooth muscle. Effects on Ca^{++} related parameters can be direct or indirect. For example, agents blocking the norepinephrine receptor site or K^+-Na^+ exchange reactions act indirectly to alter Ca^{++} movement and/or accumulation. Agents affecting Ca^{++} more specifically would include: a) those inhibiting binding or release of Ca^{++} at cell surface or membrane sites, b) those blocking one or more Ca^{++} uptake pathways, and c) those altering intracellular release or accumulation of Ca^{++}. The points at which specific agents are believed to act and some of the evidence for this are summarized in Table 3.

The aminoglycoside antibiotics (e.g. neomycin, gentamicin) appear to act at the most superficial Ca^{++} sites. These therapeutic agents do not penetrate the cell membrane (6), and strongly inhibit the initial rate of induction of K^+-induced contractile responses (measured as the half-time of rise to maximum contractile response) in vascular muscles (8). The inhibition of contraction by aminoglycosides was readily reversed by removal of agent or elevation of extracellular Ca^{++} (1). Furthermore, neomycin has been shown to decrease the equilibrated uptake but not the rate of uptake of ^{45}Ca (8) and both neomycin and gentamicin decrease the ATP-independent binding of Ca^{++} in canine aortic microsomes (19).

The voltage-sensitive Ca^{++} channel inhibitors include a rapidly growing number of agents believed to act by blocking the increased uptake of Ca^{++} accompanying depolarization. In vascular smooth muscle, methoxyverapamil (D-600) inhibited Ca^{++}-induced contractile responses in K^+-depolarized preparations at concentrations several magnitudes lower than those needed to inhibit norepinephrine-induced contractile responses (9). D-600 did not alter low affinity Ca^{++} retention in control rabbit aortic smooth muscle but did block the K^+ induced increase in Ca^{++} retention (14). A similar result was recently obtained with nitrendipine (38), a dihydropyridine voltage sensitive Ca^{++} channel blocker. Examination of the actions of a number of other voltage sensitive Ca^{++} channel blockers indicates that all agents of this type block K^+-induced tension responses and associated increases in ^{45}Ca uptake at concentrations several magnitudes lower than similar effects elicited with norepinephrine.

The effects of nitroprusside on agonist-induced tension and ^{45}Ca flux responses are readily distinguished from those of the voltage sensitive Ca^{++} channel antagonists. Nitroprusside decreased the contractile response to high concentrations of norepinephrine but had little effect on the K^+-induced contractile

TABLE 3: Agents altering Ca^{++} mobilization by affecting specific Ca^{++} sites or channels.

Agent	Ca^{++} Site or Channel	Experimental Basis or Effect
Neomycin, gentamicin	Displacement of Ca^{++} from superficial low affinity sites	Decreased microsomal Ca^{++} binding (19), K^+-induced rate of tension increase (8), and total Ca^{++} uptake (8)
D-600, nitrendipine	Blockade of voltage-sensitive Ca^{++} channel	Inhibition of K^+ induced tension (10,38) and Ca^{++} uptake (14,38)
Nitroprusside	Blockade of receptor-linked Ca^{++} channel	Inhibition of norepinephrine induced Ca^{++} uptake in K^+ depolarized muscles (14)
Hydralazine, Sr^{++}	Decreased high affinity Ca^{++} retention	Inhibition of norepinephrine induced tension and depletion of that high affinity Ca^{++} released by norepinephrine (15,38)
Antimycin A, oligomycin	Inhibits mitochondrial uptake of low affinity Ca^{++}	Decreased K^+-induced low affinity ^{45}Ca retention but not tension response (17)

responses in canine renal arteries (9) and rabbit aortic smooth muscle (Karaki and Weiss, unpublished data). Nitroprusside also inhibits the additional increment of approximately 180 nmol Ca^{++}/g of muscle obtained with 10^{-6} M norepinephrine in the presence of high K^+, but does not inhibit the increase in Ca^{++} retention elicited with high K^+. Thus, nitroprusside blocks a different specific component of ^{45}Ca uptake (a receptor-linked channel?) than does D-600.

Recent experiments comparing effects of hydralazine with those of the voltage sensitive Ca^{++} channel blocker, nitrendipine (38) indicate that the cellular basis of action of hydralazine differs from that of nitrendipine even though both agents exert strong peripheral vasodilator effects. In rabbit aortic smooth muscle, tension responses to 10^{-6} M norepinephrine were completely blocked by 2 mM hydralazine, whereas those to high K^+ were only 30%

inhibited. This was the reverse of differential inhibitory responses obtained with nitrendipine and indicated that hydralazine might have a primary effect on norepinephrine-sensitive Ca^{++} sites. Measurement of effects of hydralazine on Ca^{++} retention support this view. Hydralazine only partially inhibited the K^+ induced increase in low affinity Ca^{++} retention but did decrease the high affinity Ca^{++} retention (including that portion released by norepinephrine). Hydralazine may either directly release that high affinity Ca^{++} store important for the norepinephrine-induced response or gradually deplete these sites by preventing uptake of Ca^{++}. The result in either case is to produce vasodilation by a Ca^{++}-related mechanism which differs from that of agents such as nitrendipine.

The action of antimycin A (and oligomycin) on cellular Ca^{++} distribution appears to be a clearly delineated one (17) but does not resolve some related questions about regulation of intracellular Ca^{++} levels. Selected concentrations of antimycin A and oligomycin had only slight inhibitory effects on the K^+-induced contractile response even though both agents abolished the increase in Ca^{++} retention elicited with high K^+. Oligomycin did not block the norepinephrine-induced release of high affinity Ca^{++} but oxygen consumption in the rabbit aortic smooth muscle was decreased by antimycin A or oligomycin. In mitochondria, Ca^{++} uptake is coupled with respiratory function and both of these processes are inhibited by oligomycin or antimycin A (see 2,21). Vascular smooth muscle mitochondrial Ca^{++} uptake (24) has low affinity characteristics (as compared to the microsomal membrane fraction) and mitochondria do not appear to control free intracellular Ca^{++} levels during the physiological contraction-relaxation cycle, although they may serve as Ca^{++} storage sites during periods of excessive Ca^{++} loading. During depolarization of rabbit aortic smooth muscle cells, a large quantity of Ca^{++} enters the cell. The rapid increase in free intracellular (cytoplasmic) Ca^{++} initiates a mitochondrial accumulation of Ca^{++}. This mitochondrial Ca^{++} uptake is not seen following norepinephrine-induced Ca^{++} release. This might be attributed to a smaller increase in intracellular Ca^{++} levels even though the norepinephrine-induced contraction is not smaller than the K^+-induced one. Possibly, increased mitochondrial Ca^{++} uptake may not be proportionately related to increased intracellular Ca^{++} levels. Alternatively, the Ca^{++} released by norepinephrine may not be in a form available for mitochondrial incorporation.

The K^+-induced entry of Ca^{++} into the intracellular compartment is believed to occur because increased tension responses are elicited. If mitochondrial Ca^{++} accumulation is blocked by oligomycin or antimycin A without a simultaneous block of K^+-induced contractile responses, a question arises concerning why an increased Ca^{++} retention is no longer detected. The suggestion was offered (17) that, in the absence of mitochondrial accumulation of Ca^{++}, this

Ca^{++} fraction is readily removed from the cells. However, the precise mechanism by which this could occur is not clear.

Once Ca^{++} enters the intracellular compartment additional mechanisms become important for initiation and maintenance of vascular tone. The sequence of events involved appear to be quite complex and include more than a single pathway leading, ultimately, to activation of myosin light chain kinase. The roles of calmodulin as a regulatory calcium-dependent activator protein have been the focal point of considerable recent investigation.

Use of specific antagonists of calmodulin related processes may help elucidate intracellular Ca^{++} pathways in a manner similar to the way membrane binding and channel antagonists have helped dissociate Ca^{++} binding sites and uptake mechanisms. Recently, some of the actions of N-(6-aminohexyl)-5-chloro-1-naphthalene sulfonate (W-7), a calmodulin antagonist, on contractile responses in rabbit aortic smooth muscle have been reported (11,12). W-7 unspecifically relaxes aortic smooth muscle contractions (11) and this relaxation is only partially reversed by Ca^{++} (12). The proposal that W-7 inhibits smooth muscle contractility by selectively interacting with calmodulin (12) needs to be extended by examining effects of W-7 on Ca^{++} binding and mobilization to exclude the possibility that other non-selective effects are also present. However, W-7 does demonstrate that additional types of antagonists of Ca^{++}-related actions do exist and could provide mechanism-related information about specific physiological roles of Ca^{++} as a regulator of contractility in vascular smooth muscle.

Among other agents proposed as calmodulin inhibitors are the local anesthetics (29). These agents have long been implicated as antagonists of Ca^{++}-related actions in nerve and muscle (23). They inhibit Ca^{++} exchangeability in smooth muscle (4) and have stabilizing actions similar to Ca^{++} in a variety of excitable tissues (23). Based upon the association of similar concentrations of local anesthetics with pharmacological effects and calmodulin inhibition, it has been proposed (29) that many of the Ca^{++}-related actions of local anesthetics result from antagonism of effects of calmodulin. This possibility may help explain the basis of actions of agents acting at points in the excitation-contraction sequence subsequent to the cell membrane.

Patterns of Ca^{++} mobilization in the vascular smooth muscle cell membrane include multiple membrane sites and channels. The manner in which subsequent intracellular Ca^{++} release, binding, and distribution contributes to regulation of vascular contractility appears to be equally complex. Whether intracellular mobilization of Ca^{++} also involves multiple pathways and binding sites for Ca^{++} remains to be defined. It is entirely possible that the intracellular Ca^{++} interactions to be delineated will be as complex and heterogeneous as are those in and on the cell membrane. One of the most important current challenges in this field of research is to increase understanding of the manner in which Ca^{++} entering the intracellular compartment is coupled to the biochemical reactions directly responsible for contraction and relaxation of vascular smooth muscle systems.

ACKNOWLEDGEMENTS

The experimental work from this laboratory was supported by USPHS grants HL-14775 and HL-27145, and by a grant from Miles Laboratories. Major portions of the more recent work were performed in collaboration with Drs. H. Karaki, P. Kutsky and K. Hatano. D-600 was generously supplied by A.G. Knoll and nitrendipine by Miles Laboratories.

REFERENCES

1. Adams, H.R. & Goodman, F.R. (1975): J. Pharmacol. Exp. Ther. 193:393-402.
2. Carafoli, E. (1974): Biochem. Soc. Symp. 39:89-109.
3. Deth, R.C. (1978): Amer. J. Physiol. 243:C139-C145.
4. Feinstein, M.B. (1966): J. Pharmacol. Exp. Ther. 152:516-524.
5. Godfraind, T. (1976): J. Physiol. (Lond.) 260:21-35.
6. Goodman, F.R. (1978): Pharmacology 16:17-25.
7. Goodman, F.R. & Weiss, G.B. (1971): J. Pharmacol. Exp. Ther. 177:415-425.
8. Goodman, F.R., Weiss, G.B. & Adams, H.R. (1974): J. Pharmacol. Exp. Ther. 188:472-480.
9. Hester, R.K. & Weiss, G.B. (1981): J. Pharmacol. Exp. Ther. 216:239-246.
10. Hester, R.K., Weiss, G.B. & Fry, W.J. (1979): J. Pharmacol. Exp. Ther. 208:155-160.
11. Hidaka, H., Asano, M., Iwadare, S., Matsumoto, I., Totsuka, T. & Aoki, W. (1978): J. Pharmacol. Exp. Ther. 207:8-15.
12. Kanamori, M., Naka, M., Asano, M. & Hidaka, H. (1981): J. Pharmacol. Exp. Ther. 217:494-499.
13. Karaki, H. & Weiss, G.B. (1979): J. Pharmacol. Exp. Ther. 211:86-92.
14. Karaki, H. & Weiss, G.B. (1980): J. Pharmacol. Exp. Ther. 213:450-455.
15. Karaki, H. & Weiss, G.B. (1980): J. Pharmacol. Exp. Ther. 215:363-368.
16. Karaki, H. & Weiss, G.B. (1980): Gen. Pharmacol. 11:483-489.
17. Karaki, H. & Weiss, G.B. (1981): Blood Vessels 18:28-35.
18. Kutsky, P. & Goodman, F.R. (1978): Arch. Int. Pharmacodyn. Ther. 231:4-20.
19. Kutsky, P. & Weiss, G.B. (1981): Blood Vessels (in press).

20. Kutsky, P., Weiss, G.B. & Karaki, H. (1980): Gen. Pharmacol. 11:475-481.
21. Lehninger, A.L., Reynafarje, B., Vercesi, A. & Tew, W.P. (1978): Ann. NY Acad. Sci. 307:160-176.
22. Mekata, F. (1976): J. Physiol. (Lond.) 258:269-278.
23. Shanes, A.M. (1958): Pharmacol. Rev. 10:165-273.
24. Somlyo, A.P. (1978): IN Mechanism of Vasodilation. (eds) P.M. Vanhoutte & I. Leusen, Karger, Basel, pp. 21-29.
25. Van Breeman, C. Aaronson, P., Loutzenhiser, R. & Meisheri, K. (1980): Chest 78:157-165.
26. Van Breemen, C. & Casteels, R. (1974): Pflug. Arch. 348: 239-245.
27. Van Breemen, C., Farinas, B.R., Gerba, P. & McNaughton, E.D. (1972): Circ. Res. 30:44-54.
28. Van Breemen, C. & McNaughton, E.D. (1970): Biochem. Biophys. Res. Commun. 39:567-574 (1970).
29. Volpi, M., Sha'afi, R.I., Epstein, P.M., Andrenyak, D.M. & Feinstein, M.B. (1981): Proc. Nat. Acad. Sci. 78:795-799.
30. Weiss, G.B. (1974): Ann. Rev. Pharmacol. 14:343-354.
31. Weiss, G.B. (1977): IN Advances in General and Cellular Pharmacology, Vol.2. (eds) T. Narahashi, & C.P. Bianchi, Plenum Press, New York, pp. 71-154.
32. Weiss, G.B. (1977): IN Excitation-Contraction Coupling in Smooth Muscle. (eds) R. Casteels, T. Godfraind & J.C. Ruegg, Elsevier/North-Holland Biomedical Press, Amsterdam, pp. 253-260.
33. Weiss, G.B. (1978): IN Calcium in Drug Action. (ed) G.B. Weiss, Plenum Press, New York, pp. 57-74.
34. Weiss, G.B. (1979): Proc. West. Pharmacol. Soc. 22:333-338.
35. Weiss, G.B. (1981): IN Vasodilatation (eds) P.M. Vanhoutte & I. Leusen, Raven Press, New York, pp. 307-310.
36. Weiss, G.B. (1981): IN New Perspectives on Calcium Antagonists. (ed) G.B. Weiss, Amer. Physiol. Soc., Bethesda, pp. 83-94.
37. Weiss, G.B. & Goodman, F.R. (1969): J. Pharmacol. Exp. Ther. 169:46-55.
38. Weiss, G.B., Hatano, K. & Stull, J.T. (1981): Blood Vessels 18:230.

ROLE OF CA^{++}-CALMODULIN IN METABOLIC REGULATION IN PLANTS

M.J. Cormier, H.W. Jarrett and H. Charbonneau

Department of Biochemistry, University of Georgia
Athens, Georgia 30602, USA

INTRODUCTION

The role of Ca^{++} in physiological processes was first recognized near the turn of the century; since this time the list of Ca^{++} dependent processes in animal systems has grown rapidly. More recently, a similar list of Ca^{++} dependent processes is being recognized in plant systems as investigators become more aware of the importance of free Ca^{++} in the regulation of physiological processes in plants. The list of Ca^{++}-linked processes shown in Table 1 illustrates this point.

A better understanding of how Ca^{++} regulates so many diverse physiological processes in animal cells was not possible until the discovery of Ca^{++}-binding proteins and the realization that these proteins serve as intracellular receptors for free Ca^{++} which functions as a second messenger (for reviews see, 29,39). In smooth and non-muscle cells one of the more important targets for free Ca^{++} appears to be the ubiquitous protein, calmodulin (for reviews see, 7,28,33,44).

The recent isolation of calmodulin from fungi and several species of higher plants was the first indication that high affinity Ca^{++} binding proteins exists in these organisms (for a review see, 9). These observations suggest that Ca^{++} dependent metabolic regulation in plant cells is mediated by Ca^{++} binding proteins such as calmodulin by analogy to the known in vitro Ca^{++}-calmodulin dependent functions in animal cells.

The information provided below is intended to summarize the properties and known in vitro functions of plant calmodulin. The

TABLE 1: Examples of Ca^{++}-linked responses in plants.

Type of Response	Reference
1. Phytochrome-mediated regulation of transmembrane Ca^{++} flux	Dreyer & Weisenseel (12) Hale & Roux (14) Roux et al. (42)
2. Stimulation of chloroplast rotation in Mougeotia	Haupt (15)
3. Phytochrome-dependent depolarization of Nitella cells	Weisenseel & Ruppert (47)
4. Inhibition of cytoplasmic streaming in Nitella	Hayama et al. (16)
5. Regulation of directional growth	Jaffe et al. (20) Herth (18) Reiss & Herth (40,41)
6. Stimulation of cyclic nucleotide phosphodiesterase in Phaseolus	Brown et al. (3)
7. Inhibition of chloroplastic fructose bisphosphatase	Charles & Halliwell (6)
8. Stimulation of glutamic dehydrogenase	Kindt et al. (27)
9. Activation of NAD kinase	Anderson & Cormier (1)

implications of Ca^{++}-calmodulin dependent metabolic regulation in higher plants is also discussed.

Properties of Plant and Fungal Calmodulin

As reported earlier, many of the properties of peanut sead, pea seedling and fungal (mushroom) calmodulin are strikingly similar to their mammalian counterpart (2,5,8). The similarities include molecular weight, Stokes radii, absorption maxima, Ca^{++} dependent enhancement of tyrosine fluorescence, Ca^{++} dependent interaction with troponin I, equal potency in activating cyclic nucleotide phosphodiesterase, and Ca^{++} dependent inhibition of calmodulin function by the phenothiazine drugs. Peanut seed and rat testis calmodulin also exhibit identical behavior in a radioimmunoassay (4). These kinds of similarities have also been noted for spinach leaf calmodulin including its identical immunoreactivity in comparison to bovine brain calmodulin (45).

Table 2 compares the amino acid compositions of peanut seed, mung bean, spinach leaf, fungal and bovine brain calmodulin.

TABLE 2: Amino acid composition of calmodulin.

Amino Acid	Peanut Seed	Mung Bean	Spinach Leaf	Fungal (Basidiomycete)	Bovine Brain
	Mol (nearest integer)/16,700 g				Residues/molecule
Lysine	8	9	8	10	7
Histidine	1	1	1	1	1
Trimethyllysine	1	1	1	1	1
Arginine	4	5	5	4	6
Aspartic acid	27	25	25	25	23
Threonine	9	9	8	10	12
Serine	5	5	6	10	4
Glutamic acid	27	26	32	27	27
Proline	2	2	2	2	2
Glycine	11	11	11	12	11
Alanine	10	10	11	10	11
Half-cystine	1	1*	N.D.	0	0
Valine	6	8	5	5	7
Methionine	7	8	8	9	9
Isoleucine	5	6	5	7	8
Leucine	11	11	11	11	9
Tyrosine	1	1	1	2	2
Phenylalanine	8	9	8	8	8
Tryptophan	0	0	0	0	0

Peanut seed data from Anderson et al. (2).
Fungal data from Cormier et al (8).
Spinach leaf data from Watterson et al. (45).
Bovine brain data from Watterson et al. (46).
N.D. = not determined.
* As carboxymethylcysteine (in collaboration with Dr. T.C.Vanaman).

Considering the diversity of species from which these calmodulins have been isolated, the similarities in their amino acid compositions are noteworthy. These include the presence of trimethyllysine, a relative preponderance of negatively charged amino acids, two proline residues and the lack of tryptophan. The overall similarities in calmodulin isolated from these diverse species are striking and point out the high degree of structural conservation in these proteins. Since data from two independent laboratories agree on these basic similarities (2,8,45), it appears that the reported properties of a protein isolated from barley and a basidiomycete fungus (13) are those of calmodulin-like proteins rather than calmodulin itself. However, the possibility that these proteins represent a second example of a high affinity Ca^{++} binding protein in plants is intriguing.

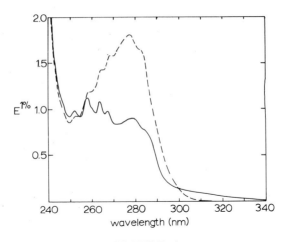

FIGURE 1

Absorption spectra of peanut seed (-) and porcine brain (--) calmodulin.

In contrast to the similarities discussed above, there are a number of differences in calmodulin isolated from plant and mammalian sources. For example, both peanut seed and spinach leaf calmodulin appear to have a lower molecular weight than bovine brain calmodulin as determined from $NaDodSO_4$ gel electrophoresis in the presence of EGTA (2,45). A lower molecular weight was also obtained for peanut seed calmodulin from sedimentation equilibrium measurements. As previously suggested, calmodulin may display anamolous behavior when subjected to these techniques (2).

Of possible importance are the differences noted in the amino acid compositions of plant and mammalian calmodulin. One interesting difference is the presence of at least one residue of cysteic acid in performic acid oxidized peanut seed calmodulin (2). Furthermore, following treatment with iodoacetate at least one residue of carboxymethylcysteine has been detected in calmodulin isolated from two leguminous plants, peanut and mung bean seeds (in collaboration with Dr. T.C. Vanaman). As the spectra in Fig. 1 show, the extinction coefficient at 276 nm of peanut seed calmodulin is approximately half that of bovine brain calmodulin. These spectral differences are explained by a phe/tyr ratio of 8 for the plant and 4 for the mammalian calmodulin (2). The presence of a single tyrosine residue has also been reported for spinach leaf calmodulin (45). However, this is not a unique property of plant calmodulin since calmodulin isolated from several species of marine invertebrates have also been found to contain a single tyrosine residue (21,34,43,50).

TABLE 3: Requirements for plant NAD kinase activity.

Additions or deletions	NAD Kinase Activity (pmol min^{-1}mg^{-1})
None	153.6 ± 8.1
−ATP	2.2 ± 1.5
−NAD$^+$	2.9 ± 2.2
−Ca^{++}	1.6 ± 1.8
−Calmodulin	1.2 ± 2.7
+Trifluoperazine (50 μM, final concentration)	7.2 ± 6.7

The complete assay medium contained 50 mM Tris, 50 mM KCl, 10 mM MgCl$_2$, 2 mM NAD, 3 mM ATP, 0.2 mM CaCl$_2$, 20 μg/ml bovine brain calmodulin, and 0.2 ml of calmodulin-sepharose purified NAD kinase. Additions and deletions from this standard assay mixture are as noted. When CaCl$_2$ was deleted it was replaced by 0.2 mM EGTA.

In addition to the differences in amino acid compositions and electrophoretic mobilities of plant and mammalian calmodulin, differences have also been noted in their peptide maps. Peptides derived from CNBr digestion of peanut seed and bovine brain calmodulin show mobility differences during gel electrophoresis (22). Similar differences have also been noted in comparing the tryptic peptide maps of spinach leaf and rabbit brain calmodulin (45). As pointed out below, some of these differences are likely to be related to the differences in some of their biological activities. Comparisons of the structures of plant and mammalian calmodulin may therefore provide valuable information on the structure-function relationships of calmodulin.

Ca^{++}-Calmodulin-Dependent Metabolic Regulation in Higher Plants

The complete dependency of plant NAD kinase activity on calmodulin (22) and μM concentrations of Ca^{++} is now well documented (1,2). As Table 3 shows, when either Ca^{++} or calmodulin are deleted from the reaction mixture, no significant level of NAD kinase activity was detected. In this respect the enzyme is analogous to myosin light chain kinase from animal sources which also exhibit complete dependence on Ca^{++}-calmodulin for its activity (for a review see, 28). Also shown in Table 3 is the inhibition of Ca^{++}-calmodulin dependent activation of NAD kinase by trifluoperazine. The inhibition of NAD kinase activation by trifluoperazine was observed with the addition of either bovine brain or peanut seed calmodulin (Table 3; ref. 2). This is not surprising since both bovine brain calmodulin (31) and plant calmodulin contain Ca^{++} dependent phenothiazine binding sites (5,45) =and the phenothiazines are known to inhibit the Ca^{++}-calmodulin-dependent stimulation of a number of mammalian enzyme activities (for a review see, 32).

Figure 2 illustrates the Ca^{++} dependent binding of plant NAD kinase to calmodulin-sepharose. This observation is consistent with a model for the activation of NAD kinase in which a Ca^{++}-calmodulin-NAD kinase complex is formed:

$$Ca^{++} + \text{calmodulin} \rightleftharpoons Ca^{++}\text{-calmodulin} \rightleftharpoons Ca^{++}\text{-calmodulin}^*$$

$$Ca^{++}\text{-calmodulin}^* + \underset{(\text{inactive})}{\text{NAD kinase}} \rightleftharpoons \underset{(\text{active})}{Ca^{++}\text{-calmodulin}^*\text{-NAD kinase}}$$

where Ca^{++}-calmodulin* indicates the active, Ca^{++} bound conformation of calmodulin. Stoichiometry is not given in the scheme above since information on that aspect of the activation is not yet available. The scheme is analogous to the calmodulin-dependent activation of several enzymes of mammalian origin (for reviews see, 7, 28, 33).

NAD kinase, extracted from pea seedlings, has recently been purified at least 4100-fold to apparent homogeneity (23). On NaDodSO$_4$ polyacrylamide gels the protein migrates as a single band of molecular weight approximately 57,000 (23). The molecular weight of the native protein is not yet known.

Throughout the purification of plant NAD kinase we noted that most of the activity was Ca^{++} dependent. In the crude extract, the Ca^{++} independent activity represented not more than 10% of the total while in all subsequent steps the Ca^{++} independent activity represented less than 3% of the total NAD kinase activity (23). This suggests that the Ca^{++} independent form of plant NAD kinase is either a minor species or nonexistent. In the presence of Ca^{++} peanut seed and porcine brain calmodulin were found to be equally active in the stimulation of porcine brain cyclic nucleotide phosphodiesterase (2,8). However, plant and mammalian calmodulins are not equally active in the activation of plant NAD kinase. This is seen in Fig. 3, which shows that peanut seed calmodulin is about seven-fold more active than porcine brain calmodulin in the activation of plant NAD kinase. These same solutions were used to compare these two calmodulins for their abilities to stimulate the red blood cell membrane (Ca^{++} + Mg^{++})-ATPase. Porcine brain calmodulin was at least as active as peanut seed calmodulin in the stimulation of the (Ca^{++} + Mg^{++})-ATPase (data not shown). Thus some of the differences noted above in the amino acid compositions and peptide maps of plant and mammalian calmodulin are functionally significant and may account for the difference in the abilities of these calmodulins to activate NAD kinase. These results are similar to the observed specificity of Tetrahymena guanylate cyclase for Tetrahymena calmodulin (25).

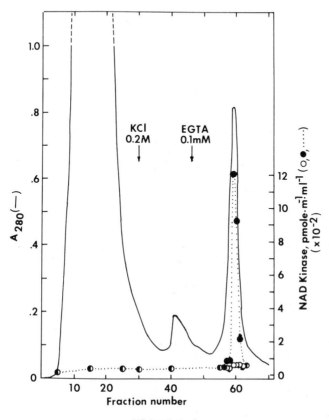

FIGURE 2

Purification of NAD kinase from peas by calmodulin-sepharose affinity chromatography. A 50 ml volume of partially purified NAD kinase was applied to a calmodulin-sepharose colume equilibrated with buffer A (50 mM Tris, 50 mM KCl, 10 mM $MgCl_2$, 0.1 mM $CaCl_2$ at pH 8.0). The arrows indicate the points at which buffer A was modified to include 0.2 M KCl (buffer B) and finally where the $CaCl_2$ in buffer B was replaced by 0.1 mM EGTA. Fractions were 6.9 ml of which duplicate 0.1 ml aliquots were used to assay NAD kinase activity for the fractions indicated. The open circles (-○-) represent the activity without and the filled circles (-●-) the activity with 20 μg of calmodulin added per ml of the assay mixture.

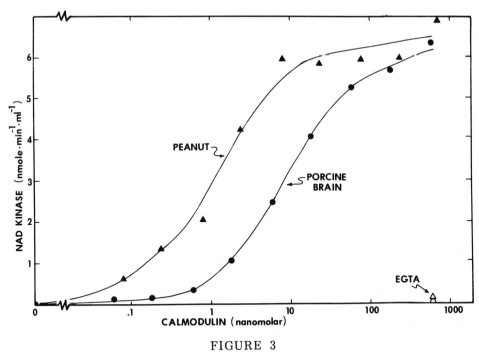

FIGURE 3

Effect of peanut seed (▲) and porcine brain (●) calmodulin on the activation of pea NAD kinase. At one saturation concentration shown, $CaCl_2$ in the assay was replaced with 0.2 mM EGTA to show the Ca^{++} dependence of calmodulin's effect.

In addition to the Ca^{++}-calmodulin-dependent activation of plant NAD kinase described above, evidence for the existence of other Ca^{++}-calmodulin-dependent plant enzymes is beginning to emerge. For example, an ATP-dependent uptake of Ca^{++} into a plasma membrane enriched microsomal fraction, prepared from a squash, was found to be stimulated by the addition of partially purified preparations of either squash or bovine brain calmodulin (10). This observation suggests the possibility that certain plant cell membranes contain a Ca^{++}-calmodulin-dependent Ca^{++} transport ATPase similar to the one observed in mammalian cells (24,30). Thus the Ca^{++}-calmodulin-dependent regulation of Ca^{++} transport could be an important component in the regulation of free Ca^{++} homeostasis in plant cells.

Other Ca^{++}-calmodulin-dependent functions in plant cells are likely to be found. For example, a number of soluble plant proteins in addition to NAD kinase bind to calmodulin-sepharose in a Ca^{++} dependent manner (22). Some of these may represent calmodulin dependent enzymes. Furthermore, calmodulin represents between 0.1 to 0.6% of the total soluble protein in crude extracts of higher plants (5,9,45) while NAD kinase was found to represent approximately 0.01% of the total soluble protein in extracts of pea seedlings (23). Therefore, there is at least a ten-fold molar excess of calmodulin over NAD kinase in extracts of pea seedlings suggesting that the bulk of it is utilized for other regulatory purposes.

If free Ca^{++} functions as a second messenger in plant cells then mechanisms must exist for maintaining low intracellular concentrations of free Ca^{++} where the targets for its second messenger functions are found. There is evidence that such mechanisms are operative in plant cells. For example, plant mitochondria can accumulate Ca^{++} in a manner similar but not identical to that found in animal mitochondria (11,19). In addition, plant cells may have an ATP-dependent Ca^{++} transport system which could function to keep the intracellular Ca^{++} concentration at a low level (10,11). Cytological studies have also indicated that several intracellular organelles (mitochondria, endoplasmic reticulum, and golgi complex) have the ability to sequester Ca^{++} and thereby maintain low cytoplasmic concentrations in the unstimulated cell (48,49).

There are also suggestions that plant cells can respond to stimuli resulting in increases in intracellular Ca^{++} concentration similar to those observed in animal cells. For example, cytoplasmic streaming in Nitella occurs normally in the unstimulated cell but is inhibited by μM concentrations of free Ca^{++} (16). This suggests that intracellular free Ca^{++} is maintained below μM in Nitella but that this level rises in response to mechanical or electrical stimulation, both of which inhibit cytoplasmic streaming. Our own data showing that plant NAD kinase, in the presence of calmodulin, is inactive in the presence of submicromolar Ca^{++} but is activated by micromolar Ca^{++} also supports the idea that intracellular free Ca^{++} in plant cells is under regulatory control (1,2).

It was not until the discovery of Ca^{++} binding proteins, and the recognition that such proteins serve as the targets for stimulus-induced changes in cytosolic free Ca^{++}, that a common molecular basis was provided for the varied roles of Ca^{++} in animal cells. Similarly, the isolation and characterization of calmodulin as the first high affinity Ca^{++} binding protein isolated from higher plant sources provides a basis for understanding the various Ca^{++} dependent physiological processes in plant cells.

In this regard, does Ca^{++}-calmodulin regulate plant NAD kinase in vivo as it does in vitro? We suggest that plant NAD kinase is regulated by Ca^{++}-calmodulin in vivo based on the evidence summarized in Table 4. Taken together, the evidence suggest the possibility that NAD kinase is regulated in vivo by light-induced changes in the levels of free Ca^{++} in plant cells. Fig. 4 illustrates

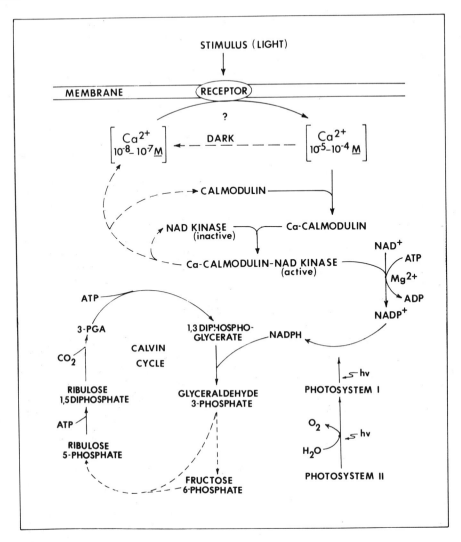

FIGURE 4

Model for the involvement of Ca^{++}-calmodulin in metabolic regulation in plants.

a model based on this concept. In this model the cell undergoes a receptor-mediated change in the levels of intracellular free Ca^{++} in response to light. Since both plant and mammalian calmodulin bind free Ca^{++} at μM levels (2,28,44), we make the assumption that the stimulus-induced changes in intracellular free Ca^{++} that may occur in plant cells approximate those observed in animal cells as indicated in Fig. 4. The stimulus (light) results in an increase

TABLE 4: Evidence that regulation of NAD kinase occurs in vivo.

Evidence	Reference
1. Illumination of green leaves or of Chlorella results in an increase of the NADP/NAD ratio in those tissues	Oh-Hama & Miyachi (36) Ogren & Krogmann (37)
2. At least 90% of all of the NAD kinase found in extracts of higher plants depends on Ca^{++}-calmodulin for activity	Jarrett et al. (23)
3. There is at least a ten-fold molar excess of calmodulin over NAD kinase in extracts of pea seedlings	Charbonneau & Cormier (5) Jarrett et al. (23)
4. The light-dependent conversion of NAD to NADP is too fast to suggest de novo synthesis of NAD kinase	Heber & Santarius (17) Muto et al. (35) Jarrett et al. (23)

TABLE 5: Evidence that regulation of NAD kinase occurs exclusively in the chloroplasts.

Evidence	Reference
1. Approximately 90% of the NAD kinase is localized in the chloroplasts of higher plants	Muto et al. (35)
2. A light induced conversion of NAD to NADP occurs in isolated chloroplasts	Muto et al. (35)
3. At least 90% of all of the NAD kinase found in higher plants depend on Ca^{++}-calmodulin for activity	Jarrett et al. (23)

in intracellular free Ca^{++}. Calmodulin could then function as a receptor for these localized changes in free Ca^{++} resulting in the activation of NAD kinase and perhaps other calmodulin-dependent enzymes. A possible candidate for the receptor is phytochrome since it has been recently implicated in the intracellular regulation of free Ca^{++} (12,14,42,47). When thought of in this way, the model could explain the involvement of phytochrome in numerous physiological responses in plants.

The observations by Ogren and Krogmann (37) and by Oh-Hama and Miyachi (36) that light increases the NADP/NAD ratio in plant cells and our own observations on the Ca^{++}-calmodulin-dependent regulation of NAD kinase suggest the possibility that the in vivo regulation of NAD kinase occurs exclusively in the chloroplast. Support for this view is given by the evidence summarized in Table 5, which includes the observation that plant NAD kinase is localized in the chloroplast.

The regulation of chloroplastic NAD kinase via Ca^{++}-calmodulin could have profound influences on the rate of photosynthesis and on the cell's metabolism. During photosynthesis, the primary electron acceptor from water is NADP. The concentration of NADP rises during illumination presumably due to an activation of NAD kinase by an increase in the level of chloroplastic free Ca^{++}. The photosynthetic increases in both NADPH and ATP concentrations presumably act to convert chloroplastic glyceraldehyde-3-phosphate dehydrogenase to the higher specific activity, monomeric NADP-specific form (38) thus facilitating the utilization of NADPH in the reductive pentose phosphate pathway. This process would also facilitate the export of reducing power to the cytoplasm via the glyceraldehyde-3-phosphate/3-phosphoglycerate shunt (26).

The model presented (Fig. 4) predicts that the rate of photosynthesis is regulated by a receptor-mediated change in the level of chloroplastic free Ca^{++} upon illumination. In the model, free Ca^{++} acts as a second messenger whose information is expressed via the formation of a Ca^{++}-calmodulin complex in the chloroplast resulting in an increased level of NAD kinase activity. Since NAD kinase is totally dependent upon Ca^{++}-calmodulin for activity in vitro (Fig. 3), the dark levels of NADP found in the chloroplast (17,35) are possibly due to a sustained low (sub-saturating) level of Ca^{++} in the dark, resulting in a much lower rate of NAD kinase activity. Upon illumination, a rise in chloroplastic free Ca^{++} would be expected to have a large effect on the rate of NAD kinase resulting in an increased NADP/NAD ratio followed by a subsequent increase in the rate of photosynthesis.

In the past there has been some tendency to think of animal and plant cell regulation in distinct terms or concepts; however, plant and animal cell regulation may be much more similar than previously recognized, particularly with regard to Ca^{++}. Free Ca^{++} may act as a second messenger in plants much as it does in animal cells and may be important in mediating some of the plant responses to stimuli such as light, auxins, gibberellins and cytokinins. Although much work remains to be done with plant systems there is an increasing body of evidence which certainly suggests that plants undergo Ca^{++} dependent regulatory events much like those in animal cells.

ACKNOWLEDGEMENTS

Supported by National Science Foundation grants PCM-79-05043 and PCM-80-07672. We wish to acknowledge the competent technical assistance of Rita Hice and Richard McCann, and Alice Harmon for helpful discussions.

REFERENCES

1. Anderson, J.M. & Cormier, M.J. (1978): Biochem. Biophys. Res. Commun. 84:595-602.
2. Anderson, J.M., Charbonneau, H., Jones, H.P., McCann, R.O. & Cormier, M.J. (1980): Biochemistry 19:3113-3120.
3. Brown, E.G., Al-Najafi, T. & Newton, R.P. (1977): Phytochemistry 16:1333-1337.
4. Chafouleas, J.G., Dedman, J.R., Munjaal, R.P. & Means, A.R. (1979): J. Biol. Chem. 254:10262-10267.
5. Charbonneau, H. & Cormier, M.J. (1979): Biochem. Biophys. Res. Commun. 90:1039-1047.
6. Charles, S.A. & Halliwell, B. (1980): Biochem. J. 188:775-779.
7. Cheung, W.Y. (1980): Science 207:19-27.
8. Cormier, M.J., Anderson, J.M., Charbonneau, H., Jones, H.P. & McCann, R.O. (1980): In Calcium and Cell Function, Vol. 1, (ed) W.Y. Cheung, Academic Press, New York, pp. 201-218.
9. Cormier, M.J., Charbonneau, H. & Jarrett, H.W. (1981): Cell Calcium 9:313-331.
10. Dieter, P. & Marme, D. (1980): Proc. Nat. Acad. Sci. 77:7311-7314.
11. Dieter, P. & Marme, D. (1980): Planta 150:1-8.
12. Dreyer, E.M. & Weisenseel, M.H. (1979): Planta 146:31-39.
13. Grand, R.J.A., Nairn, A.C. & Perry, V.S. (1980): Biochemistry 185:755-760.
14. Hale, C.C. II & Roux, S.J. (1980): Plant Physiol. 65:658-662.
15. Haupt, W. (1959): Planta 53:484-501.
16. Hayama, T., Shimmen, T. & Tazawa, M. (1979): Protoplasma 99:305-321.
17. Heber, U.W. & Santarius, K.A. (1965): Biochim. Biophys. Acta 109:390-408.

18. Herth, W. (1978): Protoplasma 96:275-282.
19. Hodges, T.K. & Hanson, J.B. (1965): Plant Physiol. 40:101-108.
20. Jaffe, L.A., Weisenseel, M.H. & Jaffe, L.F. (1975): J. Cell. Biol. 67:488-492.
21. Jamieson, G.A., Hayes, A., Blum, J.J. & Vanaman, T.C. (1980): IN Calcium Binding Proteins: Structure and Function. (ed) F.L. Siegel, Elsevier, North Holland, New York, pp. 165-172.
22. Jarrett, H.W., Charbonneau, H., Anderson, J.M, McCann, R.O. & Cormier, M.J. (1980): Ann. N.Y. Acad. Sci. 356:119-129.
23. Jarrett, H.W., DaSilva, T. & Cormier, M.J. (1982): IN Uptake and Utilization of Metals by Plants. (eds) D.A. Robb & W.S. Pierpoint, Academic Press, London (in press).
24. Jarrett, H.W. & Penniston, J.T. (1978): J. Biol. Chem. 153: 4676-4682.
25. Kakiuchi, S., Sobul, K., Yamazaki, R., Naga, S., Umeki, S., Nozawa, Y., Yazawa, M. & Yagi, K. (1981): J. Biol. Chem. 256:19-22.
26. Kelly, G.J. & Gibbs, M. (1973): Plant Physiol. 52:674-676.
27. Kindt, R., Pahlich, E. & Rasched, I. (1980): Eur. J. Biochem. 112:533-540.
28. Klee, C.B., Crouch, T.H. & Richman, P.G. (1980): Ann. Rev. Biochem. 49:489-515.
29. Kretsinger, R.H. (1976): Ann. Rev. Biochem. 45:239-266.
30. Larsen, F.L. & Vincenzi, F.F. (1979): Science 204:306-309.
31. Levin, R.M. & Weiss, B. (1977): Molec. Pharmacol. 13:690-697.
32. Levin, R.M. & Weiss, B. (1979): J. Pharmacol. Exp. Ther. 208:454-459.
33. Means, A.R. & Dedman, J.R. (1980): Nature 285:73-77.
34. Molla, A., Kilhoffer, M.C., Ferraz, C., Audemard, E., Walsh, M.P. & Demaille, J.G. (1981): J. Biol. Chem. 256:15-18.
35. Muto, S., Miyachi, S., Usuda, H. & Edwards, G.E. (1981): Plant Physiol. (in press).
36. Oh-Hama, T. & Miyachi, S. (1959): Biochim. Biophys. Acta 34:202-210.
37. Ogren, W.L. & Krogmann, D.W. (1965): J. Biol. Chem. 240: 4603-4608.
38. Pupillo, P. & Piccari, G.G. (1975): Eur. J. Biochem. 51:475-482.
39. Rasmussen, H. & Goodman, D.B.P. (1977): Physiol. Rev. 57: 421-509.
40. Reiss, H.-D. & Herth, W. (1978): Protoplasma 97:373-377.
41. Reiss, H.-D. & Herth, W. (1979): Planta 145:225-232.
42. Roux, S.J., McEntire, K., Slocum, R.D., Cedel, T.E. & Hale, C.C. (1981): Proc. Nat. Acad. Sci. 78:283-287.
43. Seamon, K.B. & Moore, B.W. (1980): J. Biol. Chem. 255: 11644-11647.
44. Wang, J.H. & Waisman, D.M. (1979): Curr. Topics Cell. Reg. 15:47-107.
45. Watterson, D.M., Iverson, D.B. & VanEldik, L.J. (1980): Biochemistry 19:5762-5768.

46. Watterson, D.M., Sharief, F. & Vanaman, T.C. (1980): J. Biol. Chem. 255:962-975.
47. Weisenseel, M.H. & Ruppert, H.K. (1977): Planta 137:225-229.
48. Wick, S.M. & Hepler, P.K. (1980): J. Cell. Biol. 86:500-513.
49. Wolniak, S.M., Hepler, P.K. & Jackson, W.T. (1980): J. Cell. Biol. 87:23-32.
50. Yazawa, M., Sakuma, M. & Yagi, K. (1980): J. Biochem. 87:1313-1320.

REGULATION OF CALMODULIN IN MAMMALIAN CELLS

A.R. Means and J.G. Chafouleas

Department of Cell Biology, Baylor College of
Medicine, Houston, Texas 77030, USA

The fact that calmodulin seems to regulate so many physiological processes suggests that alteration of the intracellular levels of this Ca^{++} receptor could be an important mechanism for control of cellular metabolism and motility. We have evaluated a variety of hormonally regulated systems to determine whether calmodulin is selectively elevated (7). The systems have included estrogen and progesterone stimulation of chick oviduct, androgen and rat prostate, corticosteroids in rat pituitary GH_1 cells, FSH and Sertoli cells, ACTH and Y1 adrenal tumor cells, TSH and rat thyroid slices, EGF and rat pituitary GH_3 cells and GnRH in rat pituitary cells. In no instance is calmodulin selectively elevated as assessed by radioimmunoassay. This is true even in the chick oviduct where estrogen results in a remarkable cytodifferentiation and the appearance of many new protein products. In the immature undifferentiated gland, during primary stimulation, withdrawal from hormone or secondary stimulation the amount of calmodulin per cell remains unchanged.

Exocytosis and Endocytosis in Hormone Action

Many peptide hormones and growth factors that interact with cell surface receptors promote the secretion of proteins, ions or steroids (32). In the past few years it has become increasingly apparent that exocytosis of many cell proteins requires packaging of the secretory product in clathrin-coated vesicles. Linden et al. (30) have revealed that calmodulin is a component of such vesicles isolated from brain and that the association is Ca^{++}-dependent and of high affinity ($K_d = 10^{-9}$ M). Since exocytosis is known to require Ca^{++}, these studies raised the possibility that calmodulin might confer the Ca^{++} sensitivity to the secretory process. Much indirect evidence exists that is compatible with this hypothesis. Anti-calmodulin compounds such as phenothiozines have been reported to inhibit glucose-mediated insulin release from pancreatic islet (27) and insulinoma cells (39), intestinal ion secretion (26), histamine release from mast cells (40), protein secretion from polymorphonuclear leukocytes (15,34), and seretonin release from platelets (47). Although the mechanism for such responses remain unknown it is likely that it does not involve changes in the total cell content of calmodulin.

All of the processes mentioned above involve changes in the net flux or distribution of Ca^{++}. Since calmodulin is a component of virtually every intracellular compartment as well as the plasma membrane, efforts have been made to determine whether cell surface acting agents promote an alteration in the distribution of calmodulin. Distinct anatomical regions of the central nervous system such as the corpus striatum contain dopamine receptors which seem to be coupled to adenylyl cyclase (21). Calmodulin has been suggested to mediate dopamine action since phosphorylation of membrane proteins promotes the apparent release of calmodulin from membrane-bound to soluble form (18). Since a soluble calmodulin-dependent phosphodiesterase exists, it has been proposed that long term stimulation of dopamine receptors is associated with an increase in the soluble calmodulin content, thereby activating PDE and decreasing receptor responsiveness. Similar data suggest interneuronal pathways also exist where opiates increase soluble calmodulin via a release of dopamine and thus act as indirect dopamine agonists (22). Smoake and Solomon (41) have reported altered calmodulin distribution in liver cells from rats with streptozotocin-induced diabetes. These authors conclude that such changes might play a role in the alteration of cAMP metabolism known to exist in such pathological states.

The difficulties with interpretation of most calmodulin distribution studies is that the protein is assayed by its ability to stimulate a calmodulin-dependent enzyme. Since all such assays are Ca^{++}-dependent and other calmodulin-binding proteins are likely to be present in each subcellular fraction it is difficult to obtain quantitative values for calmodulin. This difficulty is circumvented when a radioimmunoassay is employed since the assay can be performed in the presence of EGTA and is therefore Ca^{++}-independent (6). The radioimmunoassay has been utilized to determine the quantity and subcellular distribution of calmodulin in the rat pituitary gonadotrope before and during GnRH-induced LH release (10). Indeed, the distribution of calmodulin does change in response to GnRH. There is an initial rise in the percentage of calmodulin associated with the plasma membrane which appears concomitantly with the depletion of cytoplasmic calmodulin. These changes occur temporally in concert with secretion of LH. As the calmodulin begins to be cleared from the plasma membrane, its level increases first in the secretory granule and microsomal fractions before finally replenishing the cytoplasm. The magnitude of the changes that occur between plasma membrane and cytoplasmic content of calmodulin are related to the dose of GnRH. Calmodulin redistribution is also hormone specific since analogs such as des^1 GnRH, which has no efficacy in promoting LH secretion did not alter intracellular changes in calmodulin. Finally, a budget of calmodulin content in all subcellular fractions revealed that GnRH did not increase total calmodulin and greater than 95% of the cellular calmodulin was recovered.

The data presented above suggest that calmodulin may be important in the regulation of protein secretion but provide little information concerning the mechanism. At this juncture it is impossible to predict whether calmodulin redistribution is a cause or consequence of the secretory process. In the red blood cell (25), pancreatic islet (35) and adipocyte (36), calmodulin activated ATPases are found in the plasma membrane and, at least in the adipocyte, the enzyme appears to be hormonally regulated. Plasma membranes from islet cells also have been reported to contain a calmodulin-stimulated adenylyl cyclase activity (45). Calmodulin is also a major component of postsynaptic membranes (20,29,48), has been proposed to mediate the Ca^{++} effects on synaptic transmission (18,21,22), and, accordingly, may play a role in neurotransmitter release (14). Finally, trifluoperazine, and naphthalenesulfonamides are drugs that bind to calmodulin and inhibit many of its actions. These drugs also inhibit the receptor-mediated secretory process in a variety of systems.

Receptor-mediated endocytosis is also a Ca^{++}-dependent process and also involves clathrin-coated vesicles (19,38). Although internalization of GnRH does not appear to be required for the LH release process the gonadotrope response to this releasing hormone does include the pattern of patching, capping and internalization observed for many cell surface-mediated ligand systems (12). This receptor redistribution pattern in the gonadotrope is mimicked by changes found in calmodulin associated with the plasma membrane when assessed by indirect immunofluorescence microscopy. Recruitment of clathrin-coated vesicles to the plasma membrane of human lymphoblastoid cells occurs following stimulation with multivalent anti-IgM antibodies (38). This recruitment is inhibited by the presence of anti-calmodulin drugs and calmodulin is a component of such vesicles. Thus the appearance of calmodulin at the plasma membrane may be associated with the accumulation of coated pits involved in the receptor internalization process. Insulin, which also is internalized following cap formation, promotes the translocation of glucose transport activity from the microsomal or Golgi fractions to the plasma membrane (11,42). Actin and myosin have also been reported to co-cap with several cell surface receptors (3,16) and actin-containing matrices have been isolated from D. discoideum (9), murine tumor cells (33) and lymphocyte plasma membranes (33) associated with various receptors. Thus, the phenomenon of redistribution of new activities to the plasma membrane may be a generalized occurance for plasma membrane receptor-mediated events. This redistribution suggests a mechanism by which calmodulin-regulated events could be affected without the requirement for new protein synthesis. It is likely that calmodulin redistribution is secondary to alterations in the net flux or distribution of Ca^{++} within the cell.

Calmodulin in Transformed and Nontransformed Cells

Transformation of cells to malignancy appears to represent one general mechanism that results in a specific increase in the intracellular content of calmodulin. The initial report of such a change was by Watterson et al. (14), who suggested that transformation of chick embryo fibroblasts by Rous sarcoma virus resulted in an elevation of the activator protein of phosphodiesterase (calmodulin). In these experiments the protein was quantitated by densitometric scanning of cytoplasmic proteins distributed on polyacrylamide gels. A similar report by LaPorte et al. (28) appeared subsequently. These authors utilized an identical system but measured calmodulin by its ability to stimulate phosphodiesterase. Both reports suggested a doubling in the concentration of calmodulin in the transformed relative to the non-transformed cell. The generality of this phenomena was demonstrated by Chafouleas et al. (1,8). These authors revealed, using radioimmunoassay, that relative to the non-transformed counterparts calmodulin was elevated in SV-40 transformed 3T3 cells, normal rat kidney (NRK) cells transformed by Rous sarcoma virus, mouse mammary epithelial cells transformed in response to hormones and human bronchial epithelial cells transformed by chemical carcinogens. On the other hand bronchial cells treated in vitro with $NiSO_4$ that exhibited phenotypic changes characteristic of transformed cells but were not malignant failed to show elevated calmodulin levels. Thus a two- to threefold increase in calmodulin appears to be a general response to transforming agents regardless of their nature.

Chafouleas et al. (8) evaluated the mechanism for the increase in calmodulin accompanying cell transformation in both the 3T3 and NRK systems. Tubulin levels have been reported to remain constant following transformation (24) but the state of microtubule polymerization is regulated by Ca^{++}-calmodulin. The experimental design was to determine rates of synthesis and degradation of calmodulin, tubulin and total protein in exponentially growing normal and transformed cells (8). Calmodulin and tubulin were quantitated by radioimmunoassay and immunoprecipitation, whereas incorporation of ^{35}S-methionine into acid-precipitable material served as an estimate of total protein synthesis (or degradation). The data revealed that the rate of synthesis of calmodulin, tubulin and total protein was elevated in the transformed cells compared to the non-transformed counterparts. In addition, the rate of degradation was, in all instances, faster in transformed compared to non-transformed cells. However, whereas the rate of calmodulin synthesis was elevated threefold, only a 50% change in the rate of degradation occurred. This resulted in a two- to 2.5-fold increase in the amount of calmodulin per cell and these values agreed closely with those determined by radioimmunoassay under steady state conditions. On the other hand, tubulin synthesis was stimulated twofold as was the rate of degradation of this protein. Thus the steady state concentration of tubulin did not vary between transformed and nontransformed cells. Together these data reveal that a twofold increase in the calmodulin to tubulin ratio exists in the transformed cell.

Such changes in the calmolulin to tubulin ratio may explain why immunofluorescent techniques have been interpreted to show a diminshed cytoplasmic microtubule complex in transformed cells (4) while biochemical procedures reveal no change in tubulin content (7,8,22). Three pieces of evidence exist to support this hypothesis. First, calmodulin mediates the Ca^{++}-dependent assembly/disassembly of microtubules in vitro (31). Secondly, using a detergent permiabilized cell system to study polymerization of microtubules from 6S tubulin, Brinkley et al. (13) have shown that SV-40 transformed 3T3 cells are attenuated in their ability to nucleate and elongate microtubules. This difference is twofold in magnitude. Tash et al. (43) have reported that the SV-40 3T3 cell can be restored to normal in this regard by preincubation with anti-calmodulin. Finally, Rubin and Warren (37) have used quantitative electron microscopy to evaluate the number of polymerized microtubules in NRK and RSV-NRK cells. These authors reveal that the transformed cells contain only 50% the number of polymerized microtubules compared to NRK cells. Together these studies suggest that the degree of polymerization of tubulin into microtubules in vivo may be dependent on the concentration of calmodulin within the cell relative to that of tubulin. The elevated levels of calmodulin in transformed cells could destabilize microtubules and result in a significant reduction in number and length of these cytoskeletal components compared to those in non-transformed cells.

Cell transformation is usually accompanied by alterations in morphology, cyclic nucleotide metabolism, metabolic rate and intracellular Ca^{++} levels. In addition, transformed cells lose the requirements for anchorage to substrate and Ca^{++} for cell proliferation. All of these processes could be calmodulin-dependent based on the enzymes known to be regulated by this ubiquitous Ca^{++} binding protein. Certainly one major Ca^{++}-calmodulin regulated system, the cytoplasmic microtubule network, seems to be altered in virally transformed cells. If calmodulin does affect the intracellular regulation of tubulin assembly, it could also be postulated to indirectly alter intracellular tubulin metabolism. Ben-Ze'er et al. (2) have proposed that an increase in the ratio of unpolymerized to polymerized 6S tubulin increases the rate of degradation of these proteins by affecting the turnover of tubulin mRNA. These observations would predict that an increase in unpolymerized tubulin in the transformed cell should result in an increase in the rate of tubulin degradation relative to what occurs in the non-transformed state. Chafouleas et al. (1,8) have shown such a situation to exist in cells transformed by both RNA and DNA oncogenic viruses. Thus calmodulin may alter the cytoskeleton not only through a direct effect on microtubule assembly, but also indirectly through regulation of the intracellular levels of tubulin.

Changes in Calmodulin During the Cell Cycle

All of the experiments performed to date concerning regulation of calmodulin levels as well as those involving tubulin have been carried out on asynchronous cell populations. It is likely that effects of altered calmodulin levels on the cytoskeleton, as well as on some of the other phenotypic changes characteristic of virally transformed cells would be cell cycle dependent. For these reasons we have undertaken an analysis of when calmodulin is synthesized during the cell cycle. Because of their relatively short cell cycle and ease of handling, Chinese hamster ovary cells, CHO-K1, were selected for the synchrony experiments. Prior to synchronization, the cell cycle parameters of G_1, S, G_2 and M were ascertained by the pulse labelled mitosis procedure. All cells were synchronized without the use of drugs by the mitotic shake procedure of Terasima and Tolmach (44). In this procedure mitotic cells were selectively shaken from a monolayer in exponential growth and retained in M by placing in medium at 4°C. Such a procedure routinely resulted in starting populations of cells with mitotic indices between 93 and 98%. Cells were then released from M into G_1 by replicate plating into 60 mm dishes containing medium at 37°C. Progression through the cell cycle and verification of synchrony was monitored by determining the percent of the cells labelled at each point with a 10 min ^{3}H TdR pulse as well as quantitation of the percent mitotic cells present. In addition, calmodulin levels were also determined in duplicate samples by radioimmunoassay (6).

The initial experiments utilized CHO-K1 cells with a cell cycle of about 16 hr. Such cultures exhibited a G_1 of 5 hr, S of 8 hr, G_2 of 2 hr, and M of 1 hr. Following the mitotic shake the cells are in tight synchrony and enter S phase approximately 4-5 hr after release into G_1. Calmodulin levels at M are 150 ng/10^6 cells, but in early G_1, following cell division, the amount of calmodulin per cell is reduced to a value one half that observed in M. During late G_1, calmodulin levels increase and reach the maximum level of approximately 150 ng/10^6 cells observed in the starting M population. These values are then maintained throughout the duration of the cell cycle. In this experiment 98% of the cells were in mitosis at time 0. This value falls to 0% following release from M and remains there for at least 10 hr. By 13 hr there is a reappearance of mitotic cells representing re-entry into M, again demonstrating the synchronous progression through the cell cycle. These data suggested that calmodulin is synthesized exclusively during late G_1 or early S.

In order to further substantiate this observation a similar experiment was performed on a population of CHO-K1 cells exhibiting a cell cycle of 12 hr and a G_1 of only 2 hr. Since the G_1/S transition was of primary interest attention was focussed to only the first 7 hr of the cell cycle. Starting at about 4 hr there was a coincident temporal increase in calmodulin levels and progression into S phase, respectively. The same relationship between calmodulin levels and entry into S was observed as for the longer cycling population. However, in this cell population the increases in calmodulin occurred at approximately 1.5 hr instead of 4 hr. These data suggest that regardless of the length of G_1, calmodulin is synthesized at late G_1 and/or early S. In order to determine if there is a direct correlation between calmodulin content and progression into S the percent of cells in S was plotted against the appropriate intracellular calmodulin levels. Data from four independent experiments using cells with both 5 and 2 hr G_1 phases were analyzed by linear regression. The correlation coefficient was 0.966 which further substantiates the concept that calmodulin may be important in the G_1/S transition. If calmodulin is important in the G_1/S transition then anti-calmodulin drugs should be able to inhibit the progression of cells from G_1 into S. Two drugs synthesized by Dr. H. Hidaka, called W12 and W13 (23), were used for these experiments. While W13 binds calmodulin with high affinity and potently inhibits its biological activity, W12, the dechlorinated analog is much less effective. In addition, it should be noted that both drugs exhibit similar hydrophobicity indices. The drugs were first evaluated for effects on cell viability and morphology. Concentrations utilized were from 10^{-7} to 10^{-2} M. W12 had no effect on plating efficiency over the entire concentration range. W13 was also without effect to concentrations of 10^{-4} M. However, 2 x 10^{-4} W13 resulted in a 99% decrease in cell viability. Cells were examined by light and scanning electron microscopy as well as by indirect immunofluorescent microscopy using anti-tubulin antibody. Again W12 was without any demonstrable effect. At 1 x 10^{-4} M W13, cells became rounded and looked very similar to transformed cells at the light and SEM levels. However, this concentration of W13 did not seem to alter the number of cytoplasmic microtubules. Since calmodulin levels were also not altered, these data suggested that it was possible to dissociate changes in gross morphology from alterations in the cytoplasmic microtubule network.

The next question was to determine whether W13 at 10^{-4} M would alter the progression of cells through the cell cycle. CHO-K1 cells were plated as 7.5 x 10^5 cells per dish and grown for 18 hr at 37°C.

At this time (0 time) each dish contained 1×10^6 cells and the cells were in exponential growth. Replicate cultures were then treated with normal medium or medium containing either W12 or W13. Cultures from each group were utilized to determine total cell number at 8 and 24 hr after drug addition.

Neither W12 or W13 had demonstrable effects on cell number after the first 8 hr. After 24 hr, no change from control was observed in the W12 treated cultures. In contrast, however, treatment with W13 for 24 hr resulted in a reduction in the total number of cells in culture to a level 48% of the control cultures. It should be noted that this effect of W13 was completely reversible and the treated cells demonstrate 99% survival upon subsequent plating. These data would suggest that the reduction in cell number by W13 may be mediated through a cell cycle specific block. In order to evaluate this further the effect of these drugs on the distribution of cells in the cell cycle was determined. After 24 hr of treatment with normal media, W12 or W13, the cells were fixed and stained for DNA using the fluorescent dye propidium iodide. The distribution of cells in G_1, S and G_2/M was determined for each sample by flow cytometry. In such an analysis cell number is plotted on the Y axis against the relative fluorescence on the X axis. Since the relative fluorescence is a function of the amount of DNA, such a plot can readily discern the relative numbers of cells in G_1 and S phases. The cells used in this experiment exhibited a G_1 of 2 hr. As expected from such a short G_1 period, the control histogram revealed that a very large percentage of the cells were in S phase. The pattern obtained from cells treated with W12 was indistinguishable from that of untreated cells. In contrast, however, the analysis of the W13 treated cells showed a dramatic reduction in S phase cells with a resultant increase in the percent of cells in G_1. These data are compatible with at least one block at G_1/S. The presence of a G_1/S block in response to W13 was confirmed by an experiment which monitored the percent ^{3}H TdR-labelled cells following addition of the drug to a synchronized cell population. Cells were synchronized by mitotic shake (44) and W12 or W13 was added 1 hr after plating the mitotic cells. By 7 hr 90% of the cells were labelled in control and W12-treated cultures. However, W13-treated cells exhibited a lag in DNA labeling and only 60% of the population became labelled with TdR after 10 hr. This experiment again reveals the presence of a G_1/S block induced by the anti-calmodulin drug W13.

The next experiment was designed to evaluate the effect of W13 on the progression of cells through S phase. Cells were first treated with high levels of unlabelled TdR which is known to result in a G_1/S block (5). Cells were then released from synchrony and after 1 hr, W12 or W13 was added. The number of cells in S phase was monitored at various times using a 10 min pulse of ^{3}H-TdR to label DNA. At the time of drug addition 85% of the cells were in S.

This value increased to 92% in control and W12 treated cultures by 4 hr before declining to 40% labelled cells by 7 hr. These data are compatible with the fact that the CHO cells exhibit an S phase of 8 hr. Although the cells treated with W13 also began with 85% of the cells in S, no further increase occurred following drug addition. Moreover, by 7 hr 76% of the cells remained in S. The data indicate that W13 markedly delays S progression and suggests that the G_1/S effect noted in the earlier experiments is due to a drug induced block in early S.

The next question to address was whether W13 exhibited effects at multiple points in the cell cycle. First, drug effects on G_2 progression were evaluated. Asynchronous CHO cells in exponential growth were treated at 0 time with a 10 min ^3H TdR pulse. The label was removed and the cultures were exposed to colcemid and the test substance (W12 or W13). Colcemid arrests cells at metaphase (5). Since ^3H TdR would label those cells that pass through S, only cells in G_2 at the time of colcemid addition would be blocked in M without passing through S and becoming labelled with ^3H TdR. Thus G_2 progression is monitored by counting the number of cells not labelled by ^3H TdR. At 0 time approximately 3% of the cells were in G_2. This value increased linearly with time for 3 hr reaching a value of 12%. No differences were observed between untreated cells and those treated with W12 or W13. The data reveal that the anti-calmodulin drug does not exert any effect in G_2.

The final experiment in the cell cycle series was to determine whether W13 caused an alteration in the progression of cells through M. The mitotic shake procedure (44) was again used to synchronize cells in M. Cells were released into medium at 37°C in the presence of W12, W13, or no addition. The mitotic index was then determined at 20 min intervals between 0 and 100 min. At 0 time 92% of the cells were in M. This value decreased to 0% at 60 min, which is consistent with the 1 hr M phase of these CHO cells. Again no differences were observed in cell progression through M regardless of treatment demonstrating that W13 does not affect mitosis.

These data reveal that calmodulin is synthesized entirely during the late G_1 and/or early S phases of the cell cycle in CHO-K1 cells. Treatment of cells in exponential growth with the anti-calmodulin drug W13 results in a single cell cycle block in early S phase. The block is not due to an inhibition of calmodulin synthesis and indicates that the twofold elevation in the intracellular concentration of calmodulin that occurs in late G_1/early S may be required for DNA synthesis. A highly significant correlation exists between the increase in calmodulin levels and the length of G_1. Virally transformed cells exhibit a greatly reduced G_1 period (49) and a two- to threefold elevation in calmodulin concentration (7,8) compared to non-transformed counterparts. These results may help to explain why transformed cells multiply more rapidly than cells not infected by oncogenic viruses. Furthermore, it has been established that production of virus in SV-40 infected cells is dependent upon a host cell event that occurs during the G_1/S transition period (17). Calmodulin synthesis may represent at least one such event. This finding raises the intriguing possibility that anti-calmodulin drugs such as W13 might inhibit SV-40 replication in infected cells.

CONCLUSIONS

Although the activity of calmodulin is generally controlled by changes in the intracellular free Ca^{++} concentration, alterations in calmodulin levels can also represent an important regulatory mechanism which governs specific Ca^{++}-mediated events. One example of this latter kind of control appears to occur during the growth cycle of eukaryotic cells. Calmodulin is synthesized entirely at the G_1/S boundary and anti-calmodulin drugs specifically and reversibly prevent cells from progressing through S. Since the major event that occurs in S is DNA replication, it is tempting to speculate that calmodulin is important in the regulation of DNA synthesis. Unfortunately, this is an extremely difficult theory to prove with the present technology since no adequate assay exists for replicative DNA synthesis in eukaryotic cells. However, the importance of twofold elevation of calmodulin at G_1/S for progression through S can be predicted from a consideration of the effects of the anti-calmodulin drugs. Each cell maintains a steady state level of calmodulin that can vary by more than tenfold between cell types. However, regardless of this G_1 level, the intracellular concentration of the protein only doubles at G_1/S. Similarly, cells differ in their sensitivity to the anti-calmodulin drugs (W compounds). Chinese hamster ovary cells can accommodate 10^{-4} M (30 µg/ml) W13 without any sign of cytotoxicity. The sole difficulty seems to be in progression through S. Increasing the drug concentration by only 5 µg/ml results in 40% of the cells being killed and an additional increase of 5 µg/ml results in 99% cell death. This narrow window of drug concentration over which cells survive suggests that one can prevent the action of the increased calmodulin that occurs at G_1/S without interferring with the normal G_1 level (1 x) of the protein. As soon as sufficient drug is accumulated to bind a fraction of the 1 x calmodulin level, cells begin to die. Taken together, these observations support the contention that calmodulin is an intracellular Ca^{++} receptor and that its presence is vital for cell survival due to its pleiotypic nature. Such a dramatic requirement for cell growth and survival again points to the ubiquity of calmodulin and strengthens the argument that this protein represents the primary physiological receptor for Ca^{++} in eukaryotic cells.

REFERENCES

1. Barranco, S.C. & Bolton, W.E. (1977): Cancer Res. 37:2589-2591.
2. Ben-Ze'er, A., Farmer, S.R. & Penman, S. (1979): Cell 17:319-325.
3. Bourguignon, L.Y.W., Tokuyasu, K.T. & Singer, S.J. (1978): J. Cell Physiol. 95:239-258.
4. Brinkley, B.R. Fuller, G.M. & Highfield, D.P. (1975): Proc. Nat. Acad. Sci. 72:4981-4985.
5. Brinkley, B.R., Stubblefield, E. & Hsu, T.C. (1967): J. Ultrastruct. Res. 19:1-18.
6. Chafouleas, J.G., Dedman, J.R., Munjaal, R.P. & Means, A.R. (1979): J. Biol. Chem. 254:10262-10267.
7. Chafouleas, J.G., Pardue, R.L., Brinkley, B.R., Dedman, J.R. & Means, A.R. (1980): In Calcium-Binding Proteins: Structure and Function. (eds) F.L. Siegel, E. Carafoli, R.H. Kretsinger, D.H. MacLennan & R.H. Wasserman, Elsevier, Amsterdam, pp. 189-196.
8. Chafouleas, J.G., Pardue, R.L., Brinkley, B.R., Dedman, J.R. & Means, A.R. (1981): Proc. Nat. Acad. Sci. 78:996-1000.
9. Condeelis, J.S. (1979): J. Cell Biol. 80:751-758.
10. Conn, P.M., Chafouleas, J.G., Rogers, D. & Means, A.R.(1981): Nature 292:264-265.
11. Cushman, S.W. & Wardzala, L.J. (1980): J. Biol. Chem. 255: 4758-4762.
12. Conn, P.M., Marian, J., McMillian, M. & Rogers, D. (1980): Cell Calcium 1:7-20.
13. Dabrowska, R., Sherry, J.M.F., Aromatorio, D.K. & Hartshorne, D.J. (1978): Biochemistry 17:253-258.
14. DeLorenzo. J.R., Freedman, S.D., Yohe, W. & Maurer, S.C. (1980): 6:1838-1842.
15. Elferink, J.G.R. (1979): Biochem. Pharmacol. 28:965-968.
16. Flanagan, J. & Koch, G.L.E. (1978): Nature 273:278-281.
17. Gershey, E.L. (1979): J. Virol. 30:76-83.
18. Gnegy, M.E. & Lau, Y.S. (1980): Neuropharmacology 19:319-323.
19. Goldstein, J.L., Anderson, R.G.W. & Brown, M.S. (1979): Nature 279:679-685.
20. Grab, D.J., Berzins, K., Cohen, R.S. & Siekevitz, P. (1979): J. Biol. Chem. 254:8690-8696.
21. Hanbauer, I., Gimble, J. & Lovenberg, W. (1979): Neuropharmacology 18:851-857.
22. Hanbauer, I., Gimble, J., Sankaran, K. & Sherard, R. (1979): Neuropharmacology 18:859-864.
23. Hidaka, H., Yamaki, T., Totsuka, T. & Asano, M. (1979): Mol. Pharmacol. 15:49-59.
24. Hiller, G. & Weber, K. (1978): Cell 14:795-804.
25. Hinds, T.R., Larsen, F.L. & Vincenzi, F.F. (1978): Biochem. Biophys. Res. Commun. 81:455-461.
26. Ilundain, A. & Naftalin, R.J. (1979): Nature 279:446-448.

27. Krauz, Y., Willheim, C.B., Siegel, E. & Sharp, G.W.G. (1980): J. Clin. Invest. 66:603-607.
28. Laporte, D.C., Gidwitz, S., Weber, M.J. & Storm, D.R. (1979): Biochem. Biophys. Res. Commun. 86:1169-1177.
29. Lin, C.T., Dedman, J.R., Brinkley, B.R. & Means, A.R. (1980): J. Cell Biol. 85:473-480.
30. Linden, C.D., Dedman, J.R., Chafouleas, J.G., Means, A.R. & Roth, T.F. (1981): Proc. Nat. Acad. Sci. 78:308-312.
31. Marcum, J.M., Dedman, J.R., Brinkley, B.R. & Means, A.R. (1978): Proc. Nat. Acad. Sci. 75:3771-3775.
32. Means, A.R., Dedman, J.R., Tash, J.S., Tindall, D.J., Van Sickle, M. & Welsh, M.J. (1980): Ann. Rev. Physiol. 42:59-70.
33. Mescher, M.F., Jose, M.J.L. & Balk, S.P. (1981): Nature 289:139-144.
34. Naccache, P.H., Molski, T.F.P., Alobaidi, T., Becker, E.L., Showell, H.J. & Sha'afi, R.I. (1980): Biochem. Biophys. Res. Commun. 97:62-68.
35. Pershadsingh, H.A., Landt, M. & McDonald, J.M. (1980): J. Biol. Chem. 255:8983-8986.
36. Pershadsingh, H.A., McDaniel, M.L., Landt, M., Bry, C.G., Lacy, P.E. & McDonald, J.M. (1980): Nature 288:492-495.
37. Rubin, R.W. & Warren, R.H. (1978): J. Cell Biol. 79:279.
38. Salisbury, J.L., Condeelis, J.S. & Satir, P. (1980): J. Cell Biol. 87:132-141.
39. Schubart, U.K., Fleischer, N. & Erlichman, J. (1980): J. Biol. Chem. 255:11063-11066.
40. Sieghart, W., Theoharides, T.C., Alper, S.L., Douglas, W.W. & Greengard, P. (1978): Nature 275:329-331.
41. Smoake, J.A. & Solomon, S.S. (1980): Biochem. Biophys. Res. Commun. 94:424-430.
42. Suzuki, K. & Kono, T. (1980): J. Biol. Chem. 77:2542-2545.
43. Tash, J.S., Means, A.R., Brinkley, B.R., Dedman, J.R. & Cox, S.M. (1980): IN Microtubules and Microtubule Inhibitors. (eds) M. de Brabander & J. DeMey, North Holland/Elsevier, pp. 269-279.
44. Terasima, T. & Tolmach, L.J. (1961): Nature 190:1210-1211.
45. Valverde, I., Vandermeers, A., Anjaneyula, R. & Malaisse, W.J. (1979): Science 206:225-227.
46. Watterson, D.M., Van Eldik, L.J., Smith, R.E. & Vanaman, T.C. (1976): Proc. Nat. Acad. Sci. 73:2711-2715.
47. White, G.C. II & Raynor, S.T. (1980): Thrombosis Res. 18:279-284.
48. Wood, J.G., Wallace, R.W., Whitaker, J.N. & Cheung, W.Y. (1980): J. Cell Biol. 84:66-76.
49. Zucker, R.M., Tershakovec, A., D'Alisa, R.M. & Gershey, E.L. (1979): Exp. Cell Res. 122:15-22.

CALCIUM AND CALMODULIN DEPENDENT REGULATION OF MICROTUBULE ASSEMBLY

H. Kumagai, E. Nishida and H. Sakai

Department of Biophysics and Biochemistry,
Faculty of Science, The University of Tokyo
Bunkyo-ku, Tokyo 113, Japan

Calmodulin, which activates several enzymes in a Ca^{++} dependent manner, inhibits microtubule (MT) assembly in vitro in the presence of Ca^{++}. In this report, we propose the mechanism of calmodulin-induced inhibition of MT assembly as described below and present quantitative analysis of the association between calmodulin and the tubulin dimer.

$$Ca^{++} + calmodulin \rightleftharpoons Ca^{++}\text{-calmodulin}$$

$$Ca^{++}\text{-calmodulin} + tubulin \rightleftharpoons Ca^{++}\text{-calmodulin-tubulin}$$
$$\text{(polymerizable)} \qquad \text{(non-polymerizable)}$$

INTRODUCTION

MTs are ubiquitous organelles ranging from protozoa to mammals, which participate in a variety of cellular functions such as motility, transport, maintenance of cell shape and nerve excitation (6,22-24).

MTs prepared from brain extracts consist of tubulin and its associated proteins termed "microtubule associated proteins" (MAPs), which are required for MT assembly under physiological conditions (25,39). Since the first assembly experiment for MTs in vitro (40), many reports on MTs have been focussed on the regulatory mechanism of the assembly and disassembly (35), which seems to play an important role in MT related functions. Several factors influencing MT assembly have been identified, including polyanions such as RNA and DNA (1,38), protein factors that inhibit binding of colchicine to tubulin (18,37), an inhibitory factor in the sea urchin egg cortex (26), and 94 K protein from porcine brains that modulate MT assembly in a Ca^{++} and Mg^{++} dependent manner (29).

In addition, it was recently demonstrated that the phosphorylation of MAPs suppresses MT assembly (7). Furthermore, differences in solution conditions during MT assembly, such as pH, ionic strength, nucleotides and divalent cations have also been known to give rise to differences in assembly processes (13,27,32).

Ca^{++} ion seems to play critical roles in many cellular functions. It is known as an inhibitor of MT assembly in vivo as well as in vitro (4,27,36). An earlier finding by Nishida and Sakai (28) suggested that porcine brain contains Ca^{++}-sensitizing factors which regulate MT formation in vivo in the presence of µM levels of free Ca^{++}.

On the other hand, calmodulin, which has been known as a Ca^{++} dependent regulator of cyclic nucleotide phosphodiesterase (PDE) (2,8) and several other enzymes (3,10,24), was reported previously as a Ca^{++} dependent modulator of MT assembly and disassembly by Marcum et al. (20) and by us (11,12,30). This paper summarizes studies of the assembly modulation by calmodulin and further discusses characteristics on association of calmodulin with tubulin.

METHODS

Preparation of proteins : Fresh porcine brains were obtained from Teikoku Zoki Company immediately after slaughter. Calmodulin and calmodulin-depleted cyclic nucleotide PDE were prepared from the porcine brains by a procedure described previously (11,30). Purified microtubule proteins (PMP) were obtained by two cycles of temperature-dependent assembly-disassembly as described previously (30). Briefly, porcine brains were homogenized in a buffer solution containing 10 mM 2-(N-morpholino)-ethanesulfonic acid (MES), pH 6.8, 0.5 mM $MgCl_2$, 50 mM KCl and 1 mM ATP. The homogenate was centrifuged at 30,000 g for 1 hr at 4°C and the supernatant solution was adjusted to 4 M glycerol, 1 mM EGTA and 1 mM ATP and incubated at 35°C for 45 min. The assembled MTs were precipitated by centrifugation at 70,000 g and 30-35°C for 45 min. The precipitated MTs were depolymerized in buffer solution containing 10 mM MES, pH 6.8, 0.5 mM $MgCl_2$, 50 mM KCl and 0.5 mM GTP at 0°C and centrifuged at 100,000 g and 4°C for 45 min. The supernatant was adjusted to 4 M glycerol and the assembly and disassembly cycle was repeated to obtain two cycled MT proteins. Tubulin and MAPs were purified from PMP by the phosphocellulose column chromatography.

Preparation of tubulin-sepharose 4B and MAPs-sepharose 4B. Tubulin or MAPs were incubated with cyanogen bromide-activated sepharose 4B (Pharmacia Company) for 2 hr at room temperature in 20 mM MES, pH 6.8, 0.5 mM $MgCl_2$ and 0.2 M NaCl, and additional 0.05 mM GTP in case of coupling tubulin. After washing with the above solution, the gel was suspended in 0.5 M Tris-HCl (pH 8.3) or 0.5 M monoethanolamine (pH 8.0) for 2 hr to block the remaining cyanogen bromide activated sepharose. The gel was then washed with 10 mM MES, pH 6.8, 0.5 mM $MgCl_2$, 1 mM EGTA and 1 M NaCl to remove unbound protein and stored at 4°C until use.

Assay of PDE and calmodulin. Assay of PDE and calmodulin activity were described previously (11). It was confirmed that PMP fraction does not influence the ability of calmodulin to activate cAMP PDE (11).

Assembly assay. MT assembly was monitored by turbidimetry or viscometry as described previously (30).

Others. Free calcium ion concentration was calculated from the apparent association constant of EGTA with Ca^{++} (31). Protein concentration was determined by the method of Lowry et al. (19). SDS polyacrylamide gel electrophoresis was carried out by the method of Laemmli (15).

Regulation of MT Assembly by Calmodulin and its Binding to Tubulin

Effect of calmodulin on MT assembly-disassembly. Calmodulin inhibits MT assembly significantly in the presence of micromolar free Ca^{++} (Fig. 1A). This inhibition is dependent on the concentration of calmodulin, suggesting that the effect is not catalytic but stoichiometric. The action of calmodulin is shown to be reversibly dependent on the concentration of Ca^{++}. When free Ca^{++} concentration is decreased by the addition of EGTA, the viscosity increases (Fig. 1A). Therefore, MT assembly is reversibly regulated by a change in the concentration of either calmodulin or Ca^{++}.

On the other hand, after assembly of MTs in the presence of micromolar calcium ions which do not inhibit assembly by themselves, addition of calmodulin induces depolymerization. However, in the absence of Ca^{++}, calmodulin does not induce depolymerization of the pre-assembled MTs. When calmodulin is added to MT protein solution in the absence of Ca^{++}, MTs can be assembled to the same extent as control (Fig. 1B). Further addition of micromolar Ca^{++} then induces depolymerization of the pre-assembled MTs. Ionic strength affects the inhibitory effect of calmodulin. With increasing KCl concentration, calmodulin-induced inhibition of MT assembly is enhanced (data not shown).

Association of calmodulin with tubulin. The above results suggest that calmodulin interacts with tubulin or MAPs only in the presence of free Ca^{++}. Previous results show that calmodulin does

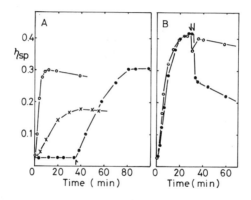

FIGURE 1

Effect of Ca^{++} and calmodulin on microtubule (MT) assembly. The extent of MT assembly at 35°C in the presence (A) or the absence (B) of 10 μM Ca^{++} was monitored by viscometry. The protein solutions used were 2.5 mg/ml PMP (○), PMP + 0.7 mg/ml calmodulin (X) and PMP + 2.7 mg/ml calmodulin (●). In A, EGTA was added at the time shown by the arrow (final concentration, 1 μM). In B, when the viscosity plateau was attained, $CaCl_2$ was added at a final free Ca^{++} concentration of 3 μM.

not bind to MAPs but to tubulin in a Ca^{++}-dependent manner. This is demonstrated by gel filtration column chromatography (Fig. 2). When a mixture of calmodulin and MT proteins is applied to a column for high speed liquid chromatography (Toyo Soda Manufacturing Co., HLC803, 3000SW columns) in the presence or the absence of Ca^{++} (Fig. 2A), a definite separation of each protein is shown. In this figure, fractions I, II and III represent MAPs, tubulin and calmodulin, respectively. In the absence of Ca^{++}, calmodulin activity is found only in the calmodulin fraction. However, in the presence of Ca^{++}, a portion of calmodulin activity shifts into tubulin fraction. In contrast, calmodulin activity is not detected in MAPs fraction even in the presence of Ca^{++}. Moreover, when a mixture of calmodulin and MAPs separated from tubulin by phosphocellulose column chromatography is subjected to the high speed gel filtration, little calmodulin activity is detected in the MAPs fraction, whereas the whole activity being recovered is in the calmodulin fraction. This result demonstrates that in a Ca^{++} dependent fashion calmodulin binds to tubulin.

An affinity column chromatography using tubulin-sepharose 4B conjugates facilitates to confirm this conclusion further (Fig. 2B). In this chromatographic procedure, calmodulin is adsorbed to tubulin-sepharose 4B only in the presence of Ca^{++}, and the

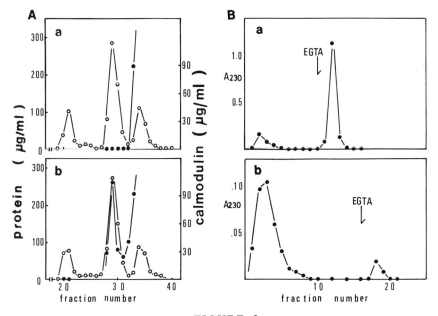

FIGURE 2

Binding of calmodulin to tubulin in the presence of Ca^{++}. A = a mixture of PMP and calmodulin in a buffer solution (10 mM MES, pH 6.8, 0.5 mM $MgCl_2$, 200 mM NaCl) containing 1.2 mM EGTA (a) or 50 µM Ca^{++} (b) was injected into a gel filtration column of high speed liquid chromatography. (○) = protein; (●) = calmodulin activity; B = calmodulin in a solution (10 mM MES, pH 6.8, 0.5 mM $MgCl_2$, 50 mM KCl) containing 2 mM $CaCl_2$ was applied to a tubulin-sepharose 4B (a) or to a MAPs-sepharose 4B (b) column equilibrated with the same buffer. At the arrow, bound calmodulin was eluted with the same buffer solution except for containing 2 mM EGTA in place of $CaCl_2$.

adsorbed calmodulin is eluted by EGTA. In contrast, calmodulin is not adsorbed to a column of MAPs-sepharose 4B even in the presence of Ca^{++}. The amount of calmodulin adsorbed to the tubulin-sepharose decreases with increasing ionic strength. However, it should be noted that the ability of calmodulin to bind to the tubulin-sepharose remains even when the column is equilibrated with 1 M KCl.

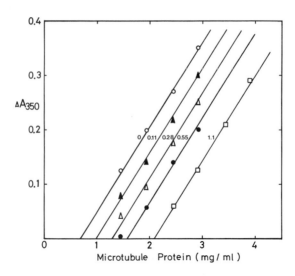

FIGURE 3

Effect of calmodulin on the critical concentration of MT assembly. The assembly medium consisted of 10 µM Ca^{++}, 10 mM MES, pH 6.8, 0.5 mM $MgCl_2$, 120 mM KCl and 0.5 mM GTP. In the presence of 0.11, 0.28, 0.55 and 1.1 mg/ml calmodulin, MT assembly was monitored at 35°C by turbidimetry.

Mechanism of Calmodulin-Induced Inhibition of MT Assembly

On the basis of these results, we have proposed the mechanism of calmodulin-induced inhibition of MT assembly as:

$$Ca^{++} + calmodulin \rightleftharpoons Ca^{++}\text{-calmodulin, and}$$

$$Ca^{++}\text{-calmodulin} + \underset{\text{(polymerizable)}}{\text{tubulin}} \rightleftharpoons \underset{\text{(non-polymerizable)}}{Ca^{++}\text{-calmodulin-tubulin}}.$$

Several examples are described for the justification of the proposed model.

Effect of calmodulin on critical concentration of MT assembly. MT assembly has been explained as a process of condensation polymerization similar to that described for G-actin to F-actin conversion (33). Therefore, it is suggested that addition of calmodulin into the MT assembly system changes tubulin concentrations available for polymerization reaction, affecting critical tubulin concentration for assembly. Figure 3 shows that calmodulin indeed causes the critical concentration of MT assembly to increase. In the MT assembly system containing micromolar Ca^{++}, if the amount of calmodulin increases, Ca^{++}-calmodulin-tubulin complex should increase, so that the critical concentration would increase because of decrease in the amount of polymerizable tubulin. This result also suggests that the calmodulin effect is stoichiometric.

Effect of calmodulin on Ca^{++}-sensitivity of MT assembly. Our model shows that the extent of inhibition is proportional to the concentration of the Ca^{++}-calmodulin complex. That is, lower Ca^{++} concentration is sufficient to suppress MT assembly with increasing calmodulin concentration; Fig. 4 shows that this is the case. Upon increasing the concentration of calmodulin, Ca^{++} concentration required for the half-maximal inhibition decreases indicating that calmodulin enhances Ca^{++} sensitivity of the MT assembly system. This result also supports the idea that Ca^{++} preferentially binds to calmodulin to form a Ca^{++}-calmodulin complex, which in turn interacts with tubulin, thereby inhibiting MT assembly.

Effect of Ca^{++} and calmodulin on MT assembly from purified tubulin alone. Calmodulin inhibits MT assembly by forming Ca^{++}-calmodulin-tubulin complex. Therefore, calmodulin shoud inhibit MT reconstitution from tubulin purified by phosphocellulose column chromatography. In the assembly system using glycerol and high concentration of Mg (16), addition of Ca^{++} or calmodulin alone shows only weak inhibitory effect on tubulin polymerization. In contrast, addition of both Ca^{++} and calmodulin inhibits polymerization significantly (Table 1). Moreover, calmodulin is shown to suppress MT elongation which can be initiated by Tetrahymena ciliary outer fibers (data not shown). This evidence indicates that the binding site of Ca^{++}-calmodulin is tubulin.

Effect of trifluoperazine on MT assembly in the presence or absence of calmodulin. Trifluoperazine (TFP) has been known as an antipsychotic reagent. It inhibits the ability of calmodulin to modulate cAMP PDE through its binding to calmodulin in the presence of Ca^{++} (17). This characteristic also facilitates investigations on the reaction between calmodulin and tubulin through analyzing the effect of TFP on MT assembly in the presence or absence of calmodulin.

A remarkable effect of TFP on MT assembly is shown in Fig. 5. In the absence of calmodulin, TFP itself inhibits MT assembly in a concentration dependent fashion (Fig. 5A). In contrast, when TFP is added to the MT assembly system to which calmodulin is supplemented, the extent of MT assembly reversely increases in a manner dependent on TFP concentration until the concentration of TFP reaches a certain level. Figure 5C summarizes these results and the scheme of reaction with TFP, calmodulin and tubulin is given as:

TABLE 1: Calmodulin-induced inhibition of MT assembly from tubulin (2.3 mg/ml) purified by phosphocellulose column chromatography.

Conditions	Percent $1/t_{\frac{1}{2}}$	
	$-Ca^{++}$	$+Ca^{++}$
Tubulin (2.3 mg/ml)	100	72
Tubulin + calmodulin (2.2 mg/ml)	83	35

The assembly buffer consisted of 3 M glycerol, 5.7 mM $MgCl_2$, 80 mM MES, pH 6.8, 0.8 mM GTP and 12 mM KCl with or without 5 μM Ca^{++}. $t_{\frac{1}{2}}$ is the time required to attain half maximal viscosity level.

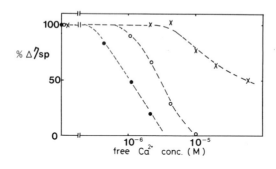

FIGURE 4

Effect of calmodulin on Ca^{++} sensitivity of MT assembly. Protein solutions contained 2.6 mg/ml PMP (X), PMP + 2.0 mg/ml calmodulin (○) and PMP + 3.6 mg/ml calmodulin (●). Assembly was monitored viscometrically. The data was expressed as percent viscosity increase. The increase of the viscosity in the absence of Ca^{++} was regarded as 100%.

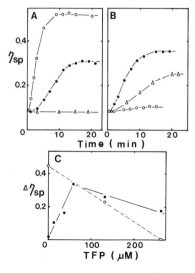

FIGURE 5

Effect of varying amounts of trifluoperazine (TFP) on MT assembly in the presence or absence of calmodulin. A = time course of MT assembly (2.7 mg/ml PMP) was monitored in the presence of 0 (○), 130 μM (●) and 250 μM TFP (△). The assembly medium contained 10 mM MES, pH 6.8, 1 mM $MgCl_2$, 10 μM Ca^{++} and 120 mM KCl. B = as for A, except that the medium contained 0.8 mg/ml calmodulin; C = summary of the effect of TFP on MT assembly in the presence (●) or absence (○) of calmodulin (from data in A and B).

Ca^{++}-calmodulin + tubulin $\rightleftharpoons$ Ca^{++}-calmodulin-tubulin,

Ca^{++}-calmodulin + tubulin + TFP $\rightleftharpoons$ Ca^{++}-calmodulin-TFP + tubulin, and

Ca^{++}-calmodulin + tubulin + TFP $\rightleftharpoons$ Ca^{++}-calmodulin-TFP + tubulin-TFP.

In the presence of TFP, calmodulin, and tubulin, TFP preferentially binds to calmodulin with a dissociation constant of approximately 1 μM and inhibits the binding of calmodulin to tubulin. Ca^{++}-calmodulin concentration in the system then decreases and assembly occurs. Upon increasing the concentration of TFP, calmodulin will be totally combined with TFP and an excess of TFP will bind to tubulin, thereby inhibiting MT assembly.

All the results described above support the idea that Ca^{++} preferentially binds to calmodulin to form Ca^{++}-calmodulin, which in turn interacts with tubulin and inhibits MT assembly.

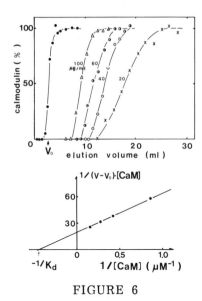

FIGURE 6

Quantitative analysis of the binding between calmodulin and tubulin. A = various concentrations of calmodulin (in 10 mM MES, pH 6.8, 0.5 mM $MgCl_2$, 50 mM KCl and 0.5 mM $CaCl_2$) were applied continuously to a tubulin-sepharose 4B column pre-equilibrated with the above Ca^{++}-containing buffer. To obtain V_o, calmodulin was chromatographed in the absence of Ca^{++}. Calmodulin content was monitored by measuring A_{230} of each fraction or calmodulin-induced activation of cAMP PDE. B = a plot of $1/(calmodulin)_o (V-V_o)$ vs $1/(calmodulin)_o$.

Quantitative Analysis of the Binding between Calmodulin and Tubulin

Quantitative analysis of the association between calmodulin and tubulin is necessary for further understanding of the reaction. The frontal analysis using tubulin-sepharose column chromatography favors this approach. This method was originally developed by Kasai and Ishii (9).

A known quantity of calmodulin, $(calmodulin)_o$, is applied continuously to the tubulin-sepharose column. When specific interaction between calmodulin and tubulin is suppressed, calmodulin is detected in a fraction designated as V_o. In contrast, under the condition that specific interaction occurs, that is, in the presence of Ca^{++}, calmodulin is eluted after a delay, the elution volume of which is designated as V. The value $V-V_o$ obtained represents the amount of calmodulin adsorbed to the tubulin-sepharose column.

Using the values V, V_o and (calmodulin)$_o$, one can obtain the amount of calmodulin bound to the tubulin column, that is, (calmodulin)$_o$(V-V_o). At equilibrium, the dissociation constant, K_d, can be calculated from

$$K_d = (tubulin)_t/(V-V_o) - (calmodulin)_o,$$

where (tubulin)$_t$ represents the total amount of tubulin coupled to sepharose 4B. Rearranging this equation, we obtain 1/(calmodulin)$_o$ (V-V_o) = K_d/(tubulin)$_t$(calmodulin)$_o$ + 1/(tubulin)$_t$. By the plot of 1/(tubulin)$_o$(V-V_o) vs 1/(calmodulin)$_o$, we can determine K_d from the intercept on the abscissa. Various concentrations of calmodulin supplied continuously into the column in the presence of Ca^{++} allows us to obtain the respective elution volume, V. Under the condition described in the legend to Fig. 6, K_d value is 2.1 μM.

DISCUSSION

We report here that Ca^{++} and calmodulin reversibly control the extent of MT assembly <u>in vitro</u>. Since MTs consist of tubulin and MAPs, it is necessary to determine whether the site of calmodulin action is tubulin or MAPs. The results presented here provide evidence that calmodulin does not bind to MAPs but to tubulin in a Ca^{++} dependent manner. The initial evidence is that when a mixture of MT proteins and calmodulin is subjected to gel filtration (Fig. 2), a portion of calmodulin activity shifts only into tubulin fraction in the presence of Ca^{++}. Secondly, calmodulin is adsorbed to tubulin-sepharose in a Ca^{++} dependent manner (Fig. 2). In the third case, Ca^{++} and calmodulin affect MT assembly from purified tubulin alone (Table 1). We therefore propose that the mechanism of calmodulin-induced modulation of MT assembly is that calmodulin preferentially takes Ca^{++} to form Ca^{++}-calmodulin complex, which interacts with tubulin and thereby inhibits MT assembly.

Our model shows that calmodulin causes a decrease in the amount of polymerizable tubulin. This is supported by the result that the critical concentration of tubulin for polymerization increases with increasing calmodulin concentration. Margolis and Wilson (21) previously reported that colchicine-tubulin complex caps the plus end of MTs, thereby inducing depolymerization of MTs. Our preliminary results indicate that calmodulin does not form cap structures. Instead, addition of Ca^{++}-calmodulin makes the equilibrium between MTs and tubulin shift into the tubulin side, resulting in depolymerization of MTs.

It was found that calmodulin decreases the amount of the ring structure which consists of tubulin and MAPs at low temperature. This suggests that either calmodulin is capable of binding to the tubulin-MAPs aggregates to dissociate the ring, or the formation of the Ca^{++}-calmodulin-tubulin complex causes tubulin concentration to decrease, resulting in shift of the equilibrium between the rings and tubulin dimers. It is most probable that once tubulin is trapped with Ca^{++}-calmodulin, it cannot participate in formation of the rings. In addition, calmodulin activity is never found in the remaining rings separated by gel filtration.

Increase in KCl concentration sensitizes MT assembly from PMP to calmodulin-induced inhibition, as described previously (30). On the other hand, in the MT assembly system from purified tubulin alone, the same increase in ionic strength reversely weakens calmolulin-induced inhibition of tubulin polymerization. Such a difference in the calmodulin effect is due to whether MAPs are contained in the assembly systems. Preliminary experiments indicate that K_d for the calmodulin-tubulin complex increases with increasing KCl concentration. This is consistent with the above finding, that for tubulin polymerization the increase in KCl concentration weakens the calmodulin action. However, evaluation of MAPs for affecting the binding reaction must await further studies.

Calmodulin is shown to adsorb to tubulin-sepharose in the presence of Ca^{++}. The use of this affinity chromatography actually facilitates isolation of calmodulin from Tetrahymena, starfish eggs and others (14). This affinity column chromatography also allows determination of the K_d value (2-5 µM) for the tubulin-calmodulin complex. In comparison with apparent K_d value of cAMP PDE for calmodulin (10 nM), K_d for the tubulin-calmodulin complex is considerably large. The low affinity of tubulin to calmodulin explains the inability of tubulin to inhibit the calmodulin action in activation of cAMP PDE. Therefore, it is valid to assume that tubulin is not responsible for inhibition of cAMP PDE as one of the calmodulin-binding proteins isolated previously (11).

It was impossible to determine the stoichiometry for calmodulin binding to tubulin using tubulin-sepharose on the quantitative basis. Instead, we measured apparent molecular weight of the calmodulin-tubulin complex by the equilibrium gel filtration method described by Hummel and Dreyer (5), using sephadex G-200 gel filtration column chromatography. By gel filtration of a mixture of tubulin and Ca^{++}-calmodulin on a column pre-equilibrated with calmodulin-containing buffer solution, apparent molecular weight of the calmodulin-tubulin complex was estimated to be in the range of 150,000-180,000. Therefore, two or more moles of calmodulin seem to bind to one mole tubulin.

In mammalian brain, the calmodulin content is unusually high as compared with the amount of calmodulin-related enzymes and calmodulin-binding proteins. It is known that the brain also contains tubulin in quite large an amount. Therefore, it is reasonable to suppose that a large amount of calmodulin is available for modulating functions of tubulin in the nervous system.

REFERENCES

1. Bryan, J., Nagle, B.W. & Doenges, K.H. (1975): Proc. Nat. Acad. Sci. 72:3570-3574.
2. Cheung, W.Y. (1970): Biochem. Biophys. Res. Commun. 38: 533-538.
3. Cheung, W.Y. (1980): Science 207:19-27.
4. Haga, T., Abe, T. & Kurokawa, M. (1974): FEBS Lett. 39: 291-295.
5. Hummel, J.P. & Dreyer, W.J. (1962): Biochim. Biophys. Acta 63:530-532.
6. Inoue, S. & Sato, H. (1967): J. Gen. Physiol. (Suppl.) 50: 259-292.
7. Jameson, L., Frey, T., Zeeberg, B., Dalldorf, F. & Caplow, M. (1980): Biochemistry 19:2472-2479.
8. Kakiuchi, S. & Yamazaki, R. (1970): Biochem. Biophys. Res. Commun. 41:1104-1110.
9. Kasai, K. & Ishii, S. (1978): J. Biochem. 85:1062-1069.
10. Klee, C., Crouch, T.H. & Richman, P.G. (1980): Ann. Rev. Biochem. 49:489-515.
11. Kumagai, H. & Nishida, E. (1979): J. Biochem. 85:1267-1274.
12. Kumagai, H. & Nishida, E. (1980): Biomed. Res. 1:223-229.
13. Kumagai, H., Nishida, E. & Sakai, H. (1979): J. Biochem. 85: 495-502.
14. Kumagai, H., Nishida, E., Ishiguro, K. & Murofishi, H. (1980): J. Biochem. 87:667-670.
15. Laemmli, U.K. (1970): Nature 227:680-685.
16. Lee, J.C. & Timasheff, S.N. (1975): Biochemistry 14:5183-5187.
17. Levin, R.M. & Weiss, B. (1976): Molec. Pharmacol. 12:581-589.
18. Lockwood, A.H. (1979): Proc. Nat. Acad. Sci. 76:1184-1188.
19. Lowry, O.H., Rosebrough, N.J., Farr, A.L. & Randall, R.J. (1951): J. Biol. Chem. 193:265-275.
20. Marcum, J.M., Dedman, J.R., Brinkley, B.R. & Means, A.R. (1978): Proc. Nat. Acad. Sci. 75:3771-3775.
21. Margolis, R.L. & Wilson, L. (1977): Proc. Nat. Acad. Sci. 74: 3466-3470.
22. Matsumoto, G. & Sakai, H. (1979): J. Memb. Biol. 50:1-14.
23. Matsumoto, G. & Sakai, H. (1979): J. Memb. Biol. 50:15-22.
24. Means, A.R. & Dedman, J.R. (1980): Nature 285:73-77.
25. Murphy, D.B. & Borisy, G.G. (1975): Proc. Nat. Acad. Sci. 72:2696-2700.
26. Naruse, H. & Sakai, H. (1980): Biomed. Res. 1:151-157.
27. Nishida, E. (1978): J. Biochem. 84:507-512.
28. Nishida, E. & Sakai, H. (1977): J. Biochem. 82:303-306.
29. Nishida, E. & Sakai, H. (1980): J. Biochem. 88:1577-1586.
30. Nishida, E., Kumagai, H., Ohtsuki, I. & Sakai, H. (1979): J. Biochem. 85:1257-1266.
31. Ogawa, Y. (1968): J. Biochem. 64:255-257.
32. Olmsted, J.B. & Borisy, G.G. (1975): Biochemistry 14:2966-3005.

33. Oosawa, F. & Asakura, S. (1975): Thermodynamics of the Polymerization of Protein, Academic Press, New York.
34. Raff, E.C. (1979): Int. Rev. Cytol. $\underline{59}$:1-96.
35. Sakai, H. (1980): Biomed. Res. $\underline{1}$:359-375.
36. Schliwa, M. (1976): J. Cell Biol. $\underline{70}$:527-540.
37. Schline, P., Shiavone, K. & Brocato, S. (1979): Science $\underline{205}$: 593-595.
38. Vater, W., Muller, H. & Unger, E. (1978): Biochem. Biophys. Res. Commun. $\underline{84}$:721-726.
39. Weingarten, M.D., Lockwood, A.H., Hwo, S.-W. & Kirschner, M.W. (1975): Proc. Nat. Acad. Sci. $\underline{72}$:1858-1862.
40. Weisenberg, R.C. (1972): Science $\underline{117}$:$\underline{1}$104-1105.

CALMODULIN AND CYTOSKELETON

S. Kakiuchi, K. Sobue, M. Fujita, Y. Muramoto,
K. Kanda and K. Morimoto

Institute of Higher Nervous Activity, Osaka University
Medical School, Nakanoshima, Kita-ku, Osaka 530, Japan

Discovery of Ca^{++}-activatable phosphodiesterase (21) and subsequent demonstration of protein modulator which confers Ca^{++}-sensitivity upon this enzyme (22,23) coincided with the discovery of a protein activator of brain phosphodiesterase (3). However, it took several years before these two independent lines of the research merged when the identity of the two proteins as a Ca^{++} binding protein was finally established (24,29,49). Since then the role of this protein, which is now called calmodulin, as a Ca^{++} dependent activator of intracellular enzymes have been thoroughly investigated in a number of laboratories. On the other hand, surprisingly few studies have been done with regard to the possible role of calmodulin in the regulation of contractile or cytoskeletal systems.

The amino acid sequence of mammalian calmodulin (7,53) is homologous to those reported for fast muscle troponin-Cs (5,51), suggesting that non-muscle calmodulin and muscle troponin-Cs may have stemmed from a common ancestral protein and may perform analogous intracellular functions as four-domain calcium receptive proteins. Previously, Perry and his associates (1) have shown that calmodulin associates with skeletal muscle troponin components to form a soluble hybrid complex and neutralizes the inhibitor action of troponin-I on the actomyosin ATPase activity. A similar observation was made by Dedman et al. (6).

In this report we will summarize our recent findings concerning the cytoskeleton-related calmodulin-binding proteins in smooth muscle and non-muscle tissues. Through these calmodulin-binding proteins, calmodulin interacts with cytoskeletal systems of those tissues.

Criteria and Assay for Calmodulin Binding Protein

The present report deals with four calmodulin-binding proteins, all of which are related to the cytoskeleton. During the course of the investigation on these proteins, however, we have felt the need to set up multiple criteria for the identification of the calmodulin-binding protein. For these criteria we employed the following: a) Ca^{++} dependent formation of a protein complex between the protein and calmodulin should be demonstrated by direct means. For the separation of such protein complex from free calmodulin, four different methods were employed. For particulate form calmodulin-binding proteins, centrifugation as in (50) and/or filtration through Whatman GF/C filter paper as in (41) were employed. For soluble form (or solubilized) proteins, gel filtration column chromatography as in (40) and/or electrophoresis on polyacrylamide gels in a buffer system of 122 mM glycine-20 mM Tris (pH 8.3) as in (25, 42-45) were used. For spectrin, 6 M urea was included in this buffer system (42,43); b) the binding of [^{3}H]calmodulin to the binding protein should be displaced by an excess of unmodified calmodulin but not other proteins that include troponin C; c) the protein should be retained by and released from a calmodulin-sepharose column in a Ca^{++} dependent fashion, and d) the binding protein should preferably be acidic (or neutral) because the interaction of calmodulin, an acidic protein, with a basic protein is likely to be non-specific, as has recently been pointed out by Itano et al. (17).

All calmodulin-binding proteins reported here fulfilled these criteria. Criterion 'a' was utilized for the determination of the binding protein. Protein samples to be assayed were incubated in the presence or absence (+EGTA) of Ca^{++} with [^{3}H]calmodulin prepared as described in (41) and then bound and unbound forms of [^{3}H]calmodulin were separated from each other as described in 'a'. The radioactivity of the bound [^{3}H]calmodulin was determined in a liquid scintillation spectrometer. Control values with EGTA were subtracted from corresponding experimental values with Ca^{++}. The subtracted value, representing Ca^{++} dependent bindings, agreed with that of the bound [^{3}H]calmodulin, displaceable by an excess of unlabelled calmodulin (specific binding).

Spectrin as a Calmodulin Binding Protein of Erythrocytes

The cytoplasmic surface of the erythrocyte membrane is lined with a meshwork, which is conventionally called erythrocyte cytoskeleton. This structure is conserved after the membrane is extracted with a non-ionic detergent such as Triton X-100, but is solubilized from the erythrocyte ghosts with 6 M urea. The cytoskeleton consists of several proteins that include spectrin (Bands 1 plus 2), actin (Band 5), Bands 2.1 and 4.1 (39, 57).

We found that the erythrocyte cytoskeleton which was solubilized from the ghost with 6 M urea (Fig. 1) binds calmodulin in a Ca^{++} dependent fashion. This calmodulin binding activity was purified from the urea extract by column chromatography using sepharose CL-6B and then calmodulin-sepharose 4B conjugate. The binding activity was retained on the affinity column of calmodulin-sepharose with Ca^{++} and eluted from the column with EGTA. The active peak thus obtained, which accounted for all the binding activity present in the original urea extract, contained only spectrin as the protein constituent (43). As shown in Table 1, the number of spectrin molecule per cell agrees well with the number of calmodulin found in the cell and the K_d value of spectrin for calmodulin also agrees with the average concentration of calmodulin in the cell.

Ca^{++} pump ATPase of the erythrocyte membrane is known to require calmodulin for its activity (12,19) and therefore binds calmodulin in a Ca^{++} dependent manner. However, the number of this enzyme per cell is 4,500 according to Jarrett and Kyte (18), which is less than 3% of the number of the cytoskeletal calmodulin binding protein (spectrin). K_d value for calmodulin of the Ca^{++} pump ATPase is three orders of magnitude less than that of spectrin (Table 1). This means that the Ca^{++} pump ATPase is always saturated with calmodulin as long as Ca^{++} is present, while the formation of spectrin-calmodulin complex may be subject to a delicate equilibrium between two proteins depending upon the local concentration of calmodulin and other intracellular conditions.

There is now good evidence that the shape and deformability of the erythrocyte is determined by its cytoskeleton and hence, to a great extent, by spectrin (31,35). For instance, in spherocytosis mutant mice, the erythrocyte shape change correlated with the degree of the decrease in spectrin (14,30). Therefore, the possibility arose that the calmodulin-spectrin interaction may be related to the control of the erythrocyte shape. By using calmodulin antagonists, we examined this possibility (see below). Human erythrocytes, washed with saline, were incubated with drugs (100 μM) for 30 min at room temperature and fixed with 1% glutaraldehyde and then examined with a scanning electron microscope for their shape. As shown in Table 2 (26), the degrees of the erythrocyte shape change produced by these compounds was well correlated with the potencies of the compounds in terms of the inhibition of the Ca^{++} and calmodulin dependent activation of brain phosphodiesterase. One might argue that the observed shape change may be due to the inhibition of Ca^{++} pump ATPase of the erythrocyte

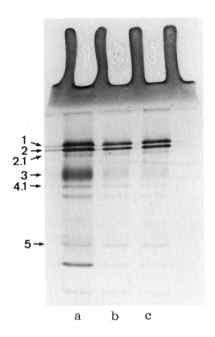

FIGURE 1

Erythrocyte cytoskeleton proteins (43). a = ghost; b = after treatment with 1% Triton X-100; c = urea extract of the Triton ghost.

TABLE 1: Calmodulin and calmodulin binding protein in human erythrocyte ghosts.

		Reference
Number of spectrin molecules per cell	200,000	(48)
Number of calmodulin molecules per cell	160,000	(43)
Number of Ca^{++}-pump ATPase per cell	4,500	(18)
K_d value of spectrin for calmodulin	2.8×10^{-6} M	(43)
K_d value of Ca^{++}-pump ATPase for calmodulin	3×10^{-9} M	(36)
Concentration of calmodulin in erythrocyte	2.5×10^{-6} M	(43)

membrane with a consequent increase in the Ca^{++} concentration in the cell. However, when the incubation was carried out in a medium containing 2 μM Ca^{++}, 100 μM trifluoperazine still produced a similar shape change. The result indicates that, although not excluding the possible contribution of the inhibition of the calmodulin dependent Ca^{++} pump to the shape change, there is another calmodulin dependent cellular process, i.e. calmodulin-spectrin interaction, whose inhibition is responsible for the shape change.

A Spectrin Like Calmodulin Binding Protein (240 K Protein) from Brain

Although calmodulin is generally regarded as a soluble protein, it was observed that, when calmodulin was prepared from a brain homogenate, inclusion of EDTA in the homogenizing buffer greatly increased the yield of this protein in the soluble fraction (52). Subsequently, work in our laboratory demonstrated the existence of calmodulin binding protein(s) in the sedimentable fraction of the homogenate (50). Through this protein a considerable amount of cytoplasmic calmodulin was tied to the particulate fraction in a Ca^{++} dependent manner. [There is another pool of the particle-associated calmodulin, which is not released into the soluble fraction by the treatment of the particulate fraction with EGTA (50). This calmodulin activity was solubilized from the particulate fraction with detergent and purified to a homogeneous protein by column chromatography. The purified protein was found to be identical to the soluble form calmodulin (46).] Thus, the particulate fraction derived from 1 g of cerebral cortex was able to bind 80-100 μg calmodulin (41,50), which accounts for about 20-25% of the total amount, i.e. 400 μg/g tissue (50), of the EGTA extractable calmodulin in the brain tissue. The amount of this particulate form calmodulin binding protein is much greater than that reported for the soluble form calmodulin binding proteins, heat-labile (calcineurin) and heat-stable proteins, in brain (38). The particulate calmodulin binding protein distributes in a variety of mammalian tissues, but its highest concentration was found in brain, followed by adrenal gland (Fig. 2). Upon subcellular fractionation of a brain homogenate, distribution of the activity of this protein paralled that of synaptic membranes (41), suggesting that this protein is implicated in the synaptic function.

We then purified this calmodulin binding protein from a microsomal fraction of the brain. The nerve ending particles (synaptosomes) had been osmotically disrupted and hence the microsomal fraction contained most of the synaptic membranes and vesicles. The calmodulin binding activity was solubilized with 6 M urea and then purified to apparent homogeneity upon polyacrylamide gel electrophoresis by successive chromatography using columns of DEAE-cellulose, calmodulin-sepharose, sepharose 4B, and calmodulin-

TABLE 2: Erythrocyte shape change produced by calmodulin antagonists (26).

Compound	Erythrocyte Shape Change	Concentration* (I_{50}, µM)
Promethazine	Stomatocyte-discocyte mixture	400
W-7	Sperocyte-stomatocyte mixture	50
Chlorpromazine	Spherocyte-stomatocyte mixture	35
Trifluoperazine	Spherocyte	8.5

*Concentration of compounds producing 50% inhibition of Ca^{++} and calmodulin dependent activation of brain phosphodiesterase.

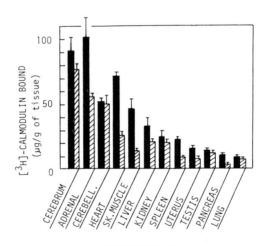

FIGURE 2

Distribution of particulate form calmodulin binding protein(s) in rat tissue (41). 105,000 x g pellets of tissue homogenates were either directly (solid bars) or after treated with 0.6 M KCl (hatched bars) subjected to the binding assay with [^{3}H]calmodulin.

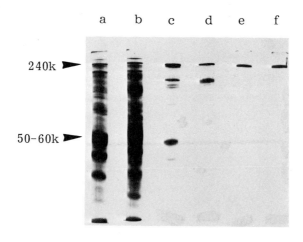

FIGURE 3

SDS-polyacrylamide gel electrophoresis of the purification steps of 240 k protein from brain (25). a = microsomal fraction (starting material); b = urea extract of (a); c = DEAE-cellulose column fraction; d = calmodulin-sepharose column fraction; e = sepharose 4B column fraction; f = after the second time calmodulin-sepharose column run.

sepharose (second time)(Fig. 3). The molecular weight of this calmodulin binding protein, as determined by SDS-polyacrylamide gel electrophoresis, was 240,000.

This 240 k protein accounts for most of the calmodulin binding activity in the microsomal fraction. Upon subcellular fractionation of a brain homogenate, we found that the 240 k protein is exclusively localized in the microsomal fraction. Indeed, a densitometric scanning of a SDS-polyacrylamide gel on which a sample from the microsomal fraction was electrophoresed showed that this protein represented 3.0% of the total protein (25). With this value the amount of the 240 k protein was calculated to be 370 mg/kg of brain.

Although this protein was obtained from the particulate fraction, it is not considered to be an intrinsic membrane protein because it was not solubilized from the particulate fraction with non-ionic detergents. Instead, it was solubilized with 6 M urea or 6 M guanidine-HCl. This fact, together with its Mr value, present a certain resemblance between the 240 k protein and spectrin. There is a good possibility that this 240 k protein associates with the structures underlying the membranes, as is the case with spectrin.

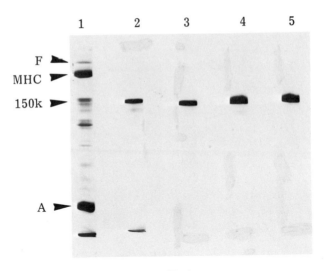

FIGURE 4

SDS-polyacrylamide gel electrophoresis of the purification steps of caldesmon from chicken gizzard muscle (44). 1 = ammonium sulfate fraction (30-50% saturation); 2 = sepharose 4B column fraction; 3 = DEAE-cellulose column fraction; 4 = calmodulin-sepharose column fraction; 5 = sepharose 4B column (second time) fraction.
F = filamin; MHC = myosin heavy chain; A = actin.

An Actin Related Calmodulin Binding Protein, Caldesmon

Recently we purified from a 0.3 M KCl extract of chicken gizzard, a calmodulin binding protein (Fig. 4). This protein, named caldesmon from calmodulin and the Greek desmos "binding," had a molecular weight of 150,000 on SDS-polyacrylamide gel electrophoresis (44,45).

Caldesmon associated with calmodulin as well as F-actin. While the interaction between caldesmon and calmodulin was Ca^{++} dependent, the interaction between caldesmon and F-actin was not (Fig. 5). Moreover, the Ca^{++} dependent association of calmodulin with caldesmon eliminated the interaction between caldesmon and F-actin. This was confirmed by two different methods, i.e. a high-speed centrifugation method (Fig. 5) and viscometry (45). Therefore, in the presence of these three protein species, the concentration of Ca^{++} acts as a flip-flop switch toward the formation of the caldesmon-calmodulin complex at the increased level and toward the formation of the caldesmon-F-actin complex at the decreased level (Fig. 6). At about 1 μM Ca^{++}, formations of both complexes were in equilibrium (45).

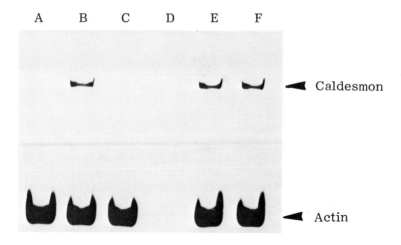

FIGURE 5

Interaction between caldesmon, calmodulin and F-actin in the presence and absence of Ca^{++} (45). Protein samples were incubated for 30 min at 30°C in the presence of either 0.2 mM $CaCl_2$ (Lanes A-D), or EGTA (E,F). At the end of the incubation, mixtures were centrifuged at 200,000 x g for 30 min at 4°C. F-actin and F-actin-associated proteins were sedimented by the centrifugation; free calmodulin and caldesmon remained in the supernatant. The sediments were dissolved in 0.1% SDS and subjected to SDS-polyacrylamide gel electrophoresis. A = F-actin; B & E = F-actin + caldesmon; C & F = F-actin + caldesmon + calmodulin; D = caldesmon.

Previously, immunofluorescent localization of calmodulin in fibroblasts and epithelial cells coincided with cellular stress fibers (actin filaments)(8). Moreover, calmodulin was found to be one of the major components of the microvillus core filament structure where actin fibers are concentrated (11,16). In spite of this indirect evidence, Howe et al (16), as well as our own results (45) found no indication of binding of calmodulin to actin. Caldesmon may be a solution to this puzzle, i.e. through this protein, calmodulin can interact with actin filaments in a Ca^{++} dependent manner. The caldesmon content in chicken gizzard muscle was estimated to be about 240 mg per 100 g of tissue (44), which accounts for about 70% of the total amount of calmodulin in this tissue taking the total amount of calmodulin to be about 40 mg/100 g (13). As shown in another chapter (Maruyama, Sobue, Kakiuchi, this volume), the Ca^{++} and calmodulin-regulated binding of caldesmon to actin did not alter the physical state of actin filaments, i.e. gel formation or shortening of actin filaments did not occur. One attractive possibility is that the interaction between caldesmon and actin filaments may modify the function of the actin filaments in relation with other proteins. We are currently investigating this possibility.

A Tubulin Related Calmodulin Binding Protein (Tau Factor) and a Ca^{++} Dependent Flip-Flop Regulation in Microtubule Assembly-Disassembly

Since Weisenberg (55) first succeeded in microtubule assembly in vitro by chelating Ca^{++} using EGTA, Ca^{++} has been thought to be a physiological regulator governing microtubule assembly-disassembly (15,33). Consequently, the possibility arose that this effect of Ca^{++} may be mediated by calmodulin. Thus, Welsh et al. (56) observed a characteristic localization of calmodulin in the chromosome-to-pole region of the mitotic apparatus visualized by immunofluorescence, and Marcum et al. (32) and Nishida et al. (34) subsequently found that calmodulin both inhibits and reverses microtubule assembly in vitro in the presence of 10^{-6} approximately 10^{-5} M Ca^{++}. Although these results strongly suggest a possible implication of calmodulin in the Ca^{++} dependent microtubule disassembly, the direct evidence, as well as its molecular mechanism, was yet to be shown.

As a first step toward this subject, we have searched the microtubule proteins for a calmodulin interacting protein. The microtubules in vivo are constituted not only of tubulin, but also several non-tubulin accessory proteins. HMr-MAPs of approximately 300,000 Mr (2) and a family of four closely related lower Mr (approximately 55,000-62,000) proteins named tau (τ) factor (54) have been characterized recently for these non-tubulin accessory proteins. Therefore, we isolated and separated these three protein species from each other by column chromatography (Fig. 7) and each protein species was examined for its ability to bind calmodulin in the presence of Ca^{++}. Ca^{++} dependent binding between protein and [^{3}H]calmodulin was studied by two different means, i.e. gel filtration column chromatography using sephadex G-100 (47) and affinity column chromatography using calmodulin-sepharose (Fig. 8). The results clearly established that only tau factor but not purified tubulin or HMr-MAPs is capable of binding to calmodulin in a Ca^{++} dependent manner. This finding is inconsistent with the reports by Kumagai and Nishida (27,28), in which the purified tubulin dimer was shown to be the calmodulin binding protein. The reason for this discrepancy is unclear at the present time.

Tau factor was discovered by Weingarten et al. (54) to be an essential factor for the assembly of the tubulin dimer into the microtubules. When the tubulin is freed of the microtubule-associated proteins by phosphocellulose column chromatography, it can no longer assemble into the microtubules under standard polymerization conditions. Addition of the tau factor back to the PC-tubulin fully

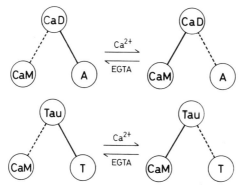

FIGURE 6

Diagrammatic presentation of the flip-flop mechanism working in the calmodulin (CaM)/caldesmon (CaD)/actin (A) system and calmodulin (CaM)/tau factor/tubulin (T) system.

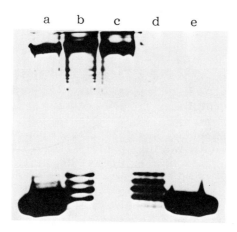

FIGURE 7

SDS-polyacrylamide gel electrophoresis on the microtubule proteins (47). a = microtubules purified by three assembly-disassembly cycles; b = crude MAPs containing both HMr-MAPs and tau factor; c = HMr-MAPs; d = tau factor; e = tubulin dimer.

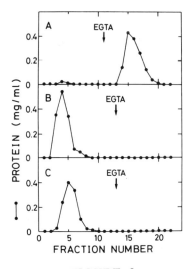

FIGURE 8

Test for the binding of microtubule proteins with a calmodulin-sepharose column (47). After the protein samples were applied to the column, the column was eluted with a medium containing approximately 10^{-5} M Ca^{++} and then, at the arrow indicated in the figure, 2 mM EGTA in place of Ca^{++} was added. A = tau factor; B = HMr-MAPs; C = tubulin dimer.

restored its polymerizability (4,9,10,37,54). Cleveland et al.(4) concluded that tau is both necessary and sufficient for nucleation and elongation of microtubules from PC-tubulin.

From the above evidence, in combination with our results that tau factor is a calmodulin binding protein, the Ca^{++} dependent microtubule disassembly is now explained by the Ca^{++} dependent association of calmodulin with tau factor thereby depriving the tubulin of the factor (tau) which confers upon the tubulin the ability to polymerize. Figure 9 shows the Ca^{++} and calmodulin dependent inhibition of the tubulin assembly in a reconstituted system consisting of the PC-tubulin and tau factor. The system was reversible depending upon the Ca^{++} concentration: 3 μM was found for the half maximum inhibition of the assembly (20). Thus, Ca^{++} acts as a flip-flop switch toward assembly and disassembly depending upon its concentration (Fig. 6).

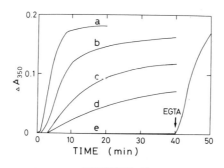

FIGURE 9

Inhibition of microtubule assembly by the addition of calmodulin (20). 1.06 mg/ml PC-tubulin was incubated at 37°C with 308 µg/ml tau factor and the increasing concentrations of calmodulin and other additions. Ca^{++} at a concentration of 1×10^{-5} M was present in all tubes except that, at the arrow indicated on curve (e), EGTA was added to decrease the concentration of Ca^{++} to approximately 1×10^{-7} M. The degree of polymerization was determined by measuring the turbidity at 350 nm. Calmodulin concentrations in µg/ml: a = 0; b = 63; c = 188; d = 375; e = 750.

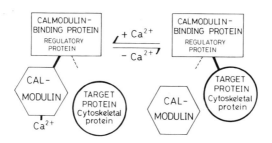

FIGURE 10

Ca^{++} and calmodulin dependent flip-flop mechanism in the regulation of cytoskeletons. Actin and tubulin represent target protein.

Ca^{++} and Calmodulin Dependent Flip-Flop Mechanism in Regulation of Cytoskeleton

The subject of this paper is four calmodulin binding proteins, spectrin from erythrocytes, 240 k protein from brain, caldesmon from chicken gizzard, and tau factor of microtubules, all of which are related to cytoskeletons. Calmodulin interacts with cytoskeletons through these calmodulin binding proteins. Of particular interest is the Ca^{++} and calmodulin dependent flip-flop mechanism as illustrated schematically in Fig. 10. In this mechanism, with calmodulin binding proteins as regulatory proteins, calmodulin is able to control functions of actin filaments or tubulin in a Ca^{++} dependent manner. Two such examples, the caldesmon/actin system, and the tau factor/tubulin system, were uncovered and are described in detail in this paper (see Fig. 6). In addition, we have recently found two other cases of this category, a 160 k protein from brain and a 140 k protein from chicken gizzard muscle, both of which interact with actin in a Ca^{++} dependent flip-flop fashion. Thus we propose that this flip-flop mechanism be extended to a general principle working in the system where calmodulin is implicated in the regulation of the cytoskeleton and contractile systems.

ACKNOWLEDGEMENTS

This research was supported in part by grants from the Scientific Research Fund of Ministry of Education, Science and Culture, Japan, and from Yamada Science Foundation and the Naito Research Grant. We are grateful to Professors S. Ebashi and M. Kurokawa for helpful discussions and encouragement at the beginning of the calmodulin work. We are also grateful for the valuable suggestions of Professor M. Endo. We also extend our appreciation to Miss A. Yoshikawa for her excellent secretarial assistance throughout this work.

REFERENCES

1. Amphlett, G.W., Vanaman, T.C. & Perry, S.V. (1976): FEBS Lett. 72:163-168.
2. Borisy, G.G., Marcum, J.M., Olmsted, J.B., Murphy, D.B. & Johson, K.A. (1975): Ann. N.Y. Acad. Sci. 253:107-132.
3. Cheung, W.Y. (1970): Biochem. Biophys. Res. Commun. 38: 533-538.
4. Cleveland, D.W., Hwo, S.Y. & Kirschner, M.W. (1977): J. Molec. Biol. 116:207-225.
5. Collins, J.H., Greaser, M.L., Potter, J.D. & Horn, M.J. (1977): 252:6356-6362.
6. Dedman, J.R., Potter, J.D. & Means, A.R. (1977): J. Biol. Chem. 252:2437-2440.

7. Dedman, J.R., Jackson, R.L., Schreiber, W.E. & Means, A.R. (1978): J. Biol. Chem. 253:343-346.
8. Dedman, J.R., Welsh, M.J. & Means, A.R. (1978): J. Biol. Chem. 253:7515-7521.
9. Fellous, A., Francon, J., Lennon, A.M. & Nunez, J. (1977): Eur. J. Biochem. 78:167-174.
10. Francon, J., Fellous, A., Lennon, A.M. & Nunez, J. (1978): Eur. J. Biochem. 85:43-53.
11. Glenney, J.R. Jr., Bretscher, A. & Weber, K. (1980): Proc. Nat. Acad. Sci. 77:6458-6462.
12. Gopinath, R.M. & Vincenzi, F.F. (1977): Biochem. Biophys. Res. Commun. 77:1203-1209.
13. Grand, R.J.A. & Perry, S.V. (1979): Biochem. J. 183:285-295.
14. Greenquist, A.C., Shohet, S.B. & Bernstein, S.E. (1978): Blood 51:1149-1155.
15. Haga, T., Abe, T. & Kurokawa, M. (1974): FEBS Lett. 39:291-295.
16. Howe, C.L., Mooseker, M.S. & Graves, T.A. (1980): J. Cell. Biol. 85:916-923.
17. Itano, T., Itano, R. & Penniston, J.T. (1980): Biochem. J. 187:455-459.
18. Jarrett, H.W. & Kyte, J. (1979): J. Biol. Chem. 254:8237-8244.
19. Jarrett, H.W. & Penniston, J.T. (1977): Biochem. Biophys. Res. Commun. 77:1210-1216.
20. Kakiuchi, S. & Sobue, K. (1981): FEBS Lett. 132:141-143.
21. Kakiuchi, S. & Yamazaki, R. (1970): Proc. Japan Acad. 46:387-392.
22. Kakiuchi, S. & Yamazaki, R. (1970): Biochem. Biophys. Res. Commun. 41:1104-1110.
23. Kakiuchi, S., Yamazaki, R. & Nakajima, H. (1970): Proc. Japan Acad. 46:587-592.
24. Kakiuchi, S., Yamazaki, R., Teshima, Y. & Uenishi, K. (1973): Proc. Nat. Acad. Sci. 70:3526-3530.
25. Kakiuchi, S., Sobue, K. & Fujita, M. (1981): FEBS Lett. 132:144-148.
26. Kidoguchi, K., Hayashi, A., Sobue, K., Kakiuchi, S. & Hidaka, H. (1982): This Volume.(Appendix)
27. Kumagai, H. & Nishida, E. (1979): J. Biochem. 85:1267-1274.
28. Kumagai, H. & Nishida, E. (1980): Biomed. Res. 1:223-229.
29. Lin, Y.M., Liu, Y.P. & Cheung, W.Y. (1974): J. Biol. Chem. 4943-4954.
30. Lux, S.E. (1979): Sem. Hematol. 16:21-51.
31. Marchesi, V.T. (1979): J. Memb. Biol. 51:101-131.
32. Marcum, J.M., Dedman, J.R., Brinkley, B.R. & Means, A.R. (1978): Proc. Nat. Acad. Sci. 75:3771-3775.
33. Nishida, E. & Sakai, H. (1977): J. Biochem. 82:303-306.
34. Nishida, E., Kumagai, H., Ohtsuki, I. & Sakai, H. (1979): J. Biochem. 85:1257-1266.
35. Palek, J. & Liu, S.-C. (1979): Sem. Hematol. 16:75-93.
36. Raess, B.U. & Vincenzi, F.F. (1980): Molec. Pharmacol. 18:253-258.

37. Sandoval, I.V. & Weber, K. (1980): J. Biol. Chem. 255:8952-8954.
38. Sharma, R.K., Desai, R., Waisman, D.M. & Wang, J.H. (1979): J. Biol. Chem. 254:4276-4282.
39. Sheetz, M.P. & Sawyer, D. (1978): J. Supramolec. Struct. 8: 399-412.
40. Sobue, K. & Kakiuchi, S. (1980): Biochem. Biophys, Res. Commun. 93:850-856.
41. Sobue, K., Muramoto, Y., Yamazaki, R. & Kakiuchi, S. (1979): FEBS Lett. 105:105-109.
42. Sobue, K., Fujika, M., Muramoto, Y. & Kakiuchi, S. (1980): Biochem. Int. 1:561-566.
43. Sobue, K., Muramoto, Y., Fujita, M. & Kakiuchi, S. (1981): Biochem. Biophys. Res. Commun. 100:1063-1070.
44. Sobue, K., Muramoto, Y., Fujita, M. & Kakiuchi, S. (1981): Biochem. Int. 2:469-475.
45. Sobue, K., Muramoto, Y., Fujita, M. & Kakiuchi, S. (1981): Proc. Nat. Acad. Sci. 78:5652-5655.
46. Sobue, K., Yamazaki, R., Yasuda, S. & Kakiuchi, S. (1981): FEBS Lett. 129:215-219.
47. Sobue, K., Fujita, M., Muramoto, Y. & Kakiuchi, S. (1981): FEBS. Lett. 132:137-140.
48. Steck, T.L. (1974): J. Cell Biol. 62:1-19.
49. Teo, T.S. & Wang, J.H. (1973): J. Biol. Chem. 248:5950-5955.
50. Teshima, Y. & Kakiuchi, S. (1978): J. Cyclic Nucleotide Res. 4:219-231.
51. Van Eerd, J.-P. & Takahashi, K. (1976): Biochemistry 15:1171-1180.
52. Watterson, D.M., Harrelson, W.G., Keller, P.M., Sharief, F. & Vanaman, T.C. (1976): J. Biol. Chem. 251:4501-4513.
53. Watterson, D.M., Sharief, F. & Vanaman, T.C. (1980): J. Biol. Chem. 255:962-975.
54. Weingarten, M.D., Lockwood, A.H., Hwo, S.Y. & Kirschner, M.W. (1975): Proc. Nat. Acad. Sci. 72:1858-1862.
55. Weisenberg, R.C. (1972): Science 177:1104-1105.
56. Welsh, M.J., Dedman, J.R., Brinkley, B.R. & Means, A.R. (1978): Proc. Nat. Acad. Sci. 75:1867-1871.
57. Yu, J., Fischman, A. & Steck, T.L. (1973): J. Supramolec. Struct. 1:233-248.

EFFECT OF THE CALMODULIN-CALDESMON SYSTEM ON THE

PHYSICAL STATE OF ACTIN FILAMENTS

K. Maruyama, K. Sobue* and S. Kakiuchi*

Department of Biology, Faculty of Science, Chiba
University, Chiba 260, Japan, and
*Institute of Higher Nervous Activity, Osaka University
Medical School, Osaka 530, Japan

F-actin solution was measured at 30°C for flow birefringence and low-shear viscosity in the presence of calmodulin and caldesmon. The results indicate that caldesmon was bound to actin filament when Ca^{++} was absent. However, in the presence of Ca^{++}, caldesmon was not bound to actin filaments. This Ca^{++}-regulated association and dissociation of caldesmon with actin filament was not influenced by tropomyosin at all. Gel formation or shortening of actin filaments was not evoked by the interaction of caldesmon with actin filaments.

INTRODUCTION

Several proteins are now known that interact with actin filaments and regulate the physical state of actin filaments in a Ca^{++}-dependent manner. Thus gelsolin from macrophage (17,18), fragmin from plasmodium (3), and villin from brush border (1) were shown to cut actin filaments to shorter pieces in the presence of micromolar concentrations of Ca^{++}; whereas actinogelin from ascites tumor cells (11,13) enhanced gelation of actin filaments in the absence of Ca^{++}. Implication of these proteins in the regulation of cell motility has been strongly suggested.

Sobue et al. (14,15) have recently purified, from chicken gizzard, a new actin-binding protein called caldesmon. Its interaction with actin filaments was regulated by Ca^{++}/calmodulin in a flip-flop fashion: presence or absence of Ca^{++} leads to the formation of protein complexes either calmodulin-caldesmon or caldesmon-actin, respectively, with the consequent association or dissociation, respectively, between caldesmon and actin filaments (14,15). The purpose of the present experiments is to see whether the calmodulin-caldesmon system affects the physical state of actin filaments or not.

MATERIALS AND METHODS

Caldesmon was purified from chicken gizzard as reported (14, 15) and used within several days after preparation. Calmodulin was prepared from bovine brain according to Kakiuchi et al. (4), and actin was purified from rabbit skeletal muscle by the method of Spudich and Watt (16).

Degrees of extinction angle and flow birefringence were determined in a Edsall-type rotating apparatus as described earlier (5) Structural viscosity was determined in a Low-Shear 100 Rheometer (Contraves, Zurich)(8).

RESULTS

Flow birefringence of a mixture of F-actin, calmodulin and caldesmon was compared with F-actin solution (Fig. 1). The results indicate that caldesmon does bind to actin filaments in the absence of Ca^{++} as evidenced by a slight but consistent increase in the flow birefringence value at all the velocity gradient examined. The results were reproducible at any given velocity gradient. There was no sign of gel formation of actin filaments. The values of extinction angle at any velocity gradient were the same as those of control within an experimental error (Fig. 1), suggesting that neither gelation nor shortening of actin filaments occurred (cf. 7).

In the presence of Ca^{++} there was no change in degree of either birefringence and extinction angle as compared with actin filaments alone (Fig. 1). This shows that caldesmon does not bind to actin filaments in the presence of calmodulin and Ca^{++}. Caldesmon alone increased the degree of birefringence of actin by approximately 10% regardless of the presence or absence of Ca^{++}.

Actin filaments are fragmented into short pieces under a sonic field, and the fragments are quickly associated by end-to-end binding to reform long filaments (10). The re-annealing process after sonication can be easily followed by measuring the increase in the degree of birefringence at a given velocity gradient. If there is an ending factor, such as muscle β-actinin that binds to the arrow-pointing end of actin filaments, the re-association is inhibited (9,10) and the birefringence value should stay at low level as represented by (---) in Fig. 2. However, both in the presence and absence of Ca^{++}, the calmodulin-caldesmon-actin mixture showed a rapid recovery after sonication (Fig. 2). Therefore, the binding of caldesmon to the ends of actin filaments is not likely.

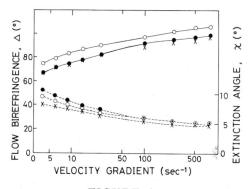

FIGURE 1

Calcium-dependent effects of calmodulin-caldesmon on flow birefringence properties of F-actin. The reaction mixture contained F-actin (0.2 mg/ml), calmodulin (0.014 mg/ml), caldesmon (0.012 mg/ml) in 0.1 M KCl, 0.3 mM ATP, 0.1 mM DTT and 10 mM Tris-HCl, pH 7.6. Either 1 mM EGTA or 10^{-5} M $CaCl_2$ was added. Flow birefringence was measured at 30°C. X = F-actin alone; O = calmodulin + caldesmon - Ca^{++}; ● = calmodulin + caldesmon + Ca^{++}. _____ = degree of birefringence; ---- = degree of extinction angle.

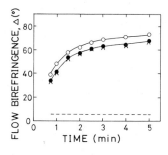

FIGURE 2

Reassociation process of actin filaments fragmented by sonic vibration under the influence of calmodulin-caldesmon. Reaction mixture (as in Fig. 1) was sonicated for 10 sec. Immediately after the sonication the degree of birefringence was measured at 100 sec^{-1} at 30°C. X = F-actin alone; O = calmodulin + caldesmon - Ca^{++}; ● = calmodulin + caldesmon + Ca^{++}; ---- = β-actin (0.05 mg/ml) was added in place of calmodulin and caldesmon.

It is well established that actin filaments in solution is weakly thyxotropic, that is, weak external force destroys the network of actin filaments (8). When an F-actin solution is put into a cell of a Low-Shear Rheometer, and then very slowly rotated, actin network is gradually formed to exhibit structural viscosity (8). The effect of the calmodulin-caldesmon system on the network formation was only minimal at velocity gradients ranging 0.01 to 1 sec^{-1}. As shown in Fig. 3, in the absence of Ca^{++}, there was an appreciable increase in the rate of network formation. In the presence of Ca^{++} no change was observed as compared to control actin filaments.

It is known that tropomyosin is bound to the grooves of actin filaments. Tropomyosin inhibited the binding of α-actinin (2) and of actin-binding protein (filamin)(12). Both proteins result in gelation of actin filaments. We tested the effect of tropomyosin on the interaction of caldesmon with actin filaments. As summarized in Table 1, the increase in the degree of birefringence of actin filaments by caldesmon, as regulated by calmodulin/Ca^{++}, was not affected by tropomyosin at all. Tropomyosin alone results in a marked increase in the degree of birefringence of actin filaments (6). The extinction angle values were not changed in any cases we observed. These results suggest that the binding of caldesmon to actin filaments is independent of the binding of tropomyosin.

DISCUSSION

The present study shows that the effect of the caldesmon-calmodulin system on the rheological properties of actin filaments is dependent upon micromolar concentrations of calcium ion. The elevation of the degree of flow birefringence, and increase in the rate of network formation of actin filaments can be explained by the binding of caldesmon to actin filaments. The bound caldesmon must be released from the actin filaments by calmodulin in the presence of Ca^{++}. It must be mentioned that caldesmon alone, but not calmodulin, was bound to actin filaments (Table 1). These observations confirm the previous results obtained by viscosity measurements and sedimentation methods (14,15). It is of interest to note that the calcium-regulated caldesmon binding to actin filaments was not affected by tropomyosin. This observation suggests that the binding sites of caldesmon on actin filaments are not the grooves of double-stranded actin filaments where tropomyosin is located.

The present study provides evidence indicating that the Ca^{++}- and calmodulin-regulated binding of caldesmon to actin filaments does not change the physical state of actin filaments, i.e. gelation or severing was not observed. This is in contrast to the cases of gelsolin (18), fragmin (3), actinogelin (11), and villin (1), where calmodulin was not required for the Ca^{++} sensitivity of the system. We have obtained preliminary results using an electron microscope

TABLE 1: Effect of calmodulin-caldesmon on the flow birefringence properties of tropomyosin-bound actin filaments.

ADDITIONS	DEGREES OF	
	BIREFRINGENCE	EXTINCTION ANGLE
F-actin	40°	9.5°
+ CalD	45	9.0
+ CalM	40	9.5
+ TM	60	8.0
+ TM + CalD + CalM - Ca^{++}	67	8.0
+ TM + CalD + CalM + Ca^{++}	61	9.0

The reaction mixture contained F-actin (0.2 mg/ml); tropomyosin (0.1 mg/ml); caldesmon (0.02 mg/ml); calmodulin (0.02 mg/ml) in 0.1 M KCl, 0.3 mM ATP, 0.1 DDT and 10 mM Tris-HCl buffer, pH 7.5. Either 1 mM EGTA or 10^{-5} M $CaCl_2$ was added. Flow birefringence measurements were carried out at 30°C at a velocity gradient of 4 sec^{-1}. CalD = caldesmon; CalM = calmodulin; TM = tropomyosin.

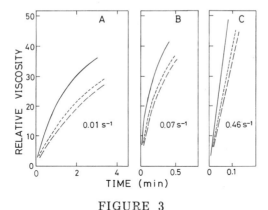

FIGURE 3

Process of network formation of actin filaments as determined by the increase in the structural viscosity under the influence of calmodulin-caldesmon. The reaction mixture contained F-actin (1.0 mg/ml); caldesmon (0.07 mg/ml); calmodulin (0.07 mg/ml) in 50 mM KCl, 0.3 mM ATP, 0.1 mM DTT, and 10 mM Tris-HCl, pH 7.6. Viscosity measurements were carried out at 30°C. The values of relative viscosity are arbitrary. Different velocity gradients were obtained as indicated in A through C. ---- = actin alone; . . . . = + calmodulin + caldesmon + Ca^{++}; ——— = + calmodulin + caldesmon - Ca^{++}.

that supports the above conclusion. However, a question then arose. What is the function of the calmodulin-caldesmon system? It may modify the function of actin in relation with other proteins. This possibility is currently being investigated.

REFERENCES

1. Bretscher, A. & Weber, K. (1980): Cell 20:839-847.
2. Drabikowski, W., Nonomura, Y. & Maruyama, K. (1968): J. Biochem. 63:761-766.
3. Hasegawa, T., Takahashi, S., Hayashi, H. & Hatano, S. (1980): Biochemistry 19:2677-2683.
4. Kakiuchi, S., Sobue, K., Yamazaki, R., Kambayashi, J., Sakon, M. & Kosaki, G. (1981): FEBS Lett. 126:203-207.
5. Maruyama, K. (1964): J. Biochem. 55:277-286.
6. Maruyama, K. (1964): Arch. Biochem. Biophys. 105:142-150.
7. Maruyama, K. (1976): Adv. Biophys. 9:157-185.
8. Maruyama, K., Kaibara, M. & Fukada, E. (1974): Biochim. Biophys. Acta 371:20-29.
9. Maruyama, K. & Kimura, S. (1981): J. Biochem. 90:563-566.
10. Maruyama, K., Kimura, S., Ishii, T., Kuroda, M., Ohashi, K. & Muramatsu, S. (1977): J. Biochem. 102:208-225.
11. Maruyama, K., Mimura, N. & Asano, A. (1981): J. Biochem. 89:317-319.
12. Maruyama, K. & Ohashi, K. (1978): J. Biochem. 84:1017-1019.
13. Mimura, N. & Asano, A. (1979): Nature 282:44-48.
14. Sobue, K., Muramoto, Y., Fujita, M. & Kakiuchi, S. (1981): Biochem. Int. 2:469-476.
15. Sobue, K., Muramoto, Y., Fujita, M. & Kakiuchi, S. (1981): Proc. Nat. Acad. Sci. (in press).
16. Spudich, J.A. & Watt, S.J. (1971): J. Biol. Chem. 245:4866-4871.
17. Yin, H.L. & Stossel, T.P. (1979): Nature 281:383-386.
18. Yin, H.L., Zaner, K.S. & Stossel, T.P. (1980): J. Biol. Chem. 255:9494-9500.

MODE OF CALCIUM BINDING TO SMOOTH MUSCLE CONTRACTILE SYSTEM

S. Ebashi, Y. Nonomura and M. Hirata*

Department of Pharmacology, Faculty of Medicine,
University of Tokyo, Hongo, Tokyo 113, Japan and
*Department of Physiology, Faculty of Dentistry,
Kyushu University, Fukuoka 812, Japan

INTRODUCTION

Troponin, the first Ca binding protein of biological importance (3,4,6), has contributed to the establishment of the concept of the "regulatory protein" (15), but it is now clear that the troponin mechanism is not common for all kinds of muscle. Although skeletal muscle of higher animals, whether deuterostomia or protostomia, are mainly dependent on this mechanism, the myosin-linked system is predominant in some lower animals, especially those of protostomia (13). The regulatory mechanism in vertebrate smooth muscle, as will be described below, has entirely different features from those in vertebrate skeletal muscle.

Characteristic Features of Smooth Muscle Regulation

Most muscle scientists had believed that the events taking place in the contractile system of vertebrate skeletal muscle should have been operating more or less in other kinds of muscle, and that the smooth muscle should have been only a dull skeletal muscle. However, it is now well understood that the pure actomyosin of smooth muscle without the regulatory system is quiescent in the presence of ATP, irrespective of the presence or absence of Ca^{++} (7), in sharp contrast with the contracting tendency of pure actomyosin of vertebrate striated muscle in response to ATP (Fig. 1). Therefore, the regulatory system of smooth muscle is an activating factor, differing distinctly from the repressive nature of the troponin system. It is interesting that the ascidian troponin system (9) has a property similar to that of the regulatory system in smooth muscle.

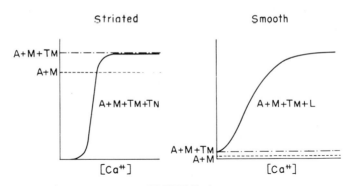

FIGURE 1

Schematic illustration of the myosin-actin-ATP interaction of vertebrate smooth muscle as a function of free Ca^{++} concentration in comparison with that of vertebrate striated muscle. A = actin; M = myosin; TM = tropomyosin; TN = troponin; L = regulatory proteins of smooth muscle (the leiotonin-tropomyosin system).

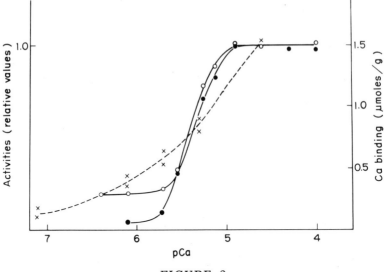

FIGURE 2

Contractile responses of aorta myosin B in comparison with Ca binding to it. Solid lines = contractile responses; open circles = superprecipitation; closed circles = ATPase activities; dashed line = Ca binding.

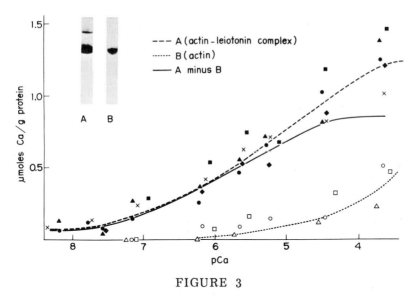

FIGURE 3

Ca binding to natural leiotonin-actin complex of aorta smooth muscle. Inserted figures show the sodium dodecyl sulfate polyacrylamide gel electrophoresis of the actin complex. In B, leiotonin was removed by α-chymotrypsin treatment.

A problem then naturally arises about what kind of regulatory mechanism is working in smooth muscle. The majority of the research group considers the Ca^{++}-dependent light chain kinase as the activating regulatory system as described in detail in another part of this volume (cf. 1,12).

Our research group is of the opinion that the regulatory system is of an actin-linked nature composed of tropomyosin (it should be a smooth muscle type of tropomyosin; skeletal tropomyosin cannot fully replace it), and a new type of Ca^{++}-dependent regulatory protein called leiotonin (cf. 4,8). The latter is composed of leiotonin A, the regulatory moiety of about 80,000 dalton in its molecular weight, and leiotonin C, the Ca-binding component of 17,000 dalton.

One important feature of the leiotonin system in comparison with the troponin system is that, although tropomyosin is absolutely required for regulation, leiotonin does not bind to tropomyosin but actin. As has been the case with various kinds of actin-modulating factors, e.g. β-actinin (14) or gelsolin (16), leiotonin also exerts its full activity with as small an amount as one-fiftieth or less of actin in its molar ratio.

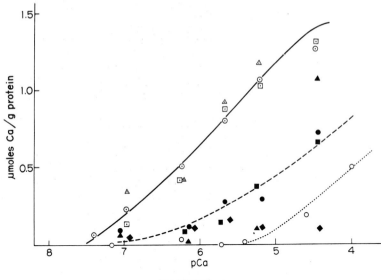

FIGURE 4

Ca binding to reconstituted actin-leiotonin-complex of aorta smooth muscle. Solid line with dotted open symbols = Ca binding to the aorta leiotonin A-actin complex with gizzard leiotonin C. Dashed line with closed symbols = Ca binding to aorta actin (without leiotonin A) with gizzard leiotonin C. Solid line with open circles = Ca binding to aorta actin. Different symbols in each group represent the results of different preparations, respectively. Note that aorta actin with gizzard leiotonin C bound occasionally much higher amount of Ca than did actin itself (the reason for this is not clear).

Ca and Sr Binding to Smooth Muscle Contractile System

Although most work concerning smooth muscle regulation has been made with gizzard, the situation in vascular smooth muscle is not essentially different from that in gizzard. Compared with gizzard, bovin aorta has advantages in that the amount of proteins other than contractile system is much smaller and the content of actin is relatively higher than myosin. Furthermore, leiotonin is so firmly bound to actin that it is not lost during various experimental procedures; leiotonin can be released from actin only by the treatment with a protease, say α-chymotrypsin (10,11).

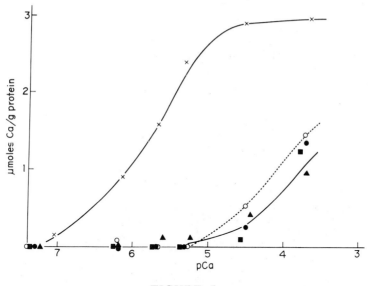

FIGURE 5

Ca binding to aorta myosin in comparison with that of rabbit skeletal myosin or scallop myosin. Solid line with filled symbols = Ca binding to aorta myosin; dashed line with open circles = rabbit skeletal myosin; solid line with crosses = ca binding to scallop striated adducter muscle myosin.

So we can compare the mode of Ca binding to the whole contractile system with that of native actin, viz., the actin preparation which still retains leiotonin. As shown in Figs. 2, 3, and 4, Ca binding to natural actomyosin, or myosin B, is well related both to its contractile responses and Ca binding to native actin and reconstituted "native" actin. Since the profile of Ca binding to myosin is entirely different from them (Fig. 5), the Ca binding to the contractile system is mainly represented by that to actin.

Owing to lower affinity of Sr^+ for the regulatory site, relatively high Sr^+ concentrations had to be used to measure its binding to contractile proteins; consequently, nonspecific Sr binding became more distinct and tended to mask specific binding (cf. Figs 3 and 4 in ref. 11). Even so the results obtained with Sr^+ (11) were essentially in accord with the above.

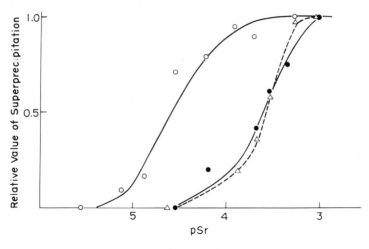

FIGURE 6

Contractile responses of aorta myosin and the aorta actin complex. Dashed line with filled circles = superprecipitation with the aorta actin-leiotonin complex; solid line with triangles = superprecipitation with the aorta actin-leiotonin A complex associated with gizzard leiotonin C; solid line with open circles = superprecipitation with the actin-leiotonin A complex with calmodulin.

The contractile responses of aorta myosin B shows much less sensitivity to Sr^{++} than phosphorylating reactions of myosin light chain (11). This will shed a light on the problem of whether leiotonin C is actually playing the major role in physiological processes or whether calmodulin would also play a substantial role. Figure 6 clearly indicates that the former is the case.

DISCUSSION

Since our old preparation of leiotonin was made when the roles of acidic proteins, calmodulin or leiotonin C were not recognized, the assay system did not contain such an acidic protein, and, consequently, it was possible that the preparation would contain the kinase unaccompanied by calmodulin. Indeed, this was the case in most of the leiotonin preparations. Although this does not affect our previous conclusion that the leiotonin system could induce the activation of the actomyosin without light chain phosphorylation, because such a preparation certainly activates the myosin-actin-ATP interaction with phosphorylation (Fig. 7) it is urgently required to establish the method of preparing leiotonin completely free of light chain kinase.

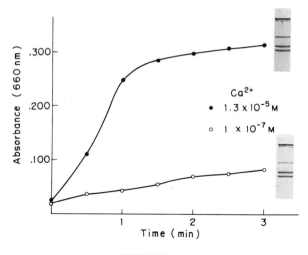

FIGURE 7

Activation of gizzard actomyosin without phosphorylation. Gizzard actin, myosin and tropomyosin together with crude leiotonin, which has been subject to Ultrogel (AC44), responded to MgATP with pronounced superprecipitation, but no phosphorylation of myosin light chain took place as indicated in the attached electrophoresis pattern. The addition of calmodulin to this system in the presence of Ca^{++} (1.3×10^{-5} M) could not markedly accelerate the superprecipitation, but induced a considerable degree of phosphorylation of the light chain (cf. Fig. 2 in ref. 5), indicating the independence of activation and phosphorylation from each other. For experimental conditions see the legend to Fig. 3 in ref. 5.

TABLE 1: Degrees of activation of actin-myosin-ATP interaction, determined by superprecipitation, relative to those of phosphorylation of myosin light chain by ammonium sulfate fractions of a chicken gizzard extract (for preparation of the extract, see ref. 5).

Range of Ammonium Sulfate Fractionation (M)	Activation/Phosphorylation	
	No.1	No2
3.1 - 2.2	0.1	0.8
2.2 - 1.8	7	0.9
1.8 - 1.4	24	17
1.4 - 0	15	2.7

Another question is whether phosphorylation is playing an active role in contractile processes. We have already presented some evidence against this hypothesis (2). It is certainly true that if we use a single preparation containing the kinase, some beautiful correlation between the degrees of phosphorylation and contractile responses will be observed. However, if we compare different kinds of preparations, no clear relationship can be observed (Table 1) supporting our previous proposal.

CONCLUSION

Ca binding to the actomyosin system of smooth muscle is represented mainly by its binding to the actin filament side. This further supports the concept that the leiotonin-tropomyosin system, an actin-linked activating system, is the main principle for the regulation of smooth muscle contraction.

ACKNOWLEDGEMENTS

The work cited in this chapter is supported in part by grants from the Muscular Dystrophy Association, the Ministry of Education, Science and Culture, Japan, the Ministry of Health and Welfare, Japan, and the Iatrochemical Foundation.

REFERENCES

1. Adelstein, R.S., Pato, M.D. & Conti, M.A. (1980): IN Muscle Contraction, Its Regulatory Mechanisms. (eds) S. Ebashi, K. Maruyama & M. Endo, Japan Science Soc. Press, Tokyo, Springer-Verlag, Berlin, pp. 303-313.
2. Ebashi, S. (1979): IN Advances in Pharmacology and Therapeutics. (ed) J.C. Stoclet, Pergamon Press, Oxford, Vol.3, pp.81-98.
3. Ebashi, S. (1980): In The Croonian Lecture, 1979; Proc. Roy. Soc. Lond. B 207:259-286.
4. Ebashi, S. & Endo, M. (1968): Prog. Biophys. Molec. Biol. 18:123-183.
5. Ebashi, S. & Nakasone, H. (1981): Proc. Japan Acad. 57:217-221.
6. Ebashi, S., Ebashi, F. & Kodama, A. (1967): J. Biochem. 62: 137-138.
7. Ebashi, S., Nonomura, Y., Toyo-oka, T. & Katayama, E. (1976): IN Calcium in Biological Systems. Symp. Soc. Exp. Biol. No.30, pp.349-360.
8. Ebashi, S., Nonomura, Y., Mikawa, T., Hirata, M. & Saida, K. (1979): IN Cell Motility: Molecule and Organization. (eds) T.Hatano, H. Sato & H. Ishikawa, Tokyo University Press, pp.225-237.
9. Endo, T. & Obinata, T. (1981): J. Biochem. 89:1599-1608.

10. Hirata, M., Mikawa, T., Nonomura, Y. & Ebashi, S. (1977): J. Biochem. 82:1793-1796.
11. Hirata, M., Mikawa, T., Nonomura, Y. & Ebashi, S. (1980): J. Biochem. 87:369-378.
12. Hartshorne, D.J., Siemankowski, R.F. (1980): IN Muscle Contraction, Its Regulatory Mechanisms. (eds) S. Ebashi, K. Maruyama & M. Endo, Japan Science Societies Press, Tokyo, Springer-Verlag, Berlin, pp. 287-301.
13. Lehman, W. & Szent-Gyorgyi, A.G. (1975): J. Gen. Physiol. 66:1-30.
14. Maruyama, K., Kimura, S., Ishii, T., Kuroda, M., Ohashi, K. & Muramatsu, S. (1977): J. Biochem. 81:215-232.
15. Maruyama, K. & Ebashi, S. (1970): IN The Physiological and Biochemical Muscle as a Food. (eds) E.J. Briskey, R.G. Cassens & B.B. Marsh, University of Wisconsin Press, Madison, pp.373-381.
16. Yin, H.L. & Stossel, T.P. (1970): Nature 281:583-586.

KINETIC STUDIES OF THE ACTIVATION OF CYCLIC NUCLEOTIDE

PHOSPHODIESTERASE BY Ca^{++} AND CALMODULIN

V. Chau, C.Y. Huang, P.B. Chock, J.H. Wang* and
R.K. Sharma*

Laboratory of Biochemistry, National Heart, Lung and
Blood Institutes, National Institutes of Health,
Bethesda, Maryland, USA, and
*Department of Biochemistry, University of Manitoba
Winnipeg, Manitoba R3E OW3, Canada

INTRODUCTION

The elucidation of the Ca^{++}-modulatory role of calmodulin originated from the studies of mammalian cyclic nucleotide phosphodiesterase. In a series of publications during the late 60's and early 70's, Cheung (2-4) found the existence in various mammalian tissues of a heat-stable and acidic protein which was capable of activating partially purified cyclic nucleotide phosphodiesterase. Shortly afterwards, Kakiuchi and co-workers (11,12) separated two forms of rat brain cyclic nucleotide phosphodiesterase; one of these was stimulated by Ca^{++}, the other was Ca^{++} independent. The stimulation of the former enzyme by Ca^{++} was enhanced by an endogenous protein which appeared similar to Cheung's activator protein (11). In 1973, Teo et al. (23) purified the activator protein from bovine heart to homogeneity and demonstrated that it was a Ca^{++} binding protein (24). Since the activation of the phosphodiesterase exhibited an absolute dependence on the simultaneous presence of both activator protein and Ca^{++} (13,24), it seemed clear that Ca^{++} activated the phosphodiesterase by binding to the activator protein-calmodulin. These and other studies during the mid 1970's have, therefore, firmly established the Ca^{++} modulatory role of calmodulin.

Although many other enzymes and proteins have subsequently been shown to be regulated by Ca^{++} and calmodulin (for reviews, see 5,14,30,31), the calmodulin dependent cyclic nucleotide phosphodiesterase is still most frequently used for the elucidation of the basic mechanisms of calmodulin action. During the last few years, homogeneous phosphodiesterases have been prepared from both bovine heart (16) and bovine brain (18,19,22), and the basic physicochemical properties of the enzyme are delineated. At the same time, physicochemical properties of calmodulin and the interactions between the protein and Ca^{++} have been extensively characterized by many investigators using a variety of techniques (for a review, see 14). Thus, it seems that the time is ripe for a comprehensive study of the mechanisms of activation of the phosphodiesterase. This communication is concerned with the kinetic study of the complex interactions between Ca^+, calmodulin and the enzyme; and the relationships between these interactions and the enzyme activation.

Properties of Calmodulin Dependent Cyclic Nucleotide Phosphodiesterase

The calmodulin dependent phosphodiesterase used in this study was purified from bovine brain according to a recently developed procedure (22). The protein purification resulted in a 3,500- to 4,000-fold enrichment from the crude brain extract in specific activity of the enzyme. The preparation appeared homogeneous as judged by analytical polyacrylamide gel electrophoresis, SDS-gel electrophoresis as well as gel isoelectric focussing. In all cases, single protein band was observed on the gel.

Some of the catalytic and physicochemical properties of the purified bovine brain phosphodiesterase are summarized in Table 1. The enzyme exhibited high responsiveness to Ca^{++} and calmodulin. For most preparations, the enzyme could be activated more than tenfold. The fully activated enzyme had specific activities greater than 300 µmol cAMP hydrolyzed per min per mg protein.

The molecular weight of the enzyme was deterimend to be about 120,000. The single protein band observed on SDS-slectrophoresis corresponded to a molecular weight of 58,000. From these results, it was suggested that the enzyme is a dimeric protein consisting of apparently identical subunits.

As has been observed previously with partially purified enzyme preparations (9,17,25), the phosphodiesterase binds calmodulin only in the presence of adequate concentrations of Ca^{++} in the medium. For example, Fig. 1 shows that when a mixture of the phosphodiesterase and calmodulin was subjected to sucrose-density gradient centrifugation in a medium containing 10^{-4} M EGTA, the two proteins separated into distinct peaks (Fig. 1A). In contrast, when the mixture was centrifuged in a medium containing 10^{-4} M Ca^{++}, a fraction containing both the phosphodiesterase and calmodulin was observed in the sedimentation profile (Fig. 1B). The results indicate that a protein complex of calmodulin and phosphodiesterase

TABLE 1: Properties of bovine brain phosphodiesterase*

Catalytic properties		
Maximal activation by calmodulin (fold)		∿ 12
Specific activity (μM cAMP hydrolyzed/min·mg)		300-350
Physicochemical properties		
Sedimentation constant(s)	Sucrose density centrifugation	6.9
Stokes radius (Å)	Gel filtration	44.2
Partial specific vol. (ml/g)	From amino acid composition	0.726
Molecular weight	From sedimentation constant & Stokes radius	124,000
	Sedimentation equilibrium	115,000
Subunit weight	SDS-gel electrophoresis	58,000
Isoelectric point (pH)	Isoelectric focussing	4.85
Absorbance at 278 nm for 1% solution		9.6
Physical properties of the calmodulin-phosphodiesterase complex		
Sedimentation constant(s)	Sucrose density centrifugation	8.0
Stokes radius (Å)	Gel filtration	48
Molecular weight (dalton)		159,000

*Summarized from Sharma et al. (22).

TABLE 2: Dissociation constants of Ca^{++}-calmodulin complexes

K_1	7.5×10^{-6}
K_2	2.7×10^{-6}
K_3	3.1×10^{-5}
K_4	3.1×10^{-5}

The interaction between Ca^{++} and calmodulin was determined at pH 7.0 and 25°C in 50 mM hepes, 5 mM $MgAc_2$ and 0.1 M KCl.

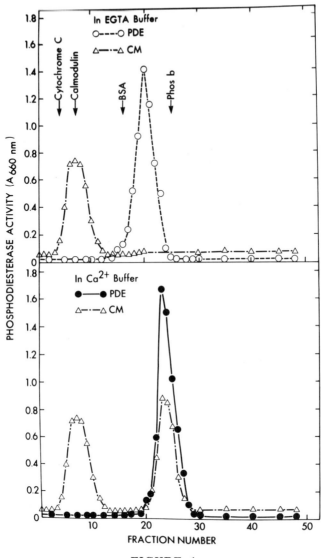

FIGURE 1

Sucrose-density gradient centrifugation of mixtures of the phosphodiesterase (19 µg) and calmodulin (72 µg) was carried out in a buffer containing either 0.1 mM EGTA (upper panel) or a buffer containing 0.1 mM Ca^{++} (lower panel).

is formed in the presence of Ca^{++}. Molecular weight of the protein complex calculated from the value of the sedimentation constants and Stoke radius was 159,000. Since calmodulin and the free enzyme have molecular weights about 17,000 and 120,000, respectively, the result suggests that the protein complex contains one molecule of the enzyme and two molecules of calmodulin. It seems reasonable to suggest further that each of the subunits of the enzyme is capable of binding of one molecule of calmodulin in the presence of Ca^{++}. Similar conclusions have been drawn by others for the interaction of calmodulin with both bovine heart and bovine brain cyclic nucleotide phosphodiesterase as examined by using the chemical cross-linking reagents (16,19).

Relationship Between Subunit Structure and Enzyme Activation

The two subunits of the cyclic nucleotide phosphodiesterase appear to act independently of each other during catalysis as well as in the enzyme activation by calmodulin. Kinetic studies using either cAMP or cGMP as the substrate revealed that the enzyme, either in its basal or fully activated states exhibited little or no homotropic interactions with respect to the nucleotide substrate. Similarly, kinetic analysis of the interaction between the enzyme and calmodulin also suggested that the binding of the two molecules of calmodulin to the enzyme were not cooperative.

Two approaches were used to study the interaction between the phosphodiesterase and calmodulin in the enzyme reactions at a saturating level of Ca^{++}. In both cases, calmodulin activation of the enzyme was used to monitor the binding of calmodulin to the enzyme. One approach involved measuring the phosphodiesterase activation at various molar ratios of calmodulin to the enzyme while the total molar concentration of the two proteins was held at a constant. The results were then analyzed according to Job's procedure to detemine the stoichiometry of the calmodulin-enzyme interaction. The results of such a study were presented previously and the stoichiometry obtained was identical to that derived from monitoring the protein-protein interaction directly, i.e. 2 mol of calmodulin to 1 mol of the phosphodiesterase (29). Therefore, it may be concluded that, at saturating Ca^{++}, the activation of the phosphodiesterase by calmodulin is directly proportional to the extent of calmodulin binding to the enzyme.

The other approach involved the analysis of the dose-dependent curve of the activation of the phosphodiesterase by calmodulin using the Scatchard plot. In this procedure, the extent of calmodulin activation of a known concentration of the phosphodiesterase was used to calculate concentrations of bound and free calmodulin. For example, if a dose-dependent curve is carried out by using 1 nM of the enzyme subunit, an enzyme activation by calmodulin to 25% of the maximal activation would indicate a concentration of enzyme-bound calmodulin of 0.25 nM. Since the total concentration of calmodulin added to the reaction is known, the concentration of free calmodulin can be readily calculated and a Scatchard plot for the interaction between calmodulin and the phosphodiesterase constructed. Such a Scatchard plot, along with the dose-dependent activation is shown in Fig. 2. The observation that the stoichiometry of calmodulin binding to the enzyme subunit was about 1 lends further support to the notion that calmodulin binding in the enzyme reaction is directly proportional to the activation of the enzyme. The linearity of the Scatchard plot indicates that the bindings of the two molecules of calmodulin to the enzyme are independent of each other.

General Scheme of the Interaction of the Enzyme, Calmodulin and Ca^{++}

Calmodulin contains four high affinity Ca^{++} binding sites. Although the relationship between the multiple Ca^{++} binding and the activation of the enzyme is not known, a general scheme may be set up to depict the various interactions between the enzyme, calmodulin and Ca^{++}.

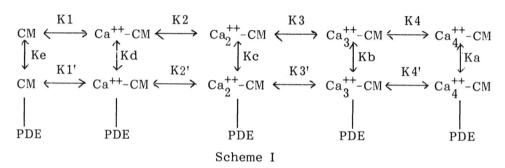

Scheme I

where CM and PDE stand for calmodulin and the enzyme, respectively and the K's represent dissociation constants for the respective reactions. The entrance of Ca^{++} and PDE into the reactions is not indicated to simplify the schematic diagram. The subunit of the enzyme is treated as the basic interacting unit since, as has been discussed in the preceding sections, the subunits in the dimeric phosphodiesterase do not exhibit cooperative interaction in calalysis or in their interaction with calmodulin. It should be noted that the schematic representation is intended to describe the degree of Ca^{++} saturation for calmodulin and the calmodulin and the enzyme complex as dictated by the individual dissociation constants, rather than implying a sequential Ca^{++} binding.

Some of the dissociation constants in Scheme I have been separately determined. Table 2 shows the values for the dissociation constants of Ca^{++}-calmodulin complexes at pH 7.0 in a solution similar

TABLE 3: Rate constants of Ca^{++} dissociation.

Complex	Monitoring Method	Rate Constant sec^{-1}	Comment
Ca^{++}_4/CM	Tyrosine fluorescence	12.1	Single exponenial curve
Ca^{++}_4/CM	Ca^{++} specific indicator	12.6	Slowest dissociating Ca^{++} was monitored
Ca^{++}_4/CM/ PDE	Ca^{++} specific indicator	3.5-4.6	Estimated for the initial rapid Ca^{++} release in a multi-exponential curve
Ca^{++}_4/CM/ PDE	Enzyme deactivation	4.5	Single exponential curve

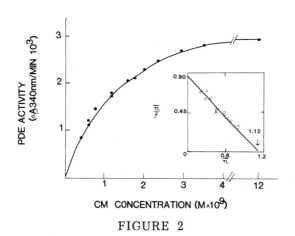

FIGURE 2

Dose dependence curve for the activation of cyclic nucleotide phosphodiesterase by calmodulin and the Scatchard plot (inset) constructed from the data (see text for details).

to that used for the steady state kinetic studies of the phosphodiesterase reaction (see next section). Two procedures were used for the measurement of the binding of Ca^{++} to calmodulin. In one procedure, the Ca^{++}-induced change in tyrosine fluorescence of calmodulin was used to monitor the concentration of bound Ca^{++}, and the concentrations of free Ca^{++} was maintained by using EGTA-Ca^{++} buffer. In the other procedure, the dansylated tropinin C was used as a Ca^{++} binding dye to measure the concentration of Ca^{++}. The values for these dissociation constants (Table 3) are comparable to, but different from those determined by Crouch and Klee (8). The slight difference may be due to the different concentration of Mg^{++} and ionic strength used in the two studies. The observation of both positive and negative cooperativities in the binding of Ca^{++} to calmodulin originally made by Crouch and Klee (8), is also evident in the values listed in Table 2.

The dissociation constant for the complex of phosphodiesterase and the fully liganded calmodulin, K_a, was determined in the phosphodiesterase reaction in the presence of saturating concentration of Ca^{++}. In one method, the Scatchard plot derived from the dose-dependent curve for the activation of the phosphodiesterase by calmodulin was used to calculate the dissociation constant (Fig. 2). The value obtained from the experiment in Fig. 2 is about 1.2 nM. The dissociation constant, however, appeared to vary among different enzyme preparations and/or to depend on the age of the preparation, the values obtained ranged from 0.2 to 1.2 nM. In another method, the rate constant of the association of the fully liganded calmodulin and the phosphodiesterase and that of the dissociation of the protein complex were determined, then these rate constants were used to calculate the dissociation constant (29). In these studies, a continuous phosphodiesterase assay was used, which involved the use of a series of three coupling enzymes: myokinase, pyruvate kinase and lactate dehydrogenase. The time course of the transition of the basal rate to the calmodulin activated rate of the enzyme reaction upon the addition of calmodulin was monitored to obtain the second order kinetic constant of the protein association. For the rate constant of the protein dissociation, the time course of inactivation of the calmodulin-activated phosphodiesterase reaction by calcineurin was followed. Calcineurin was originally discovered as an inhibitor of calmodulin activated phosphodiesterase (15,27); it inhibits the enzyme by binding to calmodulin (15,28). Thus, when excess calcineurin is added to a calmodulin activated phosphodiesterase reaction, the rate of the enzyme inactivation may correspond to the rate of dissociation of calmodulin from the enzyme (Fig. 3). The value of the dissociation constant K_a (Scheme I) calculated from the rate constants of the reactions was similar to those obtained from the dose dependence curve of the phosphodiesterase activation.

Steady State Kinetic Studies of the Enzyme Activation

Huang and co-workers (10) have recently carried out steady state kinetic studies of the activation of the phosphodiesterase by Ca^{++} and calmodulin. The rate equation was derived on the basis of Scheme I, with the following assumption and conditions: a) the activated enzyme species is the one containing fully liganded calmodulin, b) total concentration of calmodulin is in great excess over that of the enzyme, and c) free Ca^{++} concentration is maintained by using EGTA-Ca^{++} buffer. The rate equation has the following form:

$$\frac{1}{\Delta \nu} = \frac{1}{\Delta V} \left(\frac{\phi_1}{\phi_2} + \frac{\phi_3 K_e}{\phi_2 [CM]_{total}} \right) \quad [1]$$

Where $\Delta \nu$ is the activated initial rate of the enzyme reaction: observed initial rate minus basal rate. ΔV max is the maximally activated initial rate, ϕ_1, ϕ_2 and ϕ_3 are functions of free Ca^{++} concentration with the following expression:

$$\phi_1 = 1 + \frac{[Ca^{++}]}{K_1'} + \frac{[Ca^{++}]^2}{K_1' K_2'} + \frac{[Ca^{++}]^3}{K_1' K_2' K_3'} + \frac{[Ca^{++}]^4}{K_1' K_2' K_3' K_4'}$$

$$\phi_2 = \frac{[Ca^{++}]^4}{K_1' K_2' K_3' K_4'}$$

$$\phi_3 = 1 + \frac{[Ca^{++}]}{K_1} + \frac{[Ca^{++}]^2}{K_1 K_2} + \frac{[Ca^{++}]^3}{K_1 K_2 K_3} + \frac{[Ca^{++}]^4}{K_1 K_2 K_3 K_4}$$

Other symbols are the same as those used in Scheme I.

Initial rate measurements were carried out over a range of calmodulin concentration of 2×10^{-8} to 10^{-5} M at several constant levels of Ca^{++} concentrations from 2.3×10^{-7} to 2.1×10^{-6} M. The results were then plotted in the double reciprocal plot: $1/\Delta \nu$ vs $1/[CM]_{total}$, and a family of straight lines were obtained as was predicted by using equation [1]. From the expression for the slope of the double reciprocal plot, the following equation was obtained:

$$\log (\phi_3 / \Delta V \cdot S) = 4 \log [Ca^{++}] - \log K_e K_1' K_2' K_3' K_4' \quad [2]$$

Where S stands for the slope obtained in the initial plot, and ϕ_3 can be computed for the values of K_1, K_2, K_3 and K_4 (Table 2) and the given free Ca^{++} concentration. Equation [2] predicts that a plot of $\log (\phi_3 / \Delta V \cdot S)$ vs $\log [Ca^{++}]$ would result in a

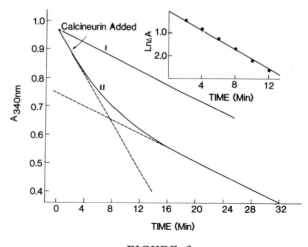

FIGURE 3

The rate of inactivation of calmodulin activated phosphodiesterase reaction by calcineurin. Basal (I) and calmodulin activated reactions (II) were monitored and then calcineurin was added to (II) to initiate the inactivation. Arrow indicates the addition of calcineurin. Inset - first order kinetic plot for the time course of the enzyme inactivation (from Ref. 29).

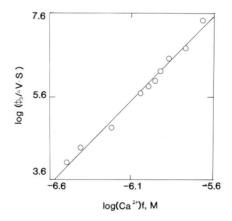

FIGURE 4

The plot of log (ϕ_3/V.S) vs log $(Ca^{++})f$. The data are plotted according to equation [2]. The slope of the line is 4.03 ± 0.15 (from Ref 10).

straight line having a slope of 4. Figure 4 reproduces such a plot showing that the prediction is fulfilled. The slope of the line shown is 4.03 ± 0.15. This result is compatible with the notion that the complex Ca_4^{++} CM-PDE is the only activated species. If, for example, both Ca_4^{++} CM-PDE and Ca_3^{++} CM-PDE are activated, the expressions for $\emptyset_2$ becomes $(Ca^{++})^3/K_1'K_2'K_3' + (Ca^{++})^4/K_1'K_2'K_3'K_4'$. The slope for a plot of log $(\emptyset_3/\Delta V.S.)$ vs log $[Ca^{++}]$ will be a function of Ca^{++} concentrations:

$$\text{slope} = \frac{3 + 4[Ca^{++}]/K_4'}{1 + [Ca^{++}]/K_4'} \quad [3]$$

The slope will be nonlinear. However, if the Ca^{++} concentration used in the experiment are much higher than the value of K_4', a straight line with slope approximately 4 will result. Thus, from the steady state kinetic study, it may be concluded that the enzyme species Ca_4^{++} CM-PDE is the predominant activated enzyme species. While the results are compatible with the suggestion that the fully liganded calmodulin-phosphodiesterase complex is the only activated enzyme species, the possibility that other species are also activated but present in negligible amounts under the experimental conditions cannot be excluded completely.

Kinetic Studies of the Dissociation of Ca^{++}-Calmodulin Complexes

To further explore the mechanisms of the interaction between Ca^{++} and calmodulin, the dissociation of Ca^{++} from the Ca_4^{++} CM complex was examined by using stopped flow fast kinetic techniques. The tyrosine fluorescence of calmodulin was used to monitor the change in Ca^{++} binding of the proteins. In one experiment, 70 μM of calmodulin in fully liganded state was mixed with 4 mM of EGTA, the time course of the tyrosine fluorescence change was monitored in a stopped flow instrument. The oscillogram showed a single exponential curve with a half-life of 57 m sec.

Since the change in tyrosine fluorescence might not correspond directly to the dissociation of Ca^{++}, a more direct method was used to monitor the release of Ca^{++} from the complex of Ca_4^{++} CM. Tsien (26) has recently synthesized a series of Ca^{++} specific indicators, one of these, 1,2 bis (O-aminophenoxy)ethane-N,N,N',N'-tetra acetic acid is shown in Fig. 5. Binding of Ca^{++} to this chelator results in a 90% decrease in its absorbance at 254 nm. Using this indicator, the release of Ca^{++} from the Ca^{++}-calmodulin complex could be monitored. Figure 6 shows the oscillogram obtained by mixing 100 μM of fully liganded calmodulin with 4 mM of the Ca^{++} specific indicator. Again, the time course followed a single exponential curve which was also indicated by the linear first-order plot (Fig. 6 inset). The half-life of the reaction was essentially the same as that obtained by monitoring the change in tyrosine fluorescence of calmodulin.

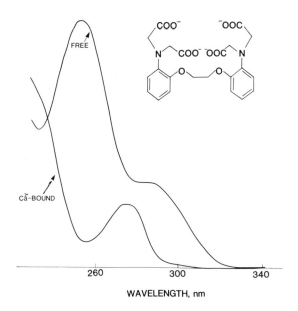

FIGURE 5

Structure and spectra of free and liganded Ca^{++} specific indicator 1,2,bis(0-aminophenoxy) ethane-N,N,N',N'-tetra acetic acid.

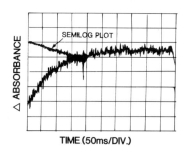

FIGURE 6

Ca^{++} release rate from calmodulin-Ca_4^{++} in a stopped flow instrument. See text for details.

While the observed single exponential time course for the Ca^{++} release from the Ca^{++}-calmodulin complex and for the change in tyrosine fluorescence of calmodulin is in agreement with a suggestion that all 4 Ca^{++} dissociate from the complex at the same rate, preliminary study indicated that one or more Ca^{++} had much higher dissociation rate than that determined. However, under the conditions used for the experiment (Fig. 6) the time resolution of the instrument was not adequate to detect the release of these Ca^{++}s. In a separate experiment when the Ca^{++} specific indicator used was lowered from 4 to 0.43 mM to slow down the process, the time course of the dissociation of Ca^{++} from the complex was found to follow a multi-exponential function. Thus, it may be concluded that the calmodulin-Ca^{++} complex contains at least two groups of Ca^{++} with distinct dissociation rates. Those having the slower dissociation rate, dissociate with a half-life of about 55 m sec. The change in calmodulin conformation which corresponds to the change in tyrosine fluorescence may be correlated with the dissociation of the slow-releasing Ca^{++}. The rate of dissociation for the fast-releasing Ca^{++} has not been determined. The number of Ca^{++} in each group in the Ca^{++}-calmodulin complex has not been established.

Kinetics of Ca^{++} Release from the Complex of Calmodulin and the Enzyme

The time course of the dissociation of Ca^{++} from the complex of the phosphodiesterase and the fully liganded calmodulin, Ca_4^{++} CM-PDE, was determined in a stopped-flow study using the Ca^{++} specific indicator. The oscillogram shown in Fig. 7 indicates that the time course is represented by multiple exponential functions. Thus, it appears that the 4 Ca^{++}s of the complex are dissociated at different rates. Although it is not clear whether each Ca^{++} has its own distinct rate of dissociation and the individual rates have not been computed, the half-life of the most rapid dissociating Ca^{++} may be estimated from the oscillogram to be in the range of 150 to 200 msec.

In order to correlate the dissociation of Ca^{++} from the enzyme-calmodulin complex with the deactivation of the enzyme, the rate of the enzyme deactivation upon the mixing of a Ca^{++} specific chelating agent with the fully liganded protein complex was examined using a three-syringe stopped-flow instrument. In this experiment, the Ca^{++}-calmodulin-phosphodiesterase complex was mixed with 10 mM EGTA and 0.2 mM ^{32}P-cAMP. The reaction was then quenched with 0.5 M HCl at various time intervals up to about 3 sec. The samples were then subjected to thin-layer chromatography to separate cAMP and 5'-AMP and the amount of cAMP hydrolyzed was determined to construct a time course of the cAMP hydrolysis. The time course was analyzed according to the following equation:

$$[AMP]/[cAMP]\text{Total} = k_B \{t + (N-1)(1-e^{-kt})/k\} \quad [4]$$

Where k_B, N and k represent basal rate of phosphodiesterase reactions, the activation factor of the enzyme by calmodulin and the rate constant of the enzyme deactivation respectively. The results obtained showed that the rate constant for the enzyme deactivation was 4.5 per sec which corresponded to a half-life of the reaction of 154 msec. Note that this value is about the same as that estimated for the dissociation of the most rapidly dissociating Ca^{++} from the protein complex (Fig. 7). Thus, it may be suggested that the loss of the most rapidly dissociating Ca^{++} results in the deactivation of the enzyme.

As was discussed in a preceding section, the results of the steady state kinetic study indicated that predominant activated enzyme species in the enzyme reactions was the enzyme-calmodulin complex containing all four Ca^{++}. The results were compatible with either of the two possibilities; the enzyme activation depending on the binding of all 4 Ca^{++} to calmodulin or the partially liganded enzyme-calmodulin complexes existing in negligible amount. The latter situation may arise under experimental conditions where the concentration of the free Ca^{++} is much greater than the dissociation constant for the most tightly bound Ca^{++} in the protein complex Ca_4^{++} CM-PDE. More detailed discussion of the steady state kinetic analysis has been provided by Huang et al. (9).

From the results of the stopped-flow study, it seems that the fast-kinetic techniques have potential for distinguishing the two possibilities. The observation that the rate of the enzyme deactivation correlated with the rapid initial dissociation of Ca^{++} from the fully liganded calmodulin phosphodiesterase complex indicates that partially liganded protein complexes are not activated. It appears likely, as may be estimated from the oscillograms of Fig. 7 that only one, or at most two, Ca^{++} dissociates from the protein complex at the fast rate. Thus, it may be suggested that all, or at least three Ca^{++} are required for calmodulin to activate cyclic nucleotide phosphodiesterase. However, since the exact number of the rapidly dissociating Ca^{++} is not established, a definitive statement on the number of Ca^{++} required for the activation of the phosphodiesterase by calmodulin cannot be made at present.

The rate constants for the dissociation of Ca^{++} from the Ca^{++}-calmodulin phosphodiesterase complexes are summarized in Table 3. It may be noted that the rate of the dissociation of the slowest Ca^{++} from the Ca^{++} calmodulin complex is about 3.5 times faster than the rate of release of the fastest dissociating Ca^{++} from the complex of Ca^{++}-calmodulin phosphodiesterase. Although the value for the rate constant of the fastest dissociating Ca^{++} of the calmodulin Ca^{++} is not known, the oscillogram of Fig. 6 suggests that it has a half-life shorter than 5 msec. Also, from the diagram of Fig. 7, it may be suggested that the half-life for the dissociation of the slowest dissociating Ca^{++} from the calmodulin phosphodiesterase complex is in the order of seconds. Thus, as an approximation, the rates of the dissociation of Ca^{++} from calmodulin are about two orders of magnitude greater than those from the complex of calmodulin and the phosphodiesterase.

DISCUSSION

The present study suggests that the activation of cyclic nucleotide phosphodiesterase is likely to require the binding of all four Ca^{++}s to calmodulin. This is at variance with a recent publication by Cox et al. (7) which suggested that the activating species of calmodulin for the phosphodiesterase contained three molecules of Ca^{++}. The discrepancy is, in part, due to the difference in experimental results obtained in these two studies. For example, Cox et al. (7) observed that calmodulin possessed three high affinity sites with K_a of 6×10^{-6} M and 1 Ca^{++} binding site with much lower affinity, and that the activation of the phosphodiesterase by calmodulin showed homotropic cooperativity. The reason for such differences are not clear. However, it may be noted that when partially purified phosphodiesterase was used, the activation of the enzyme by calmodulin may show homotropic cooperativity due to the contamination of other calmodulin binding proteins in the enzyme preparations (27).

Klee et al. (14) have made the intriguing postulate that the multiple Ca^{++} binding of calmodulin may be related to the multiple regulatory functions of the protein. Thus, calmodulin may assume different conformations depending on the degree of Ca^{++} saturation, each of these conformations may recognize and regulate a specific group of calmodulin-dependent proteins. Such a postulate would predict that different calmodulin-dependent proteins may require different numbers of calmodulin molecules for the regulation. The only other enzymes which have been examined for the requirement of Ca^{++} for activation is skeletal muscle myosin light chain kinase. Blumenthal and Stull (1) have carried out steady state kinetic studies, using a different analytical procedure, of the activation of myosin light chain kinase by Ca^{++} and calmodulin. Their results are compatible with a kinetic model indicating that this protein kinase requires the binding of four molecules of Ca^{++} per molecule of calmodulin for its activation. Thus, additional calmodulin dependent proteins have to be studied to test the possibility that the multiple Ca^{++} binding of calmodulin plays a role in the fine tuning of the multiple regulatory activities of the protein.

The significance of multiple Ca^{++} binding of calmodulin may be appreciated by considering the energy coupling of the Ca^{++} binding and phosphodiesterase binding to calmodulin. It may be seen from

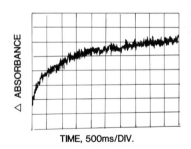

FIGURE 7

Ca^{++} release rate from calmodulin-Ca_4^{++}-PDE complex in a stopped flow instrument. See text for details.

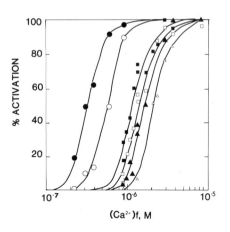

FIGURE 8

Activation of phosphodiesterase as a function of free Ca^{++} at different constant concentrations of calmodulin. The calmodulin concentrations are △ = 20.9 nM; ▲ = 58.8 nM; □ = 0.108 µM; ■ = 0.216 µM; ○ = 1.93 µM; and ● = 11.6 µM.

Scheme 1 and the values of the various dissociation constants which have been determined, the marked increase in the affinity of calmodulin for the phosphodiesterase by Ca^{++}, dissociating constant changing from 10^{-3} M to 10^{-10} M, is coupled to the change in the affinity of each of the four Ca^{++} for calmodulin by no more than two orders of magnitude. If there were only one Ca^{++} involved in the enzyme activation, the dissociation constant of Ca^{++} from the enzyme-calmodulin complex would have to be in the order of 10^{-12} M. Such a regulatory system would not respond to changes in cellular Ca^{++} concentration which is believed to always be above 10^{-8} M. Thus, for a calmodulin regulated enzyme which undergoes Ca^{++} dependent association with calmodulin, the multiple Ca^{++} binding to calmodulin is almost an essential condition.

Other advantages of having four Ca^{++} sites on calmodulin may be analyzed on the basis of the Ca^{++} concentration-dependence of the enzyme activation (Fig. 8). The activation of the phosphodiesterase by Ca^{++} is highly cooperative, the full range of the enzyme activation is achieved with a change in one order of magnitude of the Ca^{++} concentration. Thus, the presence of four Ca^{++} on calmodulin provides the cell with a very effective on/off switch of this enzyme. It is also noted in Fig. 8 that the concentration of Ca^{++} required for the activation of the phosphodiesterase depends strongly on the level of calmodulin in the reaction. Therefore, the cellular concentration of calmodulin may be considered as another determinant of the Ca^{++} sensitivity of the cells.

In a cellular regulatory system, the mechanisms of both its activation and inactivation are important. For the calmodulin-activated cyclic nucleotide phosphodiesterase, the deactivation of the enzyme may be brought about by one of the two processes; the removal of Ca_4^{++} CM from the enzyme complex or the release of Ca^{++}. From the results seen in Figs. 3 and 6, it is clear that the rate of deactivation of the enzyme by the removal of Ca^{++} is thousands-fold faster than that by the dissociation of the liganded calmodulin. Thus, the much more efficient Ca^{++} removal and the known fluctuation in the cellular-free Ca^{++} concentration together suggest strongly that the in vivo mechanisms of the deactivation of the phosphodiesterase is likely to proceed by the Ca^{++} removal.

The kinetic analysis used in this study is also applicable to other calmodulin-regulated enzymes which undergo Ca^{++} dependent reversible association. It should be stressed that some of these enzymes may not require as drastic a Ca^{++} induced change in the affinity toward calmodulin as that found for the phosphodiesterase. It is, therefore, conceivable that their interactions with calmodulin may be coupled to the binding of only two or three Ca^{++}s. An extreme case is phosphorylase kinase where calmodulin was found to be a subunit, δ subunit (6). The activation of this enzyme by Ca^{++} through the δ subunit is not coupled to the change in protein association state. It is interesting to note that phosphorylase kinase can be further activated by Ca^{++} and additional calmodulin and the enzyme activation by the second mole of calmodulin is by the mechanism of reversible Ca^{++} dependent association (6). Recently, two isozymes of skeletal muscle phosphorylase kinase were found to be regulated by calmodulin differently. While the white muscle isozyme is regulated by both the δ subunit and the additional reversibly-bound calmodulin, the red muscle isozyme is regulated solely by δ subunit (21). These observations support the notion that the regulatory functions of tightly bound calmodulin and reversibly interacting calmodulin are different.

ACKNOWLEDGEMENTS

The work from the University of Manitoba was supported by the Medical Research Council of Canada (MT-2381).

REFERENCES

1. Blumenthal, D.J. & Stull, J.T. (1980): Biochemistry 19:5608-5615.
2. Cheung, W.Y. (1967): Biochem. Biophys. Res. Commun. 29: 478-482.
3. Cheung, W.Y. (1970): Biochem. Biophys. Res. Commun. 38: 533-538.
4. Cheung, W.Y. (1971): J. Biol. Chem. 246:2858-2869.
5. Cheung, W.Y. (1980): Science 207:19-27.
6. Cohen, P., Burchell, A., Foulkes, J.G., Cohen, P.T.W., Vanaman, T.C. & Nairn, A.C. (1978): FEBS Lett. 92:287-293.
7. Cox, J.A., Malnoe, A. & Stein, E.A. (1981): J. Biol. Chem. 256:3218-3222.
8. Crouch, T.H. & Klee, C.B. (1980): Biochemistry 19:3692-3698.
9. Ho, H.C., Wirch, E., Stevens, F.C. & Wang, J.H. (1977): 252: 43-50.
10. Huang, C.Y., Chau, V., Chock, P.B., Wang, J.H. & Sharma, R.K. (1981): Proc. Nat. Acad. Sci. 78:871-875.
11. Kakiuchi, S., Yamazaki, R. & Nakajima, H. (1970): Proc. Japan. Acad. 46:587-592.
12. Kakiuchi, S. & Yamazaki, R. (1970): Biochem. Biophys. Res. Commun. 41:1104-1110.
13. Kakiuchi, S., Yamazaki, R., Teshima, Y. & Unishi, M. (1973): Proc. Nat. Acad. Sci. 70:3526-3530.
14. Klee, C.B., Crouch, T.H. & Richman, P.G. (1980): Ann. Rev. Biochem. 49:489-515.
15. Klee, C.B. & Krinks, M.H. (1978): Biochemistry 17:120-126.
16. Laporte, D.C., Toscano, W.A. Jr. & Storm, D.R. (1979): Biochemistry 18:2820-2825.
17. Lin, Y.M., Liu, Y.P. & Cheung, W.Y. (1975): FEBS Lett. 49: 356-360.

18. Morill, M.E., Thompson, S.T. & Stellwagin, E. (1979): J. Biol. Chem. 254:4371-4374.
19. Richman, P.G. & Klee, C.B. (1979): J. Biol. Chem. 254:5372-5376.
20. Sharma, R.K., Desai, R., Waisman, D.M. & Wang, J.H. (1979): J. Biol. Chem. 254:4276-4282.
21. Sharma, R.K., Tam, S.T., Waisman, D.M. & Wang, J.H. (1980): J. Biol. Chem. 255:11102-11105.
22. Sharma, R.K., Wang, T.H., Wirch, E. & Wang, J.H. (1980): J. Biol. Chem. 255:5916-5923.
23. Teo, T.S., Wang, T.H. & Wang, J.H. (1973): J. Biol. Chem. 248:588-595.
24. Teo, T.S. & Wang, J.H. (1973): J. Biol. Chem. 248:5950-5955.
25. Teshima, Y. & Kakiuchi, S. (1974): Biochem. Biophys. Res. Commun. 56:489-495.
26. Tsien, R.Y. (1980): Biochemistry 19:2396-2404.
27. Wang, J.H. & Desai, R. (1976): Biochem. Biophys. Res. Commun. 72:926-932.
28. Wang, J.H. & Desai, R. (1977): J. Biol. Chem. 252:4175-4184.
29. Wang, J.H., Sharma, R.K., Huang, C.Y., Chan, V. & Chock, P.B. (1980): Ann. N.Y. Acad. Sci. 356:190-204.
30. Wang, J.H. & Waisman, D.M. (1979): Curr. Top. Cell. Regul. 15:47-107.
31. Wolff, D.J. & Brostrom, C.O. (1979): Adv. Cycl. Nucl. Res. 11:27-88.

THE REGULATION OF MYOSIN LIGHT CHAIN KINASE BY Ca^{++} AND CALMODULIN IN VITRO AND IN VIVO

J.T. Stull, D.K. Blumenthal, B.R. Botterman, G.A. Klug, D.R. Manning & P.J. Silver

Departments of Pharmacology and Cell Biology
and Moss Heart Center, University of Texas
Health Science Center at Dallas, Dallas, Texas 75235, USA

Myosin light chain kinase calalyzes the phosphorylation of a particular light chain subunit of myosin from all types of mammalian muscles. Myosin light chain kinases from skeletal and smooth muscles are dependent upon both Ca^{++} and calmodulin for activity. From an enzymatic kinetic analysis of the activation process, the following sequence of reactions are proposed: all four divalent metal binding sites in calmodulin must be filled with Ca^{++}, a single Ca_4^{++}-calmodulin complex then binds to a myosin light chain kinase catalytic subunit to form the enzymatically active holoenzyme. In addition to variations in the kinetics of Ca^{++} release and resequestration in different types of muscles, some of the important determinants governing the rate of phosphorylation of myosin light chain include the rates of activation and inactivation of myosin light chain kinase, the calmodulin concentration, the concentration and catalytic activity of myosin light chain kinase and the amount of myosin light chain phosphatase activity.

Upon tetanic electrical stimulation in fast-twitch skeletal muscle, myosin light chain is phosphorylated at a rate much slower than contraction. When skeletal muscle is stimulated at frequencies that do not result in complete fusion of twitches (1 Hz and 20 Hz), there is a time-dependent increase in the phosphate content in myosin light chain. Repetitive stimuli at low frequencies result in phosphorylation of myosin light chain, probably due to a small but significant amount of kinase activated with each stimulus, the slow rate of inactivation of myosin light chain kinase activity after Ca^{++} sequestration by sarcoplasmic reticulum, and the very slow rate of dephosphorylation by myosin light chain phosphatase. The phosphorylation of myosin light chain is correlated to potentiation of isometric twitch after both the tetanic stimulation and stimulation at low frequencies.

In isolated trachealis smooth muscle, stimulation of contraction with carbachol, a cholinergic muscarinic agonist, resulted in an increase in isometric tension that was maintained for up to 30 min. The phosphate content of myosin light chain reached a maximum value by 1 min and then slowly declined. Isoproterenol, a β-adrenergic agonist, inhibited the development of isometric tension by carbachol and also inhibited the extent of phosphorylation of myosin light chain. These results show that isometric contraction in trachealis smooth muscle is associated with a transient phosphorylation of myosin light chain. Furthermore, the relaxation response to β-adrenergic stimulation in smooth muscle may be due to inhibition of myosin light chain phosphorylation. Biochemical mechanisms that may account for the dephosphorylation of myosin light chain during maintenance of isometric tension or inhibition of phosphorylation with β-adrenergic stimulation have not yet been elucidated.

INTRODUCTION

Myosin is a hexameric protein molecule composed of two high molecular weight subunits (heavy chains) and four low molecular weight subunits (light chains). Head regions of the heavy chains of myosin project from thick filaments to bind to thin filaments of actin. These head regions contain the ATPase activity of myosin in addition to the actin-binding domain. The sliding theory of muscle contraction postulates that tension and shortening occur as a result of actin-myosin interactions, causing thick and thin filaments to move past one another. The light chain subunits are associated with the myosin heads but their exact location has not yet been determined. Recent reviews (1,16,24) present more detailed discussions of the biochemical properties of myosin.

Myosins from various mammalian tissues each have light chain subunits which are homologous. These myosin light chains range in molecular weight from 18,500 to 20,000 and are capable of binding divalent cations. In particular, the light chains in this class are capable of being phosphorylated by myosin light chain kinases and have been referred to as phosphorylatable or P-light chains (1,35). Myosin light chain kinase catalyzes the incorporation of phosphate from MgATP into a specific serine residue of each P-light chain (Fig. 1). Calcium and the ubiquitous low molecular weight calcium binding protein, calmodulin, are both required for myosin light chain kinase activity (9,41). The activation of rabbit skeletal muscle (5) and bovine aortic smooth muscle (14) myosin light chain kinase by calcium and calmodulin has been studied in detail.

Experiments were performed in which the rate of myosin light chain phosphorylation was measured at a fixed enxyme concentration and variable concentrations of calcium and calmodulin. The results of these kinetic experiments indicate that multiple calcium binding sites in calmodulin are involved in the enzyme activation. A mathematical analysis of this activation process indicated that the best fit of the experimental data was obtained when all four calcium binding sites on calmodulin were filled with calcium. That multiple sites in calmodulin must be occupied by calcium for activation of calmodulin dependent enzymes has recently also been proposed for cyclic nucleotide phosphodiesterase (8,19). Thus, activation of myosin light chain kinase appears to require formation of a holoenzyme complex consisting of one enzyme catalytic subunit and one calmodulin with all four divalent metal binding sites occupied by calcium (Fig. 1).

The dephosphorylation of P-light chains is catalyzed by another class of enzymes, myosin light chain phosphatases (Fig. 1). More information has been presented on the properties of myosin light chain phosphatase from rabbit skeletal muscle (27) than from other tissues. Its biochemical properties distinguish it from other general phosphoprotein phosphatases. Myosin light chain phosphatase did not catalyze the dephosphorylation of the phosphorylated forms of glycogen synthase, phosphorylase kinase, histones, casein, phosphorylase, or skeletal muscle troponin. Thus, myosin light chain phosphatase appear to be highly specific in the catalysis of dephosphorylation of phosphoprotein substrates. This specificity has also been noted with myosin light chain kinases. Multiple myosin light chain phosphatases have been purified recently from gizzard smooth muscle (28).

Protein phosphorylation-dephosphorylation has been implicated in the regulation of numerous intracellular processes. These include glycogenolysis, gluconeogenesis, glycolysis, lipolysis, nucleic acid transcription, protein translation, cellular transformation, ion transport, and contractile protein interactions (1,23,35). The introduction of a phosphate moiety into a protein may result in marked changes in its biochemical properties, and may comprise an important means of regulating a particular biochemical process. These general considerations can also be applied to phosphorylation of myosin. Numerous investigators (1,35) have shown that with actomyosin prepared from various types of smooth muscle there was a good correlation between the extent of ATPase activity and the extent of phosphorylation of P-light chain. Furthermore, thiophosphorylated myosin is resistant to dephosphorylation by myosin light chain phosphatase activity and irreversible activation of actomyosin ATPase activity results from thiophosphorylation of the P-light chain (33). Recent evidence (32) has been obtained that shows there is a coincident change in the ATPase activity with reversible phosphorylation-dephosphorylation and rephosphorylation of the

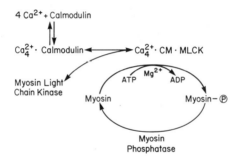

FIGURE 1

Regulation of myosin phosphorylation. This simplified scheme indicates that binding of four Ca^{++} to calmodulin is required for association of calmodulin with the myosin light chain kinase catalytic subunit to form the enzymatically active holoenzyme. Myosin is phosphorylated by myosin light chain kinase in the presence of MgATP. Dephosphorylation of myosin is catalyzed by myosin light chain phosphatase.

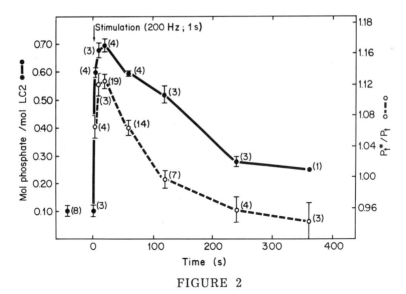

FIGURE 2

Myosin light chain 2 (P-light chain) phosphate content and potentiation of peak twitch tension following tetanic stimulation at 23°C. Rat EDL muscle preparations were tetanically stimulated for 1 sec and at times indicated were either frozen for analysis of phosphorylation or were stimulated to twitch-contract. Twitch potentiation is expressed as the ratio of post-tetanic to pre-tetanic twitch tensions (P_t^*/P_t). Pre-tetanic resting LC2 phosphate content is represented by the point prior to t = 0. Bars indicate ± 1 S.E.M. and numbers within parentheses are the number of muscles used for each point.

P-light chain in actomyosin prepared from gizzard smooth muscle. Collectively these results indicate the phosphorylation of myosin in smooth muscle may play an important role in the regulation of actomyosin interaction and hence contraction. In contrast to the results obtained from smooth muscle myosin, some investigators (27,36) have found that phosphorylation of the P-light chain of myosin from skeletal muscle had no significant effects on actomyosin ATPase activity. The ATPase activity of myosin in the presence or absence of actin was found to be unaffected by phosphorylation. One report (29), however, indicated that skeletal muscle actomyosin ATPase activity was enhanced by phosphorylation with a decrease in K_{app} of myosin ATPase activity for actin. The reason for the discrepancy is unclear at this time, but may be due to the use of unpurified kinase in the latter experiments.

In order to fully understand the regulation of myosin phosphorylation and the role of myosin phosphorylation in muscle contraction, more quantitative information on the mechanisms of regulation of purified myosin light chain kinase activity is needed. This information needs to be integrated with results obtained from investigations dealing with P-light chain phosphorylation in intact muscles. A number of factors have to be taken into account in order to fully understand this phosphorylation system. These factors include not only the kinetic and equilibrium activation properties of the myosin light chain kinase, but also such factors as calmodulin concentration, myosin light chain kinase catalytic activity and concentration, and the myosin light chain phosphatase activity in muscle in vivo. All of this information is important in elucidating the biological properties of myosin phosphorylation and the function of myosin phosphorylation in different types of muscle. Although the sequence of reactions leading to myosin phosphorylation can be described in general terms, as shown in Fig. 1, one might expect quantitative differences in the various reactions in different types of muscle because of their distinct biochemical, physiological and pharmacological properties. The results described in this report represent a preliminary attempt to integrate available information and to point out areas in which additional information is needed. The properties of phosphorylation of P-light chain in intact skeletal muscle and smooth muscle have been compared in order to emphasize important similarities and distinctions between these types of muscle which relate to myosin phosphorylation.

METHODS

Purification of proteins. P-light chain from different muscles was purified as previously described (5). Calmodulin was purified from bovine brain by the procedure of Dedman et al. (10). Myosin light chain kinase was purified by a method similar to that described by Yazawa and Yagi (41). Myosin light chain kinase activity was determined by measurement of the rates of ^{32}P incorporation into P-light chains.

Skeletal muscle preparations. Extensor digitorum longus muscles were excised from female Sprague-Dawley rats (50-90 g) and mounted vertically in oxygenated Ringer solution maintained at 23°C (25). Muscles were stimulated by a transverse electrical field and isometric tension development was recorded. Muscle lengths were adjusted so that peak twitch tensions were maximal. To obtain samples for biochemical analyses, the baths were quickly lowered and muscles were frozen by clamps precooled in liquid nitrogen. Another muscle preparation used for in situ measurements included the surgical isolation of gastrocnemius muscles of female Sprague-Dawley rats (230-270 g)(22). The gastrocnemius muscles of both legs were free from the plantaris and soleus muscles. Care was taken not to disrupt the blood or nerve supply. The calcaneous was severed from the foot and with the tendons intact attached securely to a force displacement transducer. The sciatic nerves were isolated in the lateral portion of the thigh and the entire lower leg denervated with the exception of the gastrocnemius muscle. The animal was then mounted in a specially designed apparatus which fixed the lower legs and prevented any movement from occurring. The lower legs were then submerged into a bathing chamber containing a physiological salt solution of 137 mM NaCl, 5 mM KCl, 2 mM $CaCl_2$, 1 mM $MgCl_2$, 1 mM sodium phosphate and 2 g/l of sodium bicarbonate with a pH of 7.4. The isolated sciatic nerves were placed on bipolar platinum electrodes for stimulation. Platinum electrodes were also placed into the belly of the muscle to record EMG activity. The length of the muscles were adjusted so that twitch tensions were maximal. Muscle biopsies were obtained by clamps prechilled in liquid nitrogen.

Trachealis smooth muscle preparation. Studies with bovine tracheal smooth muscle strips were performed according to previously described methods (21,34). Briefly, fresh trachea were obtained from the local abattoir and transported to the laboratory in ice-cold, balanced salt solution. The composition of the balanced salt solution was 118.5 mM NaCl, 4.74 mM KCl, 1.18 mM $MgSO_4$, 1.18 mM potassium phosphate, 24.9 mM sodium bicarbonate, 1.5 mM $CaCl_2$, 10 mM dextrose and 1 mM pyruvate. Upon arrival at the laboratory, smooth muscle was placed in balanced salt solution aerated with a 95% O_2, 5% CO_2 gas mixture at room temperature and carefully separated from cartilage and connective tissue. Transverse strips (14 mm x 1.5 mm, 15 mg wet weight) were prepared and mounted in a jacketed muscle chamber for recording isometric tension at 36°C. The muscle strips were stretched passively to a resting tension of 1.5 g which resulted in the maximum isometric tension developed following the appropriate stimulation. Following the application of passive tension, the strips were equilibrated in aerated balanced salt solution for 90 min or, alternatively, equilibrated for 120 min without passive tension followed by 45 min with passive tension. Either maneuver produced comparable experimental results. At the end of the equilibration period, passive tension was reapplied and the strips were prechallenged with 0.1 µM carbachol. Strips which did not attain 5-7 g of isometric force were eliminated from further study at this point. Following the prechallenge with 0.1 µM carbachol, the strips were rinsed 2-3 times with fresh balanced salt solution until tension returned to baseline, then 3-4 more times for an additional 20 min. After this equilibration period the strips were exposed to pharmacological agents as indicated, and the muscle strips were quick frozen at indicated times by rapidly lowering the muscle bath and immersing the strips in dichlorodifluoromethane cooled in liquid nitrogen.

Analysis of P-light chain phosphorylation. Frozen skeletal muscles were powdered by percussion at $-180°C$ and portions weighed at $-60°C$ were homogenized directly into 1 ml of 5 M guanidine hydrochloride, 50 mM potassium phosphate, pH 6.8, 1 mM EDTA, and 15 mM 2-mercaptoethanol at $-15°C$. The homogenate was warmed to $0°C$ and the light chain fraction obtained as previously described (25). The phosphate content of LC2 was determined by measuring the relative amounts of phosphorylated and nonphosphorylated forms separated by polyacrylamide gel electrophoresis in 8 M urea at pH 8.6 by densitometry. An alternative procedure was used for skeletal muscle which gave similar results to the procedure described above. This alternative procedure was also used for tracheal smooth muscle. Frozen muscle samples were homogenized in a modified sodium pyrophosphate extraction buffer (18) and centrifuged at 7,000 x g. The supernatant fraction was then applied to a pyrophosphate-agarose-polyacrylamide gel to isolate native myosin from other cellular proteins. The myosin band was then excised, homogenized in isoelectric focussing denaturing buffer and subjected to isoelectric focussing in polyacrylamide gels to separate the phosphorylated from the nonphosphorylated forms of P-light chains (13). The phosphate content of P-light chains was determined by measuring the relative amounts of phosphorylated and nonphosphorylated forms of densitometry.

RESULTS AND DISCUSSION

Studies in Intact Skeletal Muscle

P-light chain phosphorylation following tetanic stimulation has been studied with intact skeletal muscle preparations from frog (2) and rabbit (37). In these preliminary reports, it was clearly demonstrated that phosphorylation of the P-light chain could occur upon stimulation of muscle contraction. In frog muscle the temporal relationship between the extent of phosphorylation and tension developed during a tetanus was examined in more detail (3). With brief durations of tetanic stimulation, there was a net incorporation of about 0.2 mol phosphate/mol P-light chain that was rapidly dephosphorylated upon relaxation. With longer periods of tetanic stimulation, the extent of net phosphorylation increased to about 0.35 mol phosphate/mol P-light chain. However, at 30 sec after relaxation of the muscle, the extent of P-light chain phosphorylation remained elevated. Phosphorylation of P-light chain in frog skeletal muscle was also observed in overstretched fibers that developed no tension and in caffeine treated muscles. Thus, increases in sarcoplasmic calcium concentrations produced by different mechanisms result in activation of myosin light chain kinase activity.

P-light chain phosphorylation induced by tetanic stimulation of a fast-twitch skeletal muscle from rat has also been investigated (25). These studies were the first to indicate a potential physiological role for P-light chain phosphorylation in skeletal muscle. The extent of P-light chain phosphorylation was low in resting muscle with a value of 0.1 mol phosphate/mol P-light chain (Fig. 2). Following a 1 sec tetanus, the extent of phosphorylation increased to a maximum value of 0.7 mol phosphate/mol P-light chain at about 10 sec. Thus, following the 1 sec tetanic stimulation and after the muscle had relaxed, there was a rapid phase of phosphorylation of P-light chain. After a maximal value had been reached, there was a much slower rate of dephosphorylation with a $t_{\frac{1}{2}}$ of approximately 1.7 min. These data demonstrate that the rates of phosphorylation and dephosphorylation of P-light chain in rat skeletal muscle are considerably slower than the rates of contraction and relaxation. Following a tetanus, the tension generated in isometric twitches is greater than that generated before the tetanus (Fig. 2). This phenomenon has been referred to as post-tetanic potentiation of isometric twitches and has a similar time course to that shown for the phosphorylation of P-light chain in rat skeletal muscle, i.e. it consists of a rapid rising phase followed by a slower declining phase. Thus, there appears to be a good temporal correlation between the extent of myosin P-light chain phosphorylation and the extent of potentiation of the isometric twitch tension. More recent studies in rat extensor digitorum longus muscle have supported the concept that there is an important relationship between the extent of P-light chain phosphorylation and potentiation of isometric twitch tension (26). A positive correlation was obtained in the temporal relationships at 35°C as well as at 23°C. In addition, when the frequency and duration of the stimulus was varied, the positive correlation between P-light chain phosphorylation and potentiation of the isometric twitch was maintained.

The following conclusions may be drawn from these studies on intact skeletal muscles. First, myosin light chain phosphorylation occurs in the intact tissue under circumstances whereby there is an elevation of sarcoplasmic calcium concentrations. When calcium concentrations return to resting values, the P-light chain is slowly dephosphorylated, indicating the presence of myosin light chain phosphatase. However, the rates of relaxation and the rates of myosin dephosphorylation are not the same with dephosphorylation being a much slower process. As a result, P-light chain phosphorylation can be clearly dissociated from tension reponses during

TABLE 1: Effect of stimulation frequency on P-light chain phosphorylation in rat gastrocnemius muscle in situ.

	Frequency (Hz)					
	0	0.1	0.5	1.0	2.0	5.0
Mol phosphate / Mol P-light chain	0.18 ±0 03	0.27 ±0.05	0.46 ±0.02	0.52 ±0.04	0.56 ±0.01	0.71 ±0.01

Muscles were stimulated at the indicated frequencies for 100 stimuli and were quick frozen. The numbers represent means ± 1 S.E.M.

TABLE 2: Effect of carbachol on P-light chain phosphorylation, phosphorylase a formation and isometric tension in bovine tracheal smooth muscle.

Time after 1 µM Carbachol (min)	Isometric Tension Development (g)	P-Light Chain Phosphorylation (mol phosphate/ mol LC)	Phosphorylase a Activity Ratio (-5'AMP/ +5'AMP)
0	0.0	0.13 ± 0.03	0.15 ± 0.02
1	9.0	0.75 ± 0.05	0.61 ± 0.04
3	11.0	0.48 ± 0.03	0.59 ± 0.04
30	10.8	0.23 ± 0.02	0.24 ± 0.02

tetany, indicating that phosphorylation is not obligatory for contraction. Although light chain phosphorylation may not play an obligatory role in contractile element activation or force generation, it may function to modulate the contractile response, i.e. augment force generation of isometric twitches.

The relationship between isometric twitch potentiation and phosphorylation of P-chain has been examined in the rat gastrocnemius muscle in situ. It was found that very low frequencies of stimulation were sufficient to cause apparent activation of myosin light chain kinase activity (Table 1). There was a small increase in the phosphate content of P-light chain from 0.18 to 0.27 mol phosphate/mol P-light chain at a frequency of stimulation of 0.1 Hz for 100 stimuli. At 0.5 Hz this value increased to 0.46 and greater frequencies resulted in greater increases in the extent of phosphorylation of P-light chain. These data demonstrate that activation of myosin light chain kinase occurs not only with tetanic stimulation but also with repetitive single isometric twitches at appropriate frequencies and durations of stimulation. The time course of phosphorylation measured at 5 Hz was characterized by an initial rapid increase during the first 4 sec (from 0.18 to 0.43 mol phosphate/ mol P-light chain) followed by a slower rate of phosphorylation (from 0.43 to 0.71 mol phosphate/mol P-light chain over 12 sec). Potentiation of isometric twitch tension displayed a time and frequency dependence similar to that of phosphorylation. This graded increase in isometric twitch tension under these conditions is referred to as the staircase phenomenon and is well documented in the physiological literature (22). Repetitive stimulation at 5 Hz increased maximum twitch tension to 180% of the prestimulation control value. The relationship between potentiation of the isometric twitch under these conditions and the extent of phosphorylation of P-light chain was similar to that observed after a tetanic stimulation. These data demonstrate that phosphorylation of P-light chain can occur in situ with a time course and frequency dependence that would permit myosin phosphorylation to play a role in the alteration of the isometric twitch properties.

These experiments in intact skeletal muscles raise some important questions in regard to the biochemical regulation of myosin P-light chain phosphorylation. During a single twitch in intact skeletal muscle, calcium transients lasting less than 50 msec appear to be responsible for regulating the development of tension via calcium binding to troponin C. With a series of repetitive stimuli at 1 Hz, the period of time in which the muscle is contracting is only about 30 msec during the total 1 sec period between stimuli. How can this brief period of contraction result in significant phosphorylation of P-light chain? Is there a significant amount of Ca_4^{++}-calmodulin formed for activation of myosin light chain kinase? To what extent does myosin light chain kinase become activated during this brief calcium transient?

We have attempted to provide some insight into the regulation of these various processes involved in phosphorylation of P-light chain. We have used the equation and rate constants described by Potter et al. (30) to describe a typical calcium transient in the sarcoplasm of skeletal muscle during a twitch. The time course of this calcium transient is shown in the top panel of Fig. 3. In this model calcium transient, the pCa^{++} reaches a maximal value of about 5 within 5 msec. With 40 msec pCa^{++} decays to a resting value of 8. In the second panel from the top in Fig. 3, we have plotted the expected time course of Ca_4^{++}-calmodulin formation. We have assumed that the on rate for calcium binding to the four calcium binding sites is equivalent and has a value of 10^8 M^{-1} sec^{-1}. The off rate is assumed to have a value of 100 sec^{-1}. We have chosen to calculate the value of Ca_4^{++}-calmodulin formation since we have previously shown that this is probably the form of calmodulin that activates myosin light chain kinase from skeletal muscle. The values shown in Fig. 3 were solved numerically. It was assumed that the total concentration of calmodulin available for interaction with myosin light chain kinase was 3 μM (39). As can be seen in Fig. 3, about 25% of the total available calmodulin is in the form

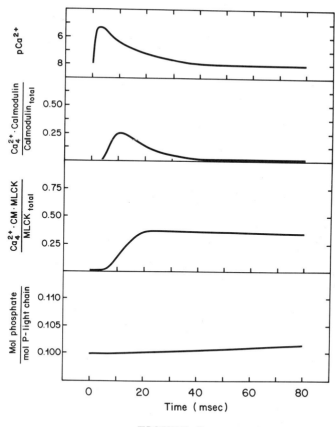

FIGURE 3

Effect of a single twitch on myosin light chain phosphorylation in skeletal muscle. The temporal relationships between the transient change in sarcoplasmic pCa^{++}, formation of Ca_4^{++}-calmodulin, active myosin light chain kinase, and the extent of phosphorylation of P-light chain were calculated as described under Results and Discussion.

of Ca_4^{++}-calmodulin by 10 msec following the onset of calcium influx. By 20 msec only about one-half of that amount remains, and by 40 msec the concentration of Ca_4^{++}-calmodulin has returned to a resting value. Thus, the formation of the Ca_4^{++}-calmodulin complex in the sarcoplasm of skeletal muscle is a rapid and transient event. The third panel from the top shows the time course of activation of myosin light chain kinase. The concentration of the active enzyme was calculated numerically, assuming an on rate constant of 10^8 M^{-1} sec^{-1} and an off rate of 0.5 sec^{-1}. The total concentration of the enzyme was estimated to be 1 μM. The determination and assumptions in these values have been previously discussed (39). The plot shown in the third panel from the top in Fig. 3 indicates that about 35% of the enzyme would be expected to be activated during a single calcium transient and that this occurs within about 20 msec after the onset of calcium influx. Because of the slow rate of decay of the active enzyme, most of the enzyme which was activated remains in that state at 50 msec, even though both the pCa^{++} value and the concentration of Ca_4^{++}-calmodulin have decayed to resting values. The bottom panel in Fig. 3 shows the time course of myosin P-light chain phosphorylation. The extent of phosphate incorporation was calculated using a value of 50 sec^{-1} as a turnover number of the active kinase and 0.007 sec^{-1} as the rate of dephosphorylation catalyzed by myosin light chain phosphatase (25,26,39). A total P-light chain concentration of 340 M and a resting value of 0.1 mol phosphate/mol P-light chain were assumed. The plot in the bottom panel indicates that by 50 msec very little phosphate incorporation has occurred, even though more than one-third of the enzyme had been activated. This result would be expected because of the low catalytic rate of the enzyme compared to the high concentration of P-light chain in skeletal muscle. Thus, several seconds would be required for significant phosphorylation of P-light chain with fully activated myosin light chain kinase.

Several of the rate constants we have employed have yet to be determined with precision. In particular, we have assumed that the rate of activation of myosin light chain kinase is diffusion controlled, i.e. 10^8 $M^{-1} sec^{-1}$, which is probably an overestimate. If an on-rate of 10^7 $M^{-1} sec^{-1}$ was assumed, maximal activation of the enzyme would occur at about 30 msec following the onset of calcium influx; about 7% of the total enzyme would become activated with a single transient. Thus, with the slow rate constant the time course of enzyme activation would be similar to that seen with the faster rate constant, but the extent of enzyme activation would be much less. Additional information is needed in regards to the kinetic constants for the activation of myosin light chain kinase in order to make critical comparisons with intact muscle data. In any event, it would appear that a significant amount of Ca_4^{++}-calmodulin can be formed with a single calcium transient and that a significant amount of myosin light chain kinase can also be activated. However, due to the low catalytic rate of the enzyme the amount of phosphorylated P-light chain formed with a single isolated twitch is probably negligible.

Because the rate of enzyme inactivation is relatively slow ($t_{\frac{1}{2}} \sim$ 1 sec) and the rate of dephosphorylation is even slower ($t_{\frac{1}{2}} \sim$ 100 sec), a train of stimuli at a frequency of approximately 1 Hz or more would be expected to result in significant phosphorylation of the P-light chain, if maintained for a long enough period of time. This would occur because the extent of enzyme activation would be maintained at a level high enough to result in significant phosphorylation of the P-light chain. Greater frequencies of stimulation would result in a greater extent of myosin light chain kinase activation. During a tetany (200 Hz) for 1 sec, myosin light chain kinase is probably fully activated. After relaxation, there would be continued phosphorylation of the P-light chain due to the slow rate of inactivation of myosin light chain kinase. This hypothesis would account for the continued phosphorylation of P-light chain after cessation of the tetanic stimulus (Fig. 2).

The information obtained on the mechanism of activation of purified skeletal muscle myosin light chain kinase by calcium and calmodulin has been useful in analyzing the properties of P-light chain phosphorylation in intact skeletal muscles. Likewise, information obtained in intact skeletal muscle on the properties of P-light chain phosphorylation has provided clues for the functional significance of myosin phosphorylation in skeletal muscle. The correlation between potentiation of isometric twitch tension and phosphorylation of P-light chain indicates that myosin phosphorylation may play a modulatory role in skeletal msucle.

Phosphorylation of P-light Chain in Intact Smooth Muscle

A number of studies by different investigators (4,11,15,20) have shown that there is in general a positive correlation between tension and the extent of phosphorylation of myosin P-light chain in different types of smooth muscles. Stimulation of smooth muscle contraction by pharmacological agonists or increases in the potassium concentration in the balanced salt solution resulted in increases in P-light chain phosphate content in addition to isometric tension. Relaxation of the muscles resulted in a return of the phosphate content to resting levels. Furthermore, elimination of calcium from the balanced salt solution lowered both the tension responses and the extent of P-light chain phosphorylation. These results indicated that P-light chain phosphorylation occurred in smooth muscle in response to agents that cause contraction, that this was a reversible phosphorylation reaction, and the the phosphorylation was dependent upon calcium. The results from these studies support the general conclusion that myosin phosphorylation may play an important role in the control of contraction of smooth muscle. This conclusion is also supported by studies utilizing phenothiazine antipsychotic drugs. These drugs inhibit the activation of myosin light chain kinase by calmodulin and cause relaxation of smooth muscle (17).

A recent study (12) has indicated that the relationship between myosin phosphorylation and contraction in smooth muscle may not be simple. When smooth muscles from swine carotid arteries were contracted in solutions containing high levels of potassium, the extent of P-light chain phosphorylation increased during the first

30 sec to a maximum value, then decreased over the next 5 min to about one-half the maximum value. The isometric tension response, however, continued to increase even though the phosphate content in P-light chain was beginning to decrease. The tension reached a maximal value that was maintained. The velocities of isotonic contractions were recorded for various tensions, and the force-velocity relationship was constructed at different times following stimulation with potassium. There was a good agreement between the maximum contraction velocity and the extent of phosphorylation of P-light chain of myosin, that is, there was first an increase in the early phase of the contractile response and then a decrease in parallel with the level of myosin phosphorylation. Thus, Murphy and his colleagues (12,15) proposed that phosphorylation of myosin is not always correlated with isometric tension and that phosphorylation of myosin may influence properties of the cyclic interaction of actin and myosin as well as the number of crossbridges. These results would indicate that vascular smooth muscle may produce a state of high tension, but low contraction velocity, and that this would result in a high economy of tension maintenance due to the decrease in cycling crossbridges and ATP hydrolysis.

We have recently obtained, in part, similar responses in bovine tracheal smooth muscle (Table 2). Incubation of isolated tracheal smooth muscle with 1 µM carbachol resulted in a marked increase in the extent of phosphorylation of P-light chain from a resting value of 0.13 mol phosphate/mol P-light chain to 0.75 mol phosphate/mol P-light chain. After this maximal response was obtained at 1 min, there was a subsequent decline in the extent of phosphorylation to a value of 0.26 mol phosphate/mol P-light chain by 30 min. The isometric tension development lagged behind the phosphorylation of P-light chain with a maximum value obtained at 3 min. This development of isometric tension was essentially the same after 30 min of incubation in the presence of 1 µM carbachol. Thus, isometric tension development was maintained for up to 30 min while the phosphate content of P-light chain was declining. The decrease in the phosphate content of P-light chain in bovine tracheal smooth muscle in the presence of pharmacological agonists is in contrast to results reported in canine tracheal smooth muscle (11). The strips of bovine tracheal smooth muscle used for studies reported in this report weighed 15 mg (wet weight) and typically developed a maximum isometric tension of 12 g. The isolated canine tracheal smooth muscle preparation normally consisted of 150-200 mg wet weight of tissue and the total isometric tension developed was 32 g (11). It is possible that in the case of the canine tracheal smooth muscle, not all the muscle cells were fully activated with the application of the pharmacological agent. The amount of phosphorylation of P-light chain for canine tracheal smooth muscle was consistently lower than the values obtained in bovine trachea. In addition, 100 µM methacholine was used to stimulate the canine tracheal smooth muscle preparation while 1 µM carbachol was used in the studies with bovine tracheal smooth muscle. Non-specific effects of the very high concentration of muscarinic agonist in addition to the large amount of tissue, could have obscured measurements of the decline in myosin P-light chain phosphorylation during maintenance of isometric tension.

We also measured phosphorylase a formation in bovine trachealis smooth muscle in response to carbachol (Table 2). Phosphorylase a formation preceded isometric tension development with a maximal value obtained at 1 min of incubation in 1 µM carbachol. This value was maintained for up to 3 min, but by 30 min had declined from a maximal value of 0.61 to a value of 0.27. Thus, even though isometric tension was maintained at 30 min there was a decline in phosphorylase a formation as well as a decline in P-light chain phosphorylation. The rate of decline of phosphorylase a formation appeared to be slower than the decline of the phosphate content of P-light chain. Phosphorylase a formation, like myosin P-light chain phosphorylation, is a calcium dependent process (Fig. 4). Phosphorylation of phosphorylase is catalyzed by phosphorylase kinase. As discussed in the preceding section, calcium activation of myosin light chain kinase is mediated by the reversible association of calcium-calmodulin with a catalytic subunit to form a holoenzyme complex. The mechanism whereby calcium stimulates phosphorylase kinase activity in smooth muscle is not known; however, results obtained by numerous investigators (6,23) for the enzyme from skeletal muscle indicates that calcium stimulation of phosphorylase kinase activity is more complex. Calcium stimulation may be mediated by calmodulin present as a tightly bound subunit in the holoenzyme complex of phosphorylase kinase and by the reversible association of phosphorylase kinase with additional calmodulin. The results obtained in isolated bovine tracheal smooth muscle are consistent with the notion that stimulation of isometric tension development with carbachol causes an increase in sarcoplasmic Ca^{++} concentrations that result in activation of both myosin light chain kinase and phosphorylase kinase activities. Subsequently, there is an increase in the phosphate content of P-light chain and phosphorylase, respectively. The decline in the phosphate content of both P-light chain and phosphorylase in the presence of carbachol suggests that the increase in the sarcoplasmic calcium concentration may be transient. However, other possibilities do exist. For example, it would not be unreasonable to suggest that the calcium concentration in the sarcoplasm may be maintained at a relatively high value in the presence of carbachol, but that there is a slow activation of myosin light chain phosphatase and phosphorylase phosphatase activities. There are certainly alternative hypotheses for these results that will require investigation.

Phosphorylation of myosin P-light chain and phosphorylase are also under regulation by cyclic AMP. It has recently been shown

TABLE 3: Effect of isoproterenol on responses to carbachol in bovine tracheal smooth muscle.

Treatment	Time	Isometric Tension Development (g)	P Light Chain Phosphorylation (mol phosphate/ mol LC)	Phosphorylase a Activity Ratio (-5'AMP/ ±5'AMP)
Carbachol (1 µM)				
	0 sec	0.0	0.13 ± 0.03	0.15 ± 0.02
	15 sec	3.5	0.25 ± 0.03	0.33 ± 0.04
	60 sec	9.0	0.75 ± 0.05	0.61 ± 0.04
Isoproterenol (5 µM) + carbachol (1 µM)				
	5 min	0.0	0.09 ± 0.03	0.40 ± 0.03
	15 sec	0.8	0.15 ± 0.02	0.50 ± 0.02
	60 sec	3.9	0.38 ± 0.02	0.56 ± 0.03

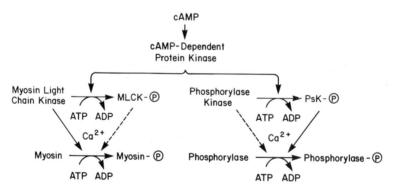

FIGURE 4

Regulation of myosin and phosphorylase phosphorylation in smooth muscle. Cyclic AMP dependent protein kinase catalyzes the phosphorylation of myosin light chain kinase and phosphorylase kinase. The phosphorylated form of myosin light chain kinase and nonphosphorylated phosphorylase kinase are the relatively less active forms of the respective enzymes (broken lines). Both the nonphosphorylated and phosphorylated forms of myosin light chain kinase and phosphorylase kinase require Ca^{++} for activity.

by Conti and Adelstein (7) that myosin light chain kinase from turkey gizzard smooth muscle may be phosphorylated by the cyclic AMP dependent protein kinase (Fig. 4). This results in a conversion of the enzyme to a form that requires a higher concentration of calcium-calmodulin for activity than the nonphosphorylated form. It should be noted that both forms of the enzyme are still dependent upon calcium and calmodulin for activity and that under saturating concentrations of calmodulin and calcium, the maximal activity obtained with both forms of the enzyme are similar. Similar results have been obtained with myosin light chain kinase from bovine aortic smooth muscle (40).

Although the phosphorylation of myosin light chain kinase by cyclic AMP dependent protein kinase results in an increase in the concentration of calmodulin required for activation, there are some difficulties in considering these results as a primary mechanism by which β-adrenergic stimulation produces relaxation of smooth muscle. The concentration of calmodulin required for half-maximal activation is approximately 1 nM with the nonphosphorylated myosin light chain kinase. After phosphorylation by cyclic AMP dependent protein kinase, this value increases to 20 nM. The apparent concentration of calmodulin in smooth muscle is about 3 µM. In addition, the concentration of myosin light chain kinase in smooth muscle is about 1 µM. The concentration of these two proteins is considerably greater than the activation constant of calmodulin even with myosin light chain kinase in the phosphorylated form. Thus, one could expect that there would always be sufficient calmodulin to cause adequate activation of myosin light chain kinase, whether it was in the phosphorylated or nonphosphorylated form. We have made some calculations on the calcium sensitivity of the nonphosphorylated and phosphorylated forms of myosin light chain kinase, assuming a concentration of calmodulin of 3 µM and myosin light chain kinase of 1 µM. There was no significant difference in the calcium sensitivity of the two forms of myosin light chain kinase when these considerations were made. These results do not necessarily indicate that phosphorylation of myosin light chain kinase is not an important regulatory mechanism in smooth muscle, but that the situation in the intact cell may be more complex and additional factors have to be taken into consideration in understanding the regulation of myosin P-light chain phosphorylation.

The effect of isoproterenol on the response of P-light chain phosphorylation and phosphorylase a formation in response to carbachol has recently been measured with the bovine tracheal smooth muscle (Table 3). Preincubation of tracheal smooth muscle with 5 µM isoproterenol, a β-adrenergic agonist, resulted in a decrease in the rate and extent of isometric tension development in response to 1 µM carbachol. This relaxant effect of isoproterenol was coincident with a decrease in the rate and extent of P-light chain phosphorylation. These results are consistent with the hypothesis that relaxation of smooth muscle produced by β-adrenergic stimulation may be due to a decrease in P-light chain phosphorylation. The mechanism by which P-light chain phosphorylation is inhibited could be the result of a number of biochemical effects. It could be proposed that myosin light chain kinase is converted to a more inactive form as a result of phosphorylation of cyclic AMP dependent protein kinase, as discussed above. In addition, it has been proposed that a decrease in sarcoplasm calcium concentrations is produced by β-adrenergic stimulation (31) which would result in inhibition of myosin light chain kinase activity.

Incubation of tracheal smooth muscle with isoproterenol alone resulted in an increase in phosphorylase a formation (Table 3). These results are analogous to observations made in skeletal muscle (38). Stimulation of cyclic AMP formation by isoproterenol may result in activation of cyclic AMP dependent protein kinase. This enzyme, in turn, catalyzes the phosphorylation of phosphorylase kinase to convert it to a more activated form that subsequently catalyzes the conversion of phosphorylase b to phosphorylase a. In skeletal muscle these effects of β-adrenergic stimulation are produced in relaxed noncontracting muscle. Thus phosphorylase a formation is not dependent upon high concentrations of calcium in the sarcoplasm that are necessary for contraction.

The addition of 1 µM carbachol to bovine tracheal smooth muscle that had been incubated for 5 min in 5 µM isoproterenol resulted in a further increase in phosphorylase a formation. This additional effect could have been the result of stimulation of calcium release by carbachol for isometric tension development that was additive to the response produced by β-adrenergic stimulation. Although nothing is known about the phosphorylation of phosphorylase kinase from smooth muscle by cyclic AMP dependent protein kinase, considerable research by Krebs and his colleagues (23) as well as by other investigators (6) has demonstrated that phosphorylase kinase is also phosphorylated by cyclic AMP dependent protein kinase. In contrast to the recent results obtained with myosin light chain kinase, phosphorylation of phosphorylase kinase results in an increase in the activity of this enzyme, that is, there is a conversion to an activated form of the enzyme that requires a lower concentration of calcium for stimulation (Fig. 4). Thus, the influence of cyclic AMP on myosin P-light chain phosphorylation is in contrast to the effects upon phosphorylase phosphorylation.

The results obtained with bovine tracheal smooth muscle are consistent with the idea that myosin P-light chain phosphorylation may play an important role in contraction. However, the regulation of phosphorylation and dephosphorylation of P-light chain appear to be more complex than the analogous reactions observed in skeletal muscle. In addition, as originally shown by Murphy and his colleagues (12,15), and supported by results obtained with bovine tracheal smooth muscle, the relationship between P-light chain phosphorylation and contraction in smooth msucle is also complex. Additional investigations with purified enzymes and with intact smooth muscles are needed to increase our understanding of these biochemical processes in relation to the physiological and pharmacological properties of smooth muscles.

ACKNOWLEDGEMENTS

The research described in this report was supported in part by USPHS grants HL-23990 and HL-26043, and from the Muscular Dystrophy Association. P.J.S. and D.K.B. are postdoctoral trainees (USPHS HL-07360).

REFERENCES

1. Adelstein, R.S. & Eisenberg, E. (1980): Ann. Rev. Biochem. 49:921-933.
2. Barany, K. & Barany, M. (1977): J. Biol. Chem. 252:4752-4754.
3. Barany, K., Barany, M., Gillis, J.M. & Kushmerick, M.J. (1979): J. Biol. Chem. 254:3617-3623.
4. Barron, J.T., Barany, M., Barany, K. & Storti, R.S. (1980): J. Biol. Chem. 255:6238-6244.
5. Blumenthal, D.K. & Stull, J.T. (1980): Biochemistry 19:5608-5614.
6. Cohen, P. (1980): Eur. J. Biochem. 111:563-574.
7. Conti, M.A. & Adelstein, R.S. (1981): J. Biol. Chem. 256:3178-3181.
8. Cox, J.A., Malnoe, A. & Stein, E.A. (1981): J. Biol. Chem. 256:3218-3222.
9. Dabrowska, R., Aromatorio, D., Sherry, J.M.F. & Hartshorne, D.J. (1977): Biochem. Biophys. Res. Commun. 78:1263-1272.
10. Dedman, J.R., Potter, J.D., Jackson, R.L., Johnson, J.D. & Means, A.R. (1977): J. Biol. Chem. 252:8415-8422.
11. deLanerolle, P. & Stull, J.T. (1980): J. Biol. Chem. 255:9993-10000.
12. Dillon, P., Aksoy, M.O., Driska, S.P. & Murphy, R.A. (1981): Science 211:495-497.
13. DiSalvo, J., Gruenstein, E. & Silver, P. (1978): Proc. Soc. Exp. Biol. Med. 158:410-414.
14. DiSalvo, J., Miller, J., Blumenthal, D. & Stull, J.T. (1981): Biophys. J. 33:276a.
15. Driska, S.P., Aksoy, M.O. & Murphy, R.A. (1981): Amer. J. Physiol. 240:C222-C233.
16. Ebashi, S. (1976): Ann. Rev. Physiol. 38:293-313.
17. Hidaka, H., Yamaki, T., Totsuka, T. & Asano, M. (1979): Molec. Pharmacol. 15:49-59.
18. Hoh, J.F.Y., McGrath, P.A. & Hale, P.T. (1978): J. Molec. Cell. Card. 10:1053-1076.

19. Huang, C.Y., Chau, V., Chock, P.B., Wang, J.H. & Sharma, R.K. (1981): Proc. Nat. Acad. Sci. 78:871-874.
20. Janis, R.A., Barany, K., Barany, M. & Sarmiento, J.G. (1981): Molec. Physiol. 1:3-11.
21. Katsuki, S. & Murad, F. (1977): Molec. Pharmacol. 13:330-341.
22. Krarup, C. (1981): J. Gen. Physiol. 311:355-372.
23. Krebs, E.G. & Beavo, J.A. (1979): Ann. Rev. Biochem. 48:923-959.
24. Mannherz, H.G. & Goody, R.S. (1976): Ann. Rev. Biochem. 45:428-465.
25. Manning, D.R. & Stull, J.T. (1979): Biochem. Biophys. Res. Commun. 90:164-170.
26. Manning, D.R. & Stull, J.T. (1982): Amer. J. Physiol. (in press).
27. Morgan, M.S., Perry, S.V. & Ottoway, J. (1976): Biochem. J. 157:687-697.
28. Pato, S. & Adelstein, R.S. (1981): Biophys. J. 33:278a.
29. Pemrick, S.M. (1980): J. Biol. Chem. 255:8836-8841.
30. Potter, J.D., Robertson, S.P., Collins, J.H. & Johnson, J.D. (1980): In Calcium-Binding Proteins: Structure and Function, (ed) F. Siegel, Elsevier/North Holland, New York, pp. 279-288.
31. Scheid, C.R., Honeyman, T.W. & Fay, F.S. (1979): Nature 277:32-36.
32. Sellers, J.R., Pato, M.D. & Adelstein, R.S. (1981): Biophys. J. 33:278a.
33. Sherry, J.M.F., Gorecka, A., Aksoy, M.O., Dabrowska, R. & Hartshorne, D.J. (1978): Biochemistry 17:4411-4418.
34. Silver, P.J., Schmidt-Silver, C.J. & DiSalvo, J. (1982): Amer. J. Physiol. (in press).
35. Stull, J.T. (1980): In Advances in Cyclic Nucleotide Research, Vol. 13 (eds) P. Greengard & G.A. Robison, Raven Press, New York, pp. 39-93.
36. Stull, J.T., Blumenthal, D.K. & Cooke, R. (1980): Biochem. Pharmacol. 29:2537-2543.
37. Stull, J.T. & High, C.W. (1977): Biochem. Biophys. Res. Commun. 77:1078-1083.
38. Stull, J.T. & Mayer, S.E. (1971): J. Biol. Chem. 246:5716-5723.
39. Stull, J.T., Sanford, C.F., Manning, D.R., Blumenthal, D.K. & High, C.W. (1982): In Protein Phosphorylation: Cold Spring Harbor Symposium on Cell Proliferation, (eds) E. Krebs & O. Rosen, Cold Spring Harbor Press, Cold Spring Harbor (in press).
40. Vallet, B., Molla, A. & DeMaille, J.G. (1981): Biochim. Biophys. Acta 674:256-264.
41. Yazawa, M. & Yagi, K. (1978): J. Biochem. (Tokyo) 84:1259-1265.

PHOSPHORYLATION AND ATPASE ACTIVITY OF SMOOTH MUSCLE MYOSIN

D.J. Hartshorne, M.P. Walsh and A. Persechini

Muscle Biology Group, Departments of Biochemistry and Nutrition and Food Science, College of Agriculture, The University of Arizona, Tucson, Arizona 85721, USA

INTRODUCTION

When the concentration of Ca^{++} within a muscle cell increases above the threshold level (about 1×10^{-6} M to 5×10^{-6} M) contraction occurs, and conversely, a reduction in the free Ca^{++} level leads to relaxation. This dependence on Ca^{++} for the activation of the contractile apparatus is similar for each muscle type. It follows therefore that for each muscle there exists a mechanism associated with the thick and thin filaments that can control cross-bridge-actin interactions in response to prevailing Ca^{++} concentrations. These components are often referred to as the regulatory proteins. Over the last few years it has become apparent that the regulatory proteins of different muscles are not identical and a variety of control mechanisms have been proposed.

The major theories proposed as regulatory mechanism in different muscles are listed in Table 1. There are two main categories of regulatory mechanisms; those associated with the thin filament, i.e. actin-linked; and those associated with the thick filament, i.e. myosin-linked. In the former belongs the classical troponin-tropomyosin mechanism and this system is operative in vertebrate skeletal and cardiac muscles. The mechanism that was next described was found in molluscs where Ca^{++} binding to the myosin light chains formed the regulatory principle (11,22). This mechanism was subsequently found to be widespread among the invertebrates and was discovered in brachiopods, echinoderms and nemertine worms. It was also discovered (12) that both myosin- and actin-linked control mechanisms could operate simultaneously in many of the invertebrates tested (including insects, most crustaceans, annelids and nematodes). The development of our knowledge on

TABLE 1: Types of regulatory mechanisms in muscle.

Category	System	Ca^{++}-Target Protein	Muscle Type	Reference
Myosin-linked	Ca^{++}-binding to myosin light chains	Two specific myosin light chains	Invertebrate muscle	(12)
	Phosphorylation of myosin light chains	Calmodulin	Vertebrate smooth muscle	(9)
Actin-linked	Troponin & tropomyosin	Troponin C	Widely distributed: vertebrate & invertebrate striated muscle	(12)
	Leiotonin & tropomyosin	Leiotonin C	Vertebrate smooth muscle	(18)
	Phosphorylation of 20,000 M$_r$ thin filament protein	Unknown	Vertebrate (vascular) smooth muscle	(13,14)

the regulatory mechanism in vertebrate smooth muscle reflected, to an extent, the historical pattern which emerged from the discovery of the two regulatory systems. Initially it was thought that a troponin-like mechanism was operative (see review: 8), although this theory is now generally not accepted. Next, it was discovered that in chicken gizzard actomyosin the control system was myosin-linked (2). However, the situation in smooth muscle was not as simple as that in molluscan muscle because it became apparent that the smooth muscle system required additional components (i.e. other than actin and myosin) which frequently were lost during purification procedures. These factors are recognized by most investigators to be the regulatory proteins in smooth muscle and will be discussed in more detail below. The consensus is that the regulation mechanism in smooth muscle is myosin-linked (see Table 1). However, this opinion is not unanimous, and Ebashi and his colleagues (18) believe that regulation is achieved via an actin-linked system termed leiotonin (Table 1). Since this system is described elsewhere in this volume it will not be discussed further here. Finally, it should be mentioned that Marston et al. (13) claim that a dual regulatory system, both myosin- and actin-linked, is present in actomyosins from pig aorta and turkey gizzard.

Requirements for Regulation in Smooth Muscle

One of the reasons why the regulatory mechanism in smooth muscle proved to be elusive was that as actin and myosin were purified the Mg^{++}-ATPase activity was lost. This contrasts to the situation in striated muscles where purified actin and myosin exhibit relatively high levels of ATPase activity, and the regulatory proteins fulfill their functions by inhibiting ATPase activity (i.e. preventing cross-bridge turnover) in the absence of Ca^{++}. In smooth muscle the situation is different: the regulatory proteins function by activating a dormant actomyosin complex but only in the presence of Ca^{++}. The requirement for an apparent activator is accepted but a controversy exists on the identity of that factor. The most accepted theory is that activation is achieved by the phosphorylation of the 20,000 M_r light chains of the myosin molecule and is therefore myosin-linked. The alternative viewpoint is that activation is caused by a thin filament based system termed leiotonin (18) and which does not involve phosphorylation of myosin.

For a myosin-linked system the simplest mode of regulation is that associated with the binding of Ca^{++} to the myosin light chains. Presumably the Ca^{++}-free light chain is inhibitory and inhibition is relieved when Ca^{++} is bound. Clearly there could be an analogy between this and the phosphorylation mechanism, and it might be proposed that the role of Ca^{++} is substituted by phosphorylation. Thus in this sense phosphorylation would be acting to derepress the actomyosin complex rather than as a true activator. With the leiotonin system the situation is more complex. Superficially the system would appear to resemble the classical troponin-tropomyosin mechanism in that it achieves its regulation only via the thin filament. However, there is a difference; the troponin-tropomyosin system effects an inhibition of ATPase activity in the absence of Ca^{++}, whereas the leiotonin system must alter the thin filament (or possibly myosin) in the presence of Ca^{++} to allow activation of ATPase activity. Mikawa (15) demonstrated that gizzard thin filaments could be frozen by cross-linking with gluteraldehyde into either "activatable" or "non-activatable" states, thus strongly suggesting that the effects of leiotonin are indeed confined to the thin filaments. Obviously the mechanism of leiotonin's action must be elucidated and it is vital to determine whether leiotonin forms an independent mechanism or one which is complementary to the phosphorylation scheme. In the case of dual regulatory mechanisms (13) the situation is not necessarily complex. It might be proposed that phosphorylation represents the primary mechanism (an on-off switch) and that a thin filament based system could attenuate the activated state. If the latter if Ca^{++}-dependent, however, that would require that the thin filament calciprotein have reduced affinity for Ca^{++} compared to calmodulin (see below).

Phosphorylation Theory of Regulation

The initial observation that led to the development of this theory was made by Sobieszek (20) who showed that chicken gizzard myosin was phosphorylated on the 20,000 dalton light chain and that this altered the actomyosin ATPase activity. Since then many reports on the phosphorylation of several smooth muscle myosins have appeared (for review, see 7) and in some instances the ATPase activity and phosphorylation were correlated (for review, see 9). One can therefore propose the following simple scheme which is consistent with a few key findings; the two 20,000 dalton light chains of myosin are phosphorylated by a Ca^{++} dependent specific enzyme, the myosin light chain kinase (MLCK) and one mole of phosphate can be incorporated per light chain; phosphorylation is assumed to be a prerequisite for the activation by actin of the Mg^{++} ATPase activity of myosin, which is thought to be equivalent to the cross-bridge cycling rate; the MLCK is composed of two subunits with a stoichiometry of 1:1 (10), one a regulatory subunit which is calmodulin (6) and a larger apoenzyme of variable molecular weight depending on the source of the kinase (for review see 7). A second enzyme(s) is also required to dephosphorylate the myosin, namely, a myosin light chain phosphatase; it is assumed that as long as the Ca^{++} concentration remains above the activation threshold then the activity of the MLCK will swamp out that of the phosphatase and the equilibrium favors the phosphorylated state. However, when the Ca^{++} level is reduced the MLCK is inactivated and then the phosphatase returns myosin to its dephosphorylated and dormant state. This scheme is depicted as

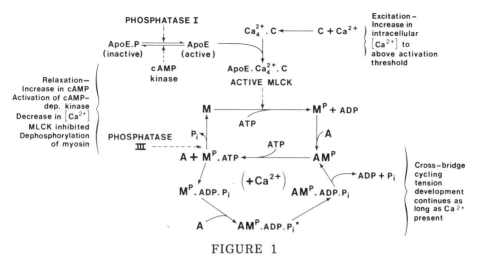

FIGURE 1

Scheme summarizing the phosphorylation theory of regulation in smooth muscle. Abbreviations: M, M^P = unphosphorylated and phosphorylated myosin, respectively; A = actin; C = calmodulin; ApoE = the larger myosin light chain kinase subunit. This scheme also illustrates the possible involvement of the cAMP dependent protein kinase in the regulatory mechanism.

a cyclic mechanism in Fig. 1. Included in Fig. 1 is also the effect of the cAMP dependent protein kinase which is thought to influence the activity of the MLCK via the phosphorylation of the larger MLCK subunit (1,5).

Despite the recent addition of several of the details given in Fig. 1 this pathway remains an oversimplification. For example, it does not consider the possible involvement of an additional regulatory system and the scheme predicts simply that the MLCK initiates the contractile event and that the phosphatase terminates the active state. From the preceding discussion it is clear that this viewpoint is not unanimous and the exact role of myosin phosphorylation must still be defined.

A comparison of substrates for the MLCK. One of the prerequisites for an understanding of any enzymatic reaction is a knowledge of some of the kinetic parameters for that system. For example, for the MLCK the V_{max} and K_m for the light chains are particularly important. These figures have been estimated (see review, 7) but in the majority of cases isolated light chains rather than whole myosin was used, and there is some concern that the two substrates may not be equivalent. This was indicated in earlier experiments (16) and it was shown that the pattern of phosphorylation for isolated light chains and the whole myosin was quite distinct (Fig. 2). The phosphorylation of the light chains occurred rapidly (approximately 12 μmol P_i incorporated min^{-1} mg^{-1} kinase) and was linear over most of the reaction. In contrast, phsophorylation of whole myosin was markedly slowed and the reaction was not linear. Under our usual assay conditions ($\mu \cong 0.06$) myosin is insoluble and it seemed possible that aggregation might reduce phosphorylation by limiting the accessibility to the MLCK. The assays were therefore repeated at higher ionic strength ($\mu \cong 0.32$) where both myosin and the isolated light chains are soluble. The two phosphorylation profiles were altered slightly (Fig. 2) but the phosphorylation of myosin remained nonlinear and significantly slower than the phosphorylation of the light chains (see later discussions).

The above experiments suggest that once the light chains are removed from the heavy chains their characteristics as a substrate for the MLCK are altered. This can be demonstrated even more dramatically using a different kinase, namely, the cAMP dependent protein kinase. It was found (17) that the isolated light chains of chicken gizzard myosin could be phosphorylated by the cAMP dependent protein kinase. The same serine residue was phosphorylated by both the MLCK and the cAMP dependent protein kinase. However, when whole myosin was tested with the cAMP dependent protein kinase (23) no phosphorylation of the light chains was observed (Fig. 3). It is interesting that the serine sites are accessible to the MLCK but not the the cAMP dependent protein kinase, as on the basis of size alone the opposite situation might be expected.

If it is accepted that the isolated light chains and the intact myosin light chains differ as substrates for the MLCK, then this raises doubts as to the physiological validity of much of the kinetic data that has been collected. Similarly, the parameters obtained for the myosin light chain phosphatase(s) may also need to be revised. Obviously the situation calls for an extensive re-examination of many of the properties of phosphorylation and dephosphorylation using whole myosin as a substrate.

<u>Phosphorylation of myosin</u>. The point that was made earlier (Fig. 2) was that the time course of phosphorylation for isolated light chains and the whole myosin were quite distinct. More recently, experiments were carried out using dephosphorylated myosin, or myosin in which the prephosphorylation levels were determined, as the substrate for the purified MLCK and measuring phosphorylation at a fixed time point (10 min) at variable MLCK concentrations (19). The results are shown in Fig. 4. Initially, at low concentrations of MLCK, the phosphorylation occurred relatively easily and an approximate linear dependence of phosphorylation on MLCK

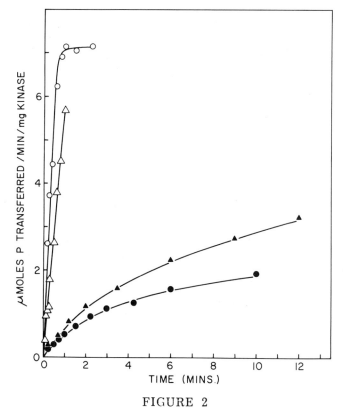

FIGURE 2

Phosphorylation of isolated myosin light chains and intact myosin. Conditions: Temperature, 25°C; 4 mM $MgCl_2$; 50 mM KCl; 1 mM [γ-^{32}P]ATP; 25 mM Tris-HCl (pH 7.6); calmodulin, 5 µg/ml; myosin light chain kinase apoenzyme, 0.92 µg/ml; myosin, 1.08 mg/ml under above assay conditions (●), and plus 0.3 M KCl (▲); 20,000 dalton light chain, 0.19 mg/ml under above assay conditions (○) and plus 0.3 M KCl (△). (From Ref. 16).

concentration was observed. When about half of the available sites were occupied subsequent phosphorylation required considerabl higher levels of MLCK. In contrast to this behavior the phosphorylation of the isolated light chains showed essentially a linear dependence on MLCK concentration. From this data, illustrated more clearly by the inset of Fig. 4, it could be suggested that the phosphorylation sites on the whole myosin can be divided into two classes. One class which is phosphorylated relatively easily and the second class which is phosphorylated with more difficulty. When the light chains are removed from the myosin molecule only one class of sites is apparent and this appears to correspond to the more easily phosphorylated class of sites.

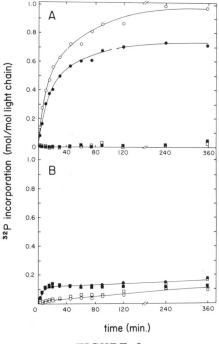

FIGURE 3

Phosphorylation of isolated myosin light chains (A) and intact myosin (B) by cAMP dependent protein kinase. Conditions: Temperature 37°C; 4 mM $MgCl_2$, 1 mM [γ-^{32}P]ATP; 25 mM Tris-HCl (pH 7.5); 1 x 10^{-4} M $CaCl_2$ (●,■) or 1 mM EGTA (○,□); in the presence (○,●) or absence (□,■) of the catalytic subunit of the cAMP dependent protein kinase (E:S = 1:100). The slight phosphorylation of myosin (B) in the absence and presence of the catalytic subunit was due to contamination by the myosin light chain kinase. (From Ref. 23).

One explanation for this type of behavior is that when one of the myosin heads is phosphorylated this will hinder the subsequent phosphorylation of the second head. This demands that the two heads are not independent and suggests that the two sites are subject to cooperative interactions. In effect this would generate an ordered phosphorylation sequence in which myosin with only one head phosphorylated would be formed initially. An alternative explanation is that the nonlinearity of the phosphorylation reaction is due to a gradual reduction in substrate concentration to the extent that the K_m of the MLCK for the light chains is approached. However, this explanation is unlikely for two reasons: at concentrations of isolated light chains similar to the concentration of light chains in whole myosin the reaction is essentially linear. This

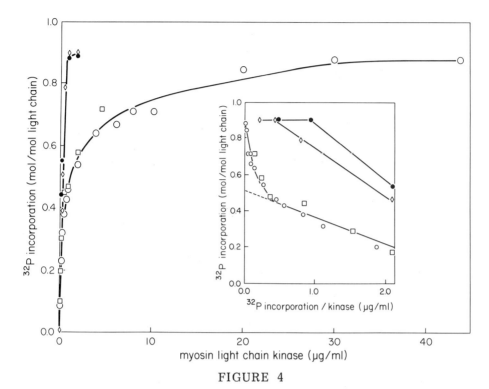

FIGURE 4

Phosphorylation of intact myosin and partially purified myosin light chains at varying concentrations of myosin light chain kinase. Assay conditions as given in Fig. 1. Myosin, 0.1 mg/ml (□) and 0.44 mg/ml (○); myosin light chains, 0.01 mg/ml (●) and 0.02 mg/ml (◊). (From Ref. 19).

assumes that the K_m values of the MLCK for isolated light chains and for whole myosin are similar, but this is not unreasonable. The second reason is that a similar profile was obtained at different myosin concentrations and thus suggests that substrate depletion is not the dominant factor.

Our opinion, therefore, is that the nonlinearity of the phosphorylation reaction is induced by cooperative interaction between the two heads of myosin. Whether or not cooperative responses can be detected in the ATPase behavior is obviously the next point to consider. However, before doing this it is worthwhile to outline the various possibilities that could occur in the relationship of phosphorylation and the activation of ATPase activity.

Relationship between phosphorylation and ATPase activity: Hypothetical. In the following discussion it is assumed that each smooth muscle myosin molecule has two heads, and that each head contains a phosphorylation site (i.e. a 20,000 dalton light chain), an ATP hydrolysis site and an actin-binding site. The ATPase activity that is relevant is the actin-activated Mg^{++} ATPase activity of myosin. There are two basic possibilities to consider for the behavior of the heads; either the heads are independent (i.e. non-cooperative), or one head is influenced by its partner (i.e. cooperative).

Non-cooperative behavior: This is the simplest situation to envisage. It assumes that each head is independent and will respond to phosphorylation irrespective of the phosphorylation state of the other head. The result being that ATPase activity (or, more correctly, activation of Mg^{++}-ATPase activity by actin) is directly proportional to phosphorylation. This situation is shown diagrammatically in Fig. 5A. A dependence of this type has been seen experimentally for gizzard actomyosin and myofibrils (21) and for stomach muscle actomyosin (3).

Cooperative behavior: This possibility covers the situation where the phosphorylation of one head influences the properties of the second head. There are two variations; one is that the phosphorylation of one head activates the ATPase activity of the second head, and the second variation is that myosin with one head phosphorylated is inactive and the phosphorylation of the second head is required for activity. In this case the unphosphorylated head can be visualized as inhibitory to the phosphorylated head. Within both of these groups (which can be thought of loosely as examples of positive and negative cooperativity, respectively) there is also the chance that the phosphorylation itself is either random or ordered. The various possibilities are depicted diagrammatically in Fig. 5B and 5C. Considering in more detail the first of these two hypothetical possibilities, ordered phosphorylation would result in full activation of ATPase activity at 50% phosphorylation, whereas for a random phosphorylation process the ATPase activity would increase rapidly at lower levels of phosphorylation but would reach 100% activity only when completely phosphorylated (shown by the dashed line in Fig. 5B). This type of situation has not been reported experimentally. Considering now the second of the two possibilities, if the phosphorylation is ordered than at 50% phosphorylation each myosin molecule would contain one phosphorylated light chain and the ATPase activity would be zero. Subsequent phosphorylation would form the "doubly" phosphorylated species and thus from 50% to 100% phosphorylation the ATPase activity would show a linear increase in activity. If, on the other hand, the phosphorylation is random the ATPase activity would still reflect the formation of myosin molecules with both heads phosphorylated but this would increase as the square of the fractional phosphorylation, i.e. the dependence of ATPase activity on phosphorylation would be nonlinear and would show slight activity at less than 50% phosphorylation and a more rapid increase of ATPase

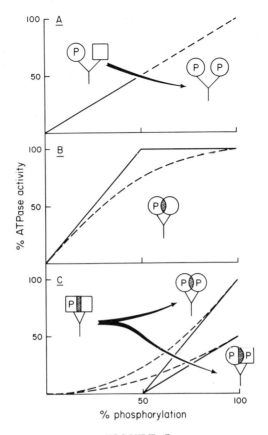

FIGURE 5

Hypothetical schemes to illustrate the various relationships between phosphorylation of myosin and the actin-activated ATPase activity of myosin. Each myosin is depicted with two heads; an active head is shown as a circle and an inactive head as a square. The cross-hatched contact areas of the myosin heads implies cooperativity. A = non-cooperative situation; B = positive cooperativity; C = negative cooperativity. In B and C the solid line illustrates the situation for ordered phosphorylation and the dashed line for random phosphorylation.

activity on subsequent phosphorylation. This is depicted as the dashed lines in Fig. 5C. An additional complication is that one or both of the myosin heads might be active, and this possibility is also considered in Fig. 5C.

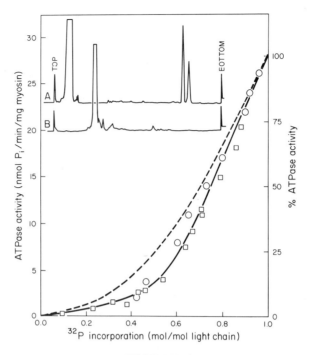

FIGURE 6

Observed relationship between phosphorylation of myosin and the actin-activated ATPase activity of myosin. Points taken from time courses of phosphorylation and ATPase activity (○) and from fixed time point assays (□). The theoretical curve for random phosphorylation shown as a dashed line. Assay conditions as given in Fig. 1. Myosin, 0.44 mg/ml; actin, 0.25 mg/ml; myosin light chain kinase varied 0-43 µg/ml; calmodulin constant, 11 µg/ml. Scans of gradient SDS-polyacrylamide gels of myosin (A) and myosin light chain kinase (B) are also shown. (From Ref. 19).

Observed. For these studies (19) we used as simple an assay system as possible. The protein components were myosin and MLCK isolated from chicken gizzard, actin from rabbit skeletal muscle and calmodulin from bull testes. The reason for choosing this defined system was to eliminate, as much as possible, effects due to contaminants and thereby to allow us to determine whether phosphorylation alone is sufficient for the activation of ATPase activity. The disadvantage of this system is that it is undoubtedly much simpler than that occurring in vivo and for the final analysis of the regulatory mechanism in smooth muscle other factors may have to be considered.

The relationship between the actin-activated ATPase activity of smooth muscle myosin and phosphorylation is shown in Fig. 6. These values were taken either from fixed time point assays (at 10 min), or were estimated from time courses of ATPase activity and phosphorylation at times between 4 and 10 min. The dependence of ATPase activity on phosphorylation is obviously not linear and can be approximated by two phases. The first phase covers phosphorylation levels up to 50% and during this phase only a slight activation of ATPase activity is observed. During the subsequent phase, the phosphorylation levels increase from 50% to 100% and this is accompanied by a marked activation of ATPase activity. The actin-activated ATPase activity of the fully phosphorylated myosin is about 28 nmol P_i liberated min^{-1} mg^{-1} myosin.

The profile shown in Fig. 6 resembles the hypothetical models depicted in Fig 5C and would, therefore, suggest that the myosin heads are subject to cooperative interactions. It can be proposed that myosin with only one head phosphorylated is activated only slightly by actin, whereas the fully phosphorylated myosin molecule shows a more marked activation of ATPase activity. An analogous situation was shown recently with respect to Ca^{++} binding to scallop heavy meromyosin (4) where it was suggested that actin-activated ATPase activity was expressed only when Ca^{++} was bound to both heads. Ca^{++} binding to the light chains of molluscan myosin presumably being equivalent to phosphorylation of the vertebrate smooth muscle myosin light chains.

For the data shown in Fig. 6, there remains the question of whether the phosphorylation was ordered or random. If the phosphorylation were completely ordered and a cooperative model of ATPase activity is assumed then the ATPase activity would remain at zero up to 50% phosphorylation. This is clearly not the case. However, the data do not conform to the random phosphorylation curve either (shown as the dashed line in Fig. 6). When this evidence is considered with that given earlier in Fig. 4 it may be tentatively concluded that the phosphorylation process is ordered, and that myosin with one head phosphorylated has a much lower activity (but not zero as in the hypothetical model) than myosin with both heads phosphorylated.

SUMMARY

These results indicate that the two heads of smooth muscle myosin may interact and generate cooperative phenomena. Cooperativity is manifest by two of the properties associated with myosin, namely phosphorylation and actin-activated ATPase activity. It is suggested that phosphorylation of one of the myosin heads somehow restricts the subsequent phosphorylation of the second head. It is also suggested that myosin with one of its two heads phosphorylated is activated only slightly by actin, whereas the fully phosphorylated myosin shows a more marked activation of ATPase activity by actin. Obviously these statements are based on only a very restricted area of observation, and many other conditions must be analyzed before cooperativity in smooth muscle myosin can be claimed as a general feature. Among the variables to be tested would be other smooth muscle myosins and variations in the assay conditions. It is possible that the linear dependence of ATPase activity and phosphorylation observed previously (3,21) could be due to differences such as these.

The cooperative behavior observed in vitro must be compared to the situation occurring in the intact muscle, and the highest priority should be given to establishing the correlation between tension development and the levels of myosin phosphorylation. It is also possible that the relationship between phosphorylation and tension development could vary during different phases of the contractile cycle and this should be investigated.

REFERENCES

1. Adelstein, R.S., Conti, M.A., Hathaway, D.R. & Klee, C.B. (1978): J. Biol. Chem. 253:8347-8350.
2. Bremel, R.D. (1974): Nature 252:405-407.
3. Chacko, S. (1981): Biochemistry 20:702-707.
4. Chantler, P.D., Sellers, J.R. & Szent-Gyorgyi, A.G. (1981): Biochemistry 20:210-216.
5. Conti, M.A. & Adelstein, R.S. (1981): J. Biol. Chem. 256: 3178-3181.
6. Dabrowska, R., Sherry, J.M.F., Aromatorio, D.K. & Hartshorne, D.J. (1978): Biochemistry 17:253-258.
7. Hartshorne, D.J. (1982): In Muscle and Non-Muscle Motility, Vol. 2, (eds) R.M. Dowben & J.W. Shay, Plenum Press, New York (in press).
8. Hartshorne, D.J. & Gorecka, A. (1980): IN Handbook of Physiology, Sect. 2: The Cardiovascular System, Vol. 2. (eds) D.F. Bohr, A.P. Somlyo & H.V. Sparks, American Physiology Society, Bethesda, Maryland, pp. 93-120.
9. Hartshorne, D.J. & Mrwa, U. (1982): Blood Vessels (in press).
10. Hartshorne, D.J., Siemankowski, R.J. & Aksoy, M.O. (1980): IN Muscle Contraction: Its Regulatory Mechanisms. (eds) S. Ebashi, K. Maruyama & M. Endo, Springer-Verlag, New York, pp. 287-301.
11. Kendrick-Jones, J., Szentkiralyi, E.M. & Szent-Gyorgyi, A.G. (1976): J. Molec. Biol. 104:747-775.
12. Lehman, W. & Szent-Gyorgyi, A.G. (1975): J. Gen. Physiol. 66:1-30.

13. Marston, S.B., Trevett, R.M. & Walters, M. (1980): Biochem. J. 185:355-365.
14. Marston, S.B. & Walters, M. (1980): Cell Biol. Int. Rep. 4:799.
15. Mikawa, T. (1979): J. Biochem. (Tokyo) 85:879-881.
16. Mrwa, U. & Hartshorne, D.J. (1980): Fed. Proc. 39:1564-1568.
17. Noiman, E.S. (1980): J. Biol. Chem. 255:11067-11070.
18. Nonomura, Y. & Ebashi, S. (1980): Biomed. Res. 1:1-14.
19. Persechini, A. & Hartshorne, D.J. (1982): Science 213:1383-1385.
20. Sobieszek, A. (1977): IN The Biochemistry of Smooth Muscle (ed) N.L. Stephens, University Park Press, Baltimore, Maryland, pp. 413-443.
21. Sobieszek, A. (1977): Eur. J. Biochem. 73:477-483.
22. Szent-Gyorgyi, A.G., Szentkiralyi, E.M. & Kendrick-Jones, J. (1973): J. Molec. Biol. 74:179-203.
23. Walsh, M.P., Persechini, A., Hinkins, S. & Hartshorne, D.J. (1981): FEBS Lett. 126:107-110.

DEMONSTRATION OF TWO TYPES OF Ca^{++} DEPENDENT PROTEIN KINASES IN THE BRAIN AND OTHER TISSUES

E. Miyamoto, K. Fukunaga and K. Matsui*

Department of Pharmacology, Kumamoto University
Medical School, Kumamoto-shi, Kumamoto 860, Japan, and
*On leave from: Department of Obstetrics and Gynecology,
Kumamoto University Medical School

The calcium dependent regulator protein (calmodulin) has been recognized to be ubiquitous in the eukaryotes, and to be widely involved in regulating a variety of cellular processes (1,9,13). Recently, we have found that a substance of calmodulin-like activity (7) exists in the prokaryote.

Cyclic AMP, cyclic GMP and Ca^{++} are considered to function as intracellular mediators in diverse biological activities. The effects of these regulators are performed largely, if not entirely, through activation of protein kinases. Cyclic AMP and cyclic GMP dependent protein kinases have been extensively studied in the wide range of the animal kingdom (11,22). Ca^{++} dependent protein kinase was first discovered in the muscle system (21) and subsequently in other muscle (2,6,25) and non-muscle tissues (3,4,26). During the course of the studies on the demonstration of Ca^{++} dependent protein kinase in the central nervous system, we found that the level of a Ca^{++} dependent and calmodulin independent protein kinase, in contrast to that of cyclic AMP dependent protein kinase, increases in the brain at a later stage of development after birth (16). Subsequently, we found another Ca^{++} dependent protein kinase which is stimulated with a small amount of calmodulin in the cytosol of the brain (14,17). This Ca^{++} dependent and calmodulin dependent enzyme is eluted at a different position on DEAE-cellulose or gel filtration column from the Ca^{++} dependent enzyme mentioned above. The results indicate that the enzymes are different entities, although the concentration of Ca^{++} required to stimulate both of the enzymes is in a physiological range. The

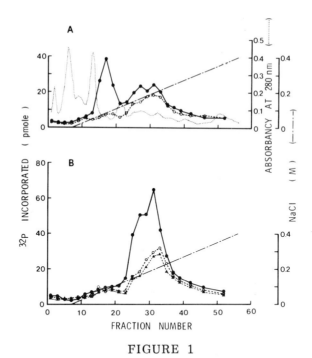

FIGURE 1

DEAE-cellulose column chromatography. The cytosol fraction (8.3 ml) containing 30.3 mg of protein was applied to a 1.6 x 5.3 cm column of DEAE-cellulose. An aliquot (0.07 ml) of the fractions was assayed for protein kinase activity with brain A (A) and chicken gizzard myosin light chains (B) in the presence of 0.1 mM Ca^{++} (●——●) or 1 mM EGTA (○---○) with calmodulin and in the presence of 0.1 mM Ca^{++} without calmodulin (▲---▲) (From Ref. 14).

present communication describes the demonstration and properties of two types of Ca^{++} dependent protein kinases in the brain and other tissues.

MATERIALS AND METHODS

Preparation of myosin and myosin light chains from chicken gizzard. Myosin from chicken gizzard was prepared by the method of Yamaguchi et al. (27). Calmodulin deficient myosin light chains from myosin were obtained using the method of Perrie and Perry (20) and Matsuda et al. (12) at the final step of DEAE-cellulose column.

Preparation of brain proteins. In principle, procedures for preparation of actomyosin from the muscle system were applied to

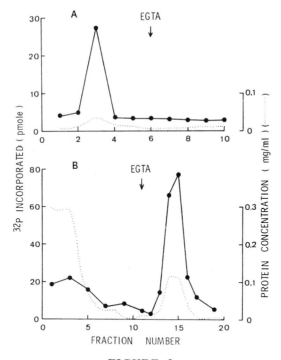

FIGURE 2

Calmodulin-affinity column chromatography. The enzyme for myosin light chains was collected in a large scale on the gel filtration column. The concentrates of active fractions for brain A (A) and myosin light chains (B) from sephadex G-200 columns containing 0.20 and 11.2 mg protein were applied to 1.1 x 5.0 cm and 1.7 x 4.4 cm columns of calmodulin-sepharose 4B, respectively. The columns were washed with 8.0 and 30 ml of 40 mM Tris-HCl buffer, pH 7.5, 50 mM NaCl, 0.2 mM Ca^{++}, 1 mM Mg^{++} and 10 mM 2-mercaptoethanol, and then eluted with 10 and 30 ml of 40 mM Tris-HCl buffer, pH 7.5, 200 mM NaCl, 1 mM Mg^{++}, 1 mM EGTA and 10 mM 2-mercaptoethanol. Fractions (2 ml and 3 ml) were each collected in (A) and (B), respectively. An aliquot (0.07 ml) was assayed for protein kinase activity in the presence of Ca^{++} and calmodulin (●——●) (From Ref. 14).

brain tissue. Whole bovine brains were extracted with 0.6 M KCl. After centrifugation the supernatant was diluted with cold distilled water to lower the ionic strength of KCl to 0.1 M and centrifuged. The dissolution and precipitation were repeated twice. The resulting precipitate was treated by the method of Perrie and Perry (20). The protein preparation obtained was tentatively referred to hereafter as "brain A."

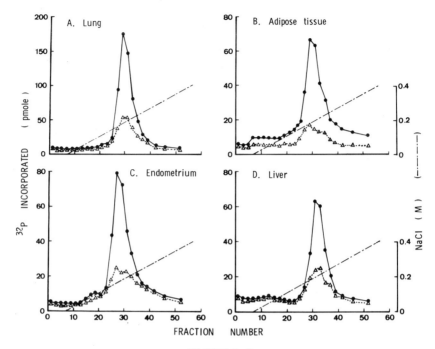

FIGURE 3

DEAE-cellulose column chromatography. The cytosol fraction of each tissue containing about 12 mg of protein was applied to a 1.1 x 3.2 cm column of DEAE-cellulose. The enzyme protein was eluted with a linear gradient of NaCl (0-0.4 M) in a total volume of 150 ml of 10 mM Tris-HCl buffer, pH 7.5, 0.05 mM EGTA, 1 mM EDTA and 0.43 mM PMSF. Fractions (3 ml) were collected. Aliquots (0.07 ml) of fractions were assayed for protein kinase activity in the presence (●—●) or the absence (△—△) of 1.1 μg calmodulin with 0.1 mM Ca^{++}. (From Ref. 17).

Preparation of calmodulin and calmodulin-binding affinity column. Calmodulin was prepared from bovine brain as described previously (15). Calmodulin-binding sepharose 4B affinity column was prepared by the method of Klee and Krinks (10).

Assay for Ca^{++} dependent protein kinase activity. The standard assay system for Ca^{++} dependent protein kinase activity was described elsewhere (14,17). The procedures for collection of TCA-insoluble protein and for count of the precipitate were as described previously (18,19). Phosphodiesterase was assayed by the method of Kakiuchi et al. (8).

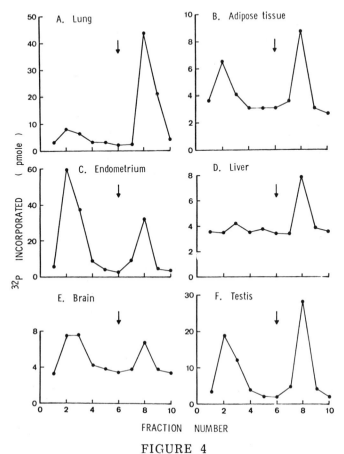

FIGURE 4

Calmodulin-affinity column chromatography. Active fractions obtained on DEAE-cellulose column were collected and pooled. The concentration was adjusted to 40 mM Tris-HCl buffer, pH 7.5, 50 mM NaCl, 0.2 mM Ca^{++}, 1 mM Mg^{++} and 10 mM 2-mercaptoethanol. The enzyme solution was applied to a 1.1 x 5.0 cm column of calmodulin-sepharose 4B. After washing the column with the buffer, the enzyme was eluted with 40 mM Tris-HCl buffer, pH 7.5, 200 mM NaCl, 1 mM Mg^{++}, 1 mM EGTA and 10 mM 2-mercaptoethanol at the positions of the arrows. Fractions (2 ml) were collected. Protein kinase activity was assayed in the presence of Ca^{++} and calmodulin (From Ref. 17).

TABLE 1: Protein kinase activity of peaks I and II from the brain with brain A and myosin light chains as substrates.

		Protein Kinase Activity			
		+EGTA		+Ca^{++}	
Enzyme	Substrate	-Cal %	+Cal %	-Cal %	+Cal %
Peak I	Brain A	42	40	91	100
	MLC	23	21	24	30
Peak II	Brain A	20	16	20	32
	MLC	9	9	6	100

Activities are expressed as 100% for peak I with brain A and for peak II with myosin light chains assayed in the presence of Ca^{++} and calmodulin. From these values, each activity was calculated as percent. Values represent the mean of duplicate determinations. MLC = myosin light chains; Cal = calmodulin (From Ref. 14).

Preparation of the cytosol fraction from tissues. The endometrium and myometrium from Japanese white rabbit, gizzard from chicken, and other tissues from Wistar rat were each homogenized with 5 vol of 0.25 M sucrose (0.32 M sucrose for the brain), 10 mM Tris-Cl buffer, pH 7.5, 4 mM EDTA, 0.05 mM EGTA and 0.43 mM phenylmethylsulfonylfluoride (PMSF) in a teflon-homogenizer or an ultraturrax (Ika-Werk). The homogenate was centrifuged at 105,000 x g for 60 min. The cytosol fraction obtained was used for the following experiments.

RESULTS

Purification of Ca^{++} dependent protein kinases from the cytosol fraction of the brain. The supernatant obtained was directly applied to a column of DEAE-cellulose. The enzyme protein was eluted with a linear gradient of NaCl (0-0.4 M) (Fig. 1). Ca^{++} dependent protein kinases which phosphorylate brain A (Fig. 1A) and chicken gizzard myosin light chains (Fig. 1B) as substrates were separately eluted from the column. The activity for brain A was resolved into two peaks, major and minor, eluted at 0.08 and 0.18 M NaCl, respectively. The first peak of Ca^{++} dependent activity was analyzed in the following experiments. The enzyme which phosphorylates myosin light chains was eluted as a single peak at 0.18 M NaCl. The activity was stimulated with calmodulin and Ca^{++}. The activity without calmodulin or Ca^{++} was a level similar to that without both calmodulin and Ca^{++}. The active fractions for brain A and for myosin light chains were separately

TABLE 2: Activity of peaks I and II in the brain and other tissues.

Tissue	Protein Kinase Activity	
	Peak I	Peak II
	(pmol)	(pmol)
Brain	31.5	12.1
Myometrium	7.7	515.0
Small intestine	4.8	93.0
Stomach	4.6	319.5
Liver	3.4	42.0
Endometrium	2.6	59.0
Ovarium	2.1	10.5
Heart	1.6	9.5
Kidney	1.1	2.0

The supernatant solution of the crude homogenate from various tissues containing similar amount of protein (about 12 mg) was applied to a 1.1 x 3.0 cm column of DEAE-cellulose. The enzyme protein was eluted with a linear gradient of NaCl (0-0.4) in 10 mM Tris-Cl buffer pH 7.5, 0.05 mM EGTA, 0.43 mM PMSF and 1 mM EDTA in a total volume of 150 ml. Fractions (3 ml) were collected. Peaks I and II represent activities of peaks eluted with about 0.08 and 0.18 M NaCl which were assayed with brain A and chicken gizzard myosin light chains as substrates, respectively, under standard conditions. Values were corrected for those determined in the presence of EGTA (Miyamoto et al., unpublished data).

collected and concentrated by Amicon ultrafiltration with a PM-10 membrane. Each concentrate was applied to a column of sephadex G-200. The molecular weights of proteins were estimated on gel filtration with γ-globulin and bovine serum albumin as standards. The enzyme which phosphorylates brain A showed a single peak of activity with a molecular weight of 88,000 on the column. The enzyme which phosphorylates myosin light chains was resolved into two peaks of activity, independent of and dependent on Ca^{++}. The first and second peaks of activity were eluted at the positions of void volume and of molecular weight 120,000, respectively. The second peak of Ca^{++} and calmodulin dependent activity was analyzed further.

The active fractions for brain A and myosin light chains were collected. The enzyme solution was applied to a column of calmodulin-sepharose 4B. The kinase activity for brain A (peak I) passed through the column (Fig. 2A). In constrast, the activity for myosin light chains (peak II) was adsorbed and eluted with a buffer containing EGTA (Fig. 2B). As described below, both

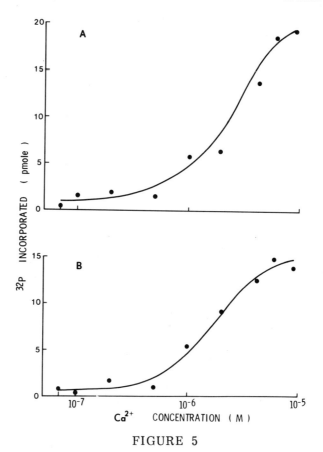

FIGURE 5

Effect of Ca^{++} concentration. Brain A (A) and chicken gizzard myosin light chains (B) as substrates with peaks I (2.5 µg protein) and II (1.2 µg protein) from the brain, respectively, were phosphorylated under standard assay conditions, except for the variation in concentration of Ca^{++}, as indicated. To obtain a given concentration of free Ca^{++}, calcium-EGTA buffers were used. Ca^{++} concentrations at pH 7.0 were calculated with the association constant of 4.83×10^6 M^{-1} for the calcium-EGTA complex (24). Activities have been corrected for values determined at 5×10^{-8} M Ca^{++}. (From Ref. 14).

enzyme preparations contained no detectable endogenous calmodulin. The results indicate that peak I had no affinity for calmodulin. In contrast, peak II had a good affinity for calmodulin.

Whether or not each enzyme preparation contains endogenous calmodulin was examined by the ability to activate brain phosphodiesterase and brain myosin light chain kinase. Both preparations had no detectable calmodulin.

Separation of Ca^{++} dependent protein kinases in the cytosol fractions of other tissues. The cytosol fraction (0.9 to 5.5 ml) of each tissue containing approximately similar amounts of protein was applied to a column of DEAE-cellulose. The typical results with chicken gizzard myosin light chains as substrates are shown in Fig. 3. A peak of Ca^{++} dependent and calmodulin dependent activity was eluted with 0.18 to 0.20 M NaCl in all tissues studied. The enzymes from the lung, adipose tissue, endometrium and liver (Fig. 3) showed a good dependency on calmodulin. On the other hand, the enzymes from the testis, ovarium and kidney contained high calmodulin independent activities (data now shown). The occurrence of the Ca^{++} independent and calmodulin independent activities may be due to proteolysis during the preparation, although PMSF and EDTA were used as protease inhibitors in the buffer.

When brain A was used as substrate, a peak of activity appeared at the position eluted with 0.08 M NaCl in all tissues studied (data not shown). The magnitude of activity varied from tissue to tissue.

The active fractions of each tissue for myosin light chains as substrates were collected from DEAE-cellulose column, and applied to a calmodulin-sepharose 4B affinity column. The relative amount of protein kinase activities which passed through or were retained on the column varied from tissue to tissue (Fig. 4). The activities which were retained on the column were dependent on addition of Ca^{++} and calmodulin. In contrast, the activities which passed through the column was independent of Ca^{++} and calmodulin. The results indicate that all non-muscle tissues studied have more or less Ca^{++} dependent and calmodulin dependent activity of enzymes.

Phosphorylation of brain A and myosin light chains with peaks I and II. The phosphorylation of brain A and myosin light chains as substrates with peaks I and II from the brain is shown in Table 1. Peaks I and II prefer brain A and myosin light chains as substrates, respectively. The phosphorylation of brain A with peak I was Ca^{++} dependent but calmodulin independent. Myosin light chains were poor substrates for peak I. Furthermore, Ca^{++} and calmodulin had little effect on the reaction by peak I. In contrast, peak II phosphorylated myosin light chains in a Ca^{++} dependent and calmodulin dependent manner. From this and above results, e.g. the molecular weight and affinity for calmodulin column, peak II may belong to a class of myosin light chain kinase. Slight phosphorylation of brain A by peak II was observed. Ca^{++} and calmodulin had only a small effect on its reaction.

The distribution of peaks I and II in other muscle and non-muscle tissues was examined by assaying the activity from DEAE-cellulose column with brain A and myosin light chains as substrates (Table 2). The ratio of activity of peak I to that of peak II was different from tissue to tissue. The brain had the highest activity of peak I. In other tissues studied, peak II activity was more predominant than peak I.

When the samples were subjected to SDS slab gel electrophoresis, the 17,000 and 13,000 dalton components in brain A were mainly phosphorylated by peak I of the brain enzyme in a Ca^{++} dependent manner. A slight incorporation of phosphate into 20,000 dalton myosin light chain by peak I was observed. On the other hand, peak II phosphorylated the 20,000 dalton myosin light chain in a Ca^{++} dependent and calmodulin dependent manner. Peak II from the brain slightly phosphorylated the 17,000 dalton component of brain A.

Properties of the enzymes: a) Effect of Ca^{++} concentration. The effect of Ca^{++} concentration on phosphorylation of brain A and myosin light chains with peaks I and II from the brain, respectively, is shown in Fig. 5. From the plots in Figs. 5A and 5B, the concentration of Ca^{++} required to give half-maximal activation was determined to be 2.4 and 1.6 µM for peaks I and II, respectively; b) Effect of calmodulin concentration. The effect of calmodulin concentration on phosphorylation of myosin light chains with peak II from the brain was examined. The concentration of calmodulin required to give half-maximal activation was determined to be 0.055 µg per 0.2 ml of reaction mixture. This corresponds to 16.7 nM, assuming that the molecular weight of bovine brain calmodulin is 16,500.

DISCUSSION

The present study indicates that two types of Ca^{++} dependent protein kinases occur in the cytosol fraction of the brain, and other non-muscle and muscle tissues. The properties of both enzymes are summarized as follows: 1) Peaks I and II are eluted with 0.08 and 0.18 M NaCl, respectively, from DEAE-cellulose column; 2) peak I has no affinity for calmodulin. In contrast, peak II has a good affinity for calmodulin, as shown in the experiments with calmodulin affinity column and its activation by calmodulin; 3) peaks I and II prefer brain A and myosin light chains as substrates, respectively.

Peak II phosphorylates chicken gizzard myosin light chains in the presence of Ca^{++} and calmodulin and, therefore, should be called myosin light chain kinase. Ca^{++} and calmodulin dependent phosphorylation of proteins in the brain has been reported for synaptic membrane (23), synaptic vesicle (5) and cytosol (26) fractions. Our recent results (unpublished data) indicate that peak II from the brain can phosphorylate several endogenous brain proteins in a Ca^{++} dependent and calmodulin dependent manner. These results suggest that there are many substrates other than myosin light chains for brain myosin light chain kinase. The search for the substrates in each tissue may lead to the elucidation of functional significance of the enzymes in respective tissues.

SUMMARY AND CONCLUSION

Two types of Ca^{++} dependent protein kinases were demonstrated in the brain, and other non-muscle and muscle tissues. The enzymes were partially purified by DEAE-cellulose, sephadex G-200 and calmodulin-affinity column chromatography, using endogenous proteins and chicken gizzard myosin light chains as substrates. One of the enzymes had no affinity for calmodulin, whereas the other had a good affinity for calmodulin with Ka value of 16.7 nM for the brain enzyme. The Ka values of the enzymes for Ca^{++} were in a range of micromolar order.

ACKNOWLEDGEMENTS

This research was supported in part by the Naito Research Grant for 1979. We are grateful to Ms. T. Muramatsu for her secretarial assistance.

REFERENCES

1. Cheung, W.Y. (1980): Science 207:19-27.
2. Dabrowska, R., Aromatorio, D., Sherry, J.M.F. & Hartshorne, D.J. (1977): Biochem. Biophys. Res. Commun. 78:1263-1272.
3. Dabrowska, R. & Hartshorne, D.J. (1978): Biochem. Biophys. Res. Commun. 85:1352-1359.
4. Daniel, J.L. & Adelstein, R.S. (1976): Biochemistry 15:2370-2377.
5. DeLorenzo, R.J., Freedman, S.D., Yohe, W.B. & Maurer, S.C. (1979): Proc. Nat. Acad. Sci. 76:1838-1842.
6. Ikebe, M., Aiba, T., Onishi, H. & Watanabe, S. (1978): J. Biochem. 83:1643-1655.
7. Iwasa, Y., Yonemitsu, K., Matsui, K., Fukunaga, K. & Miyamoto, Y. (1981): Biochem. Biophys. Res. Commun. 98: 656-660.
8. Kakiuchi, S., Yamazaki, R., Teshima, Y., Uenishi, K. & Miyamoto, E. (1975): Biochem. J. 146:109-120.
9. Klee, C.B., Crouch, T.H. & Richman, P.G. (1980): Ann. Rev. Biochem. 49:489-515.
10. Klee, C.B. & Krinks, M.H. (1978): Biochemistry 17:120-126.
11. Krebs, E.G. & Beavo, J.A. (1979): Ann. Rev. Biochem. 48: 923-959.
12. Matsuda, G., Suzuyama, Y., Maita, T. & Umegane, T. (1977): FEBS Lett. 84:53-56.
13. Means, A.R. & Dedman, J.R. (1980): Nature 285:73-77.
14. Miyamoto, E., Fukunaga, K., Matsui, K. & Iwasa, Y. (1981): J. Neurochem. 37:1324-1330.

15. Miyamoto, E., Hideshima, Y., Teshima, Y., Yamazaki, R. & Kakiuchi, S. (1977): J. Cyclic Nucl. Res. 3:85-94.
16. Miyamoto, E., Hirose, R. & Setoyama, C. (1980): Biomed. Res. 1:158-163.
17. Miyamoto, E., Matsui, K., Fukunaga, K., Nishime, S. & Iwasa, Y. (1981): Biomed. Res. 2:341-346.
18. Miyamoto, E., Miyazaki, K. & Hirose, R. (1978): J. Neurochem. 31:269-275.
19. Miyamoto, E., Petzold, G.L., Kuo, J.F. & Greengard, P. (1973): J. Biol. Chem. 248:179-189.
20. Perrie, W.T. & Perry, S.V. (1970): Biochem. J. 119:31-38.
21. Perrie, W.T., Smillie, L.B. & Perry, S.V. (1973): Biochem. J. 135:151-164.
22. Rubin, C.S. & Rosen, O.M. (1975): Ann. Rev. Biochem. 44:831-887.
23. Schulman, H. & Greengard, P. (1978): Nature 271:478-479.
24. Schwarzenbach, G. (1960): F. Enke, Stuttgart.
25. Walsh, M.P., Vallet, B., Autric, F. & Demaille, J.G. (1979): J. Biol. Chem. 254:12136-12144.
26. Yamauchi, T. & Fujisawa, H. (1980): FEBS Lett. 116:141-144.
27. Yamaguchi, M., Miyazawa, Y. & Sekine, T. (1970): Biochim. Biophys. Acta 216:411-421.

CALMDOULIN DEPENDENT PROTEIN KINASES FROM RAT BRAIN

T. Yamauchi and H. Fujisawa

Department of Biochemistry, Asahikawa Medical College
Asahikawa 078-11, Japan

Three distinct calmodulin dependent protein kinases with different substrate specificities were present in rat brain cytosol. One of them was specifically distributed in the brain and it appeared to be involved in the regulation of tyrosine 3-mono-oxygenase and tryptophan 5-mono-oxygenase. The activation of mono-oxygenases by the calmodulin dependent protein kinase required the presence of activator protein as well as Ca^{++}, calmodulin and ATP. The activation of the mono-oxygenases occurred in a two-step reaction: first, phosphorylation by the calmodulin dependent protein kinase, and then the activation by the activator protein. A possible regulatory mechanism of the biosynthesis of catecholamines and serotonin in the brain was presented.

INTRODUCTION

Tyrosine 3-mono-oxygenase and tryptophan 5-mono-oxygenase are the rate-limiting enzymes in the biosynthesis of catecholamines and serotonin in the brain, respectively, and therefore, the mechanism of their regulation is of great interest. Tyrosine 3-mono-oxygenase was demonstrated to be activated by phosphorylation by cAMP dependent protein kinase (5,12,15,20,22,24). Brain tyrosine 3-mono-oxygenase was also found to be activated by Ca^{++} as well as cAMP in the presence of ATP and Mg^{++} (26). On the other hand, tryptophan 5-mono-oxygenase was reported to be activated by Ca^{++} in the presence of ATP and Mg^{++} (4,7,9,23). Recently we demonstrated that the activation of tyrosine 3-mono-oxygenase and tryptophan 5-mono-oxygenase was mediated by a calmodulin dependent protein kinase in the presence of Ca^{++} in rat brainstem (25,26). The calmodulin dependent activation of tryptophan 5-mono-oxygenase was confirmed by Kuhn et al. (6). The present studies demonstrate that at least three distinct calmodulin dependent protein kinases are present in the rat brain cytosol; two of them are likely to correspond to phosphorylase kinase and myosin light chain kinase, respectively, in brain tissues (1,2,10). The other kinase, kinase II, appeared to be a new calmodulin dependent protein kinase. Kinase II was almost specifically distributed in the brain and involved in the activation of both tyrosine 3-mono-oxygenase and tryptophan 5-mono-oxygenase. The activation of the mono-oxygenases occurred in a two-step reaction, phosphorylation of the mono-oxygenases by kinase II, followed by the activation of activator protein.

MATERIALS AND METHODS

$[\gamma-^{32}P]$ATP was purchased from Radiochemical Center (Amersham). Rabbit antiserum against bovine adrenal tyrosine 3-mono-oxygenase was prepared as described previously (24). Tyrosine 3-mono-oxygenase from bovine adrenal medulla was prepared as described previously (22). Tyrptophan 5-mono-oxygenase was prepared from rat brainstem as described previously (27). Tryptophan 5-mono-oxygenase, which was purified about 500-fold from rat brainstem, was kindly donated by Dr. H. Nakata of our laboratory. Calmodulin dependent protein kinase (kinase II) was prepared as described previously (27). cAMP dependent protein kinase from bovine heart was prepared by the method of Rubin et al. (13) without treatment of alumina Cγ-gel. Activator protein was purifed from rat cerebral cortex as described previously (28). Rat brain calmodulin was prepared by the method of Wang and Desai (16). Light chain from chicken gizzard myosin was prepared by the method of Perrie and Perry (11).

Tyrosine 3-mono-oxygenase was assayed fluorometrically (19). Tryptophan 5-mono-oxogenase was assayed essentially according to the method of Friedman et al. (3). Activation of tyrosine 3-mono-oxygenase and tryptophan 5-mono-oxygenase by calmodulin dependent protein kinase was carried out as described previously (28). The standard preincubation mixture for the activation of tyrosine 3-mono-oxygenase contained 50 mM 4-(2-hydroxyethyl)-1-piperazineethanesulfonic acid (Hepes) buffer, pH 7.0, 0.5 mM ATP, 3 mM Mg(CH$_3$COO)$_2$, 0.1 mM ethylene glycol bis (β-aminoethyl ether)-N,N,N',N'-tetraacetic acid (EGTA), 20 mM NaF, 0.12 mM CaCl$_2$, calmodulin, activator protein, calmodulin dependent protein kinase, and a suitable amount of tyrosine 3-mono-oxygenase. The mixture was incubated at 30°C for 5 min and the activity of tyrosine 3-mono-oxygenase was then assayed as described above. For the activation of tryptophan 5-mono-oxygenase, the activation reaction was carried out simultaneously with tryptophan 5-mono-oxygenase reaction in order to prevent denaturation of the enzyme. The standard incubation mixture for the activation of tryptophan 5-mono-oxygenase contained 50 mM Hepes buffer, pH 7.0, 0.5 mM

ATP, 3 mM $Mg(CH_3COO)_2$, 0.1 mM EGTA, 0.12 mM $CaCl_2$, calmodulin, 0.4 mM tryptophan, 0.3 mM 2-amino-4-hydroxyl-6-methyl-5, 6,7,8-tetrahydropteridine (6-MPH_4), 0.1 mM Fe $(NH_4)_2(SO_4)_2$, 2 mM dithiothreitol (DTT), 50 μg of catalase, activator protein, calmodulin dependent protein kinase, and a suitable amount of tryptophan 5-mono-oxygenase. The incubation was carried out at 30°C for 20 min, and 5-hydroxy-tryptophan was determined as described above. The phosphorylation of tyrosine 3-mono-oxygenase was carried out under the conditions used for the activation of the enzyme, except that 0.5 mM ATP was replaced by 0.1 mM [γ-^{32}P]ATP. The incorporation of [^{32}P]phosphate into tyrosine 3-mono-oxygenase. was analyzed by immunoprecipitation followed by sodium dodecyl sulfate (SDS)-polyacrylamide gel electrophoresis as described previously (24). Kinase II activity was assayed on the basis of its ability to activate tryptophan 5-mono-oxygenase under the standard activation conditions as described above. One unit of the activity of kinase II is defined as 1 incremental nmol of 5-hydroxy-tryptophan produced under standard conditions over controls without ATP. Assay for calmodulin dependent protein kinase activity were carried out as described previously (21). Endogenous substrates (27), chicken gizzard myosin light chain, or casein were used as protein substrates. Phosphorylase kinase was assayed by the method of Yamamura et al. (18).

Protein concentrations were estimated by the method of Lowry et al. (8) with bovine serum albumin as the standard. SDS-polyacrylamide gel electrophoresis was carried out by the method of Weber and Osborn (17).

RESULTS

We recently reported that a calmodulin dependent protein kinase was involved in the regulation of tyrosine 3-mono-oxygenase and tryptophan 5-mono-oxygenase (25,26) and that most of the effect of Ca^{++} on the endogenous protein phosphorylation in brain cytosol might be mediated by calmodulin (21). In order to characterize the calmodulin dependent protein kinases in the brain, the elution patterns of the enzymes on sepharose CL-6B were examined using several protein substrates, as shown in Fig. 1. The enzyme activities were eluted as three peaks (peaks I, II and III), assuming the activity eluted near the void volume to be due to a heterogeneous aggregate of the kinases. The activity of phosphorylase kinase, which was eluted as peak I at a position corresponding to a molecular weight of about 1×10^6, was stimulated by the addition of Ca^{++} and calmodulin. The activating activity of tryptophan 5-mono-oxygenase, which was eluted as peak II at a position corresponding to a molecular weight of about 5×10^5, absolutely required the presence of ATP, Mg^{++}, Ca^{++} and calmodulin. Calmodulin could not be replaced by phosphatidylserine, indicating that

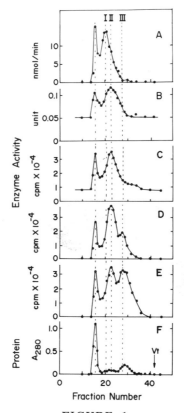

FIGURE 1

Gel filtration of calmodulin dependent protein kinases on sepharose CL-6B column. The column (1.4 x 60 cm) was equilibrated with 40 mM Tris-HCl, pH 7.6, containing 50 mM NaCl, 3 mM $Mg(CH_3COO)_2$, and 1 mM DTT and eluted with the same buffer. Each fraction was assayed for phosphorylase kinase (A), for activation of tryptophan 5-mono-oxygenase (B), for calmodulin dependent phosphorylation of cytosol proteins (C), for calmodulin dependent phosphorylation of casein (D), for calmodulin dependent phosphorylation of myosin light chain (E), and protein was determined (F).

this enzyme should be distinguished from phospholipid-sensitive Ca^{++} dependent protein kinase (14). The kinase eluted as peak II also stimulated the phosphorylation of casein, myosin light chain, and endogenous cytosol proteins and the activities were also dependent on the presence of both Ca^{++} and calmodulin. These results indicate that the kinase eluted as peak II has a broad substrate specificity. When either casein or myosin light chain was used as a substrate, the calmodulin dependent protein kinase activity, eluted as peak III, appeared at a position corresponding

TABLE 1: The activity of kinase II in various tissues of rat.

Tissue	Kinase II Activity	
	SA (units/mg protein)	RA (percent)
Cerebral cortex	10.8	100
Brainstem	9.8	91
Cerebellum	7.4	69
Adrenal gland	0.4 ± 0.4	4
Lung	0.4 ± 0.3	4
Kidney	0.3 ± 0.2	3
Testis	0.3 ± 0.3	3
Heart	0	0
Liver	0	0
Spleen	0	0
Skeletal muscle	0	0

SA = specific activity; RA = relative activity.
The tissues, as indicated above, of Wistar rats were homogenized in 4 vol of 0.32 M sucrose containing 10 mM Tris-HCl, pH 7.6, 0.01 mM EDTA, 1 mM DTT, and 1 mM phenylmethylsulfonylfluoride with a Potter Elvehjem homogenizer. Each soluble fraction was obtained from the homogenate by centrifugation at 105,000 x g for 1 hr was passed through a BioGel P-10 column in order to remove low molecular weight substances. The kinase activity of each cytosol fraction was assayed on the basis of its ability to activate tryptophan 5-mono-oxygenase, as described in Materials and Methods.

to a molecular weight of 1×10^5. Phosphorylation activity of casein and myosin light chain by the enzyme eluted as peak III also required the presence of Ca^{++} and calmodulin. However, the ratio of myosin light chain kinase activity to calmodulin dependent casein kinase activity of the enzyme eluted as peak III was twice as high as the enzyme eluted as peak II. Thus, myosin light chain served as a substrate more efficiently for the kinase eluted as peak III than for the kinase eluted as peak II, compared with casein. These results indicate that the enzymes eluted as peak I and peak III appear to be similar or identical to, phosphorylase kinase (2,10) and myosin light chain kinase (1), respectively.

Table 1 shows the tissue distribution of the activity of the kinase eluted as peak II, designated kinase II, as assayed by the activation of tryptophan 5-mono-oxygenase. The activity of kinase II was distributed in various regions of the brain, such as the cerebral cortex, the brainstem and the cerebellum. In contrast, the activity of the enzyme was considerably less or negligible in

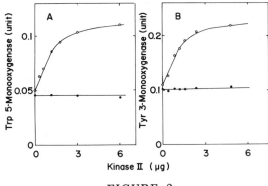

FIGURE 2

Effect of varying the concentration of kinase II on activation of tryptophan 5-mono-oxygenase and tyrosine 3-mono-oxygenase. Tryptophan 5-mono-oxygenase (224 µg protein) (A) and tyrosine 3-mono-oxygenase (125 µg cytosol protein of adrenal medulla)(B) were incubated with different amounts of kinase II under standard conditions (○) and without Ca^{++} (●).

other tissues, suggesting that kinase II was almost specifically distributed in brain tissues.

Figure 2 shows that both tyrosine 3-mono-oxygenase and tryptophan 5-mono-oxygenase were activated as a function of the concentration of kinase II. The extent of the activation of both monooxygenases by kinase II were almost equal. Thus, kinase II appeared to contribute to the activation of both mono-oxygenases to a similar extent.

When highly purified mono-oxygenase preparations were used to study the activation of mono-oxygenases by kinase II, activating activity was resolved into two distinct fractions (Fractions I and II) by gel filtration on sepharose CL-6B, as shown in Fig. 3. The activating activity of tryptophan 5-mono-oxygenase contained in Fraction I was assayed in the presence of Fraction II and that contained in Fraction II was assayed in the presence of Fraction I. Thus, Fraction I and Fraction II required the presence of the other to activate tryptophan 5-mono-oxygenase. The activity of Fraction I could be completely replaced by kinase II, indicating that the activating component in Fraction I may be identical to kinase II. Another activating component which was contained in Fraction II, designated as activator protein, was purified to apparent homogenity from a rat brain (28). The molecular weight of the activator protein was 70,000 and it was composed of two identical subunits with a molecular weight of 35,000.

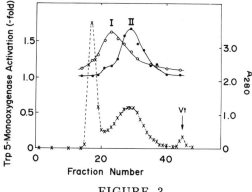

FIGURE 3

Resolution of activating activity of tryptophan 5-mono-oxygenase into two components on a sepharose CL-6B column. Cerebral cortex extract from rat brain was brought to 55% saturation with ammonium sulfate and the protein fraction precipitating in 55% saturated $(NH_4)_2SO_4$ (about 50 mg protein) was subjected to gel filtration on sepharose CL-6B column (1.4 x 60 cm) previously equilibrated with 40 mM Tris-HCl buffer, pH 7.6, containing 50 mM NaCl, 3 mM $Mg(CH_3COO)_2$, and 1 mM DTT and eluted with the same buffer at a flow rate of 6 ml/hr in fraction of 1.6 ml. Each fraction (a 30 μl aliquot) was assayed for the activation of tryptophan 5-mono-oxygenase in the presence of a 50 μl aliquot of fraction 31 (o——o) or a 50 μl aliquot of fraction 23 (●——●) and for absorbance at 280 nm (x---x).

When tryptophan 5-mono-oxygenase was incubated with Ca^{++}, ATP, calmodulin, kinase II, and activator protein, the mono-oxygenase became activated approximately two-fold, as shown in Table 2. The activation was almost completely blocked by the omission of any of these from the incubation mixture. The concentration of activator protein required for half-maximal activation of tryptophan 5-mono-oxygenase was about 1 μg protein/ml. Similar results were obtained in the activation of tyrosine 3-mono-oxygenase, indicating that the regulatory system, consisting of kinase II and activator protein, may be involved in the regulation of tyrosine 3-mono-oxygenase as well as tryptophan 5-mono-oxygenase (Table 2).

When immunoprecipitates of tyrosine 3-mono-oxygenase after incubation with kinase II in the presence of [γ-^{32}P]ATP were analyzed by SDS-polyacrylamide gel electrophoresis, a single major peak of ^{32}P radioactivity was observed, as shown in Fig. 4. The radioactivity was located at the position corresponding to that of tyrosine 3-mono-oxygenase subunit with a molecular weight of 60,000. The radioactivity was not observed in the control serum. The phosphorylation of tyrosine 3-mono-oxygenase required the

TABLE 2: Requirements for activation of tryptophan 5-mono-oxygenase and tyrosine 3-mono-oxygenase.

System	Tryptophan 5-Mono-oxygenase	Tyrosine 3-Mono-oxygenase
	Unit (-fold)	Unit (-fold)
Complete	0.057 (1.8)	0.51 (2.0)
- Kinase II	0.035 (1.2)	0.28 (1.1)
- Activator protein	0.034 (1.1)	0.31 (1.2)
- Calmodulin	0.033 (1.1)	0.25 (1.0)
- ATP	0.031 (1.0)	0.25 (1.0)
- Ca^{++}	0.031 (1.0)	0.26 (1.0)

Tryptophan 5-mono-oxygenase (5.7 μg protein) or tyrosine 3-mono-oxygenase (11 μg protein) were incubated under standard conditions and the enzyme activity was assayed as described in Materials and Methods.

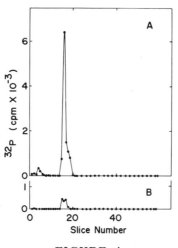

FIGURE 4

Phosphorylation of tyrosine 3-mono-oxygenase. Tyrosine 3-mono-oxygenase (28 μg protein) was incubated in 50 mM Hepes buffer, pH 7.0, 0.1 mM [γ-^{32}P]ATP, 5 mM $Mg(CH_3COO)_2$, 0.12 mM $CaCl_2$, 0.1 mM EGTA, 20 mM NaF, 3.7 μg of activator protein, 5 μg of calmodulin, 11 μg of kinase II in a final volume of 0.1 ml. After incubation for 5 min at 30°C, 20 μl of specific antiserum to tyrosine 3-mono-oxygenase (A), or control serum (B) were added to the reaction mixture and immunoprecipitation was carried out as described in Materials and Methods. The immunoprecipitate was subjected to SDS-polyacrylamide gel electrophoresis.

presence of kinase II, calmodulin and Ca^{++} did not require the presence of the activator protein, as shown in Table 3. These results indicate that the phosphorylation of tyrosine 3-mono-oxygenase by kinase II occurred in the absence of the activator protein, suggesting that the activation and the phosphorylation of the mono-oxygenase may be distinct reactions.

To examine the relationship between the phosphorylation and the activation of tyrosine 3-mono-oxygenase, separation of both reactions was attempted, as shown in Table 4. Preincubation mixtures contained the ingredients necessary for the reaction of protein phosphorylation and, therefore, EDTA was added to the assay mixture to prevent phosphorylation of the enzyme during the incubation. The preincubation of tyrosine 3-mono-oxygenase with either kinase II or activator protein alone was not effective in activating the enzyme but the preincubation with both was effective, suggesting that the phosphorylation of the enzyme by kinase II may be a necessary but not sufficient condition for the enzyme activation. When tyrosine 3-mono-oxygenase was preincubated with kinase II and activator protein was then added to the assay mixture, the enzyme was activated to the same extent as when the enzyme was preincubated with the kinase and activator protein together, indicating that activator protein may act immediately on tyrosine 3-mono-oxygenase phosphorylated by the action of kinase II. From the facts described above, it was concluded that the activation of tyrosine 3-mono-oxygenase occurred in a two-step reaction. Since the activation of tryptophan 5-mono-oxygenase required the presence of both kinase II and activator protein, it is reasonable to assume that tryptophan 5-mono-oxygenase is also activated by the two-step reaction.

Recent studies from this and other laboratories (5,12,15,20,22,24) have demonstrated that tyrosine 3-mono-oxygenase is phosphorylated by the action of a cAMP dependent protein kinase with a concomitant increase of tyrosine 3-mono-oxygenase activity. The effect of the activator protein on the activation of tyrosine 3-mono-oxygenase by cAMP dependent protein kinase was examined, as shown in Table 5. In contrast to the phosphorylation of tyrosine 3-mono-oxygenase by kinase II, the phosphorylation of the enzyme by cAMP dependent protein kinase resulted in the activation of the enzyme without activator protein. No further activation was observed on incubation with the activator protein. These results indicate that the activator protein was effective with the enzyme phosphorylated by kinase II, but not effective with the enzyme phosphorylated by cAMP dependent protein kinase. Furthermore, the effect of kinase II on the activation of tyrosine 3-mono-oxygenase was additive to that of cAMP dependent protein kinase, suggesting that tyrosine 3-mono-oxygenase may be regulated independently by calmodulin dependent protein kinase and by cAMP dependent protein kinase.

TABLE 3: Requirements for phosphorylation of tyrosine 3-mono-oxygenase.

System	^{32}P Incorporation	
	cpm	percent
Complete	6716	100
- Kinase II	1058	16
- Activator protein	6806	102
- Calmodulin	123	2
- Ca^{++}	341	5

Tyrosine 3-mono-oxygenase (28 µg protein) was incubated with 0.1 mM [γ-^{32}P]ATP as described in legend for Fig. 4. After incubation for 5 min at 30°C immunoprecipitation was carried out as described in Materials and Methods. The immunoprecipitate was subjected to SDS-polyacrylamide gel electrophoresis and ^{32}P radioactivity located at the position corresponding to that of tyrosine 3-mono-oxygenase subunit with a molecular weight of 60,000 was measured.

TABLE 4: Effect of preincubation on the activation of tyrosine 3-mono-oxygenase by kinase II and/or activator protein.

Time of Addition		Tyrosine 3-mono-oxygenase	
- 5 min	0 min	unit	(-fold)
None	None	0.484	(1.0)
Kinase	None	0.550	(1.1)
Activator	None	0.496	(1.0)
Kinase + activator	None	0.985	(2.0)
Kinase	Activator	0.968	(2.0)
Activator	Kinase	0.484	(1.0)

Tyrosine 3-mono-oxygenase (11 µg protein) was previously incubated at 30°C for 5 min in standard conditions. The phosphorylation reaction was terminated by the addition of EDTA at 0 min and the reaction of tyrosine 3-mono-oxygenase was started by the addition of the assay mixture for the enzyme at 0 min. Kinase II (11 µg protein) and activator protein (3.7 µg protein) were added at the indicated times.

TABLE 5: Effects of activator protein on the activation of tyrosine 3-mono-oxygenase by kinase II and/or cAMP dependent protein kinase.

Kinase Added	Tyrosine 3-Mono-oxygenase	
	+ Activator Unit (-fold)	- Activator Unit (-fold)
None	0.35 (1.0)	0.35 (1.0)
Kinase II	0.79 (2.3)	0.40 (1.1)
cAMP dependent	0.84 (2.4)	0.84 (2.4)
Kinase II + cAMP dependent	1.17 (3.4)	0.95 (2.7)

Tyrosine 3-mono-oxygenase (7.6 µg protein) was incubated with kinase II (11 µg protein) and/or cAMP dependent protein kinase (10 µg protein) in the presence or absence of activator protein (3 µg protein) under standard conditions. Kinase II was added to the incubation mixture together with 0.12 mM Ca^{++} and 2 µg of calmodulin. cAMP dependent protein kinase was added with 0.01 mM cAMP and 1.5 mM isobutyl-1-methylxanthine. Tyrosine 3-mono-oxygenase activity was assayed as described in Materials and Methods.

DISCUSSION

Present studies demonstrate that at least three distinct calmodulin dependent protein kinases are present in the rat brain cytosol. Although two of them appeared to correspond to the phosphorylase kinase (2,10) and myosin light chain kinase (1), respectively, the other, kinase II, with a molecular weight of about 5×10^5, is considered to be a new calmodulin dependent protein kinase. Kinase II was found to be almost specifically but widely distributed in the brain and it showed a broad substrate specificity, suggesting that kinase II might play a number of roles in the nervous system. One of these may be related to the regulation of the biosynthesis of catecholamines and serotonin, since kinase II mediated the activation of tyrosine 3-mono-oxygenase and tryptophan 5-mono-oxygenase.

The activation of tyrosine 3-mono-oxygenase and tryptophan 5-mono-oxygenase by kinase II required the presence of activator protein. The regulation of the mono-oxygenases by kinase II occurred in a two-step reaction, first, phosphorylation by kinase II and then activation of the activator protein. Since activator protein was abundantly found in the cytosol fraction of a variety of tissues, including brain tissues (28), the actual regulatory step may be the phosphorylation step through the action of kinase II. In contrast to the kinase system, the activation of tyrosine 3-

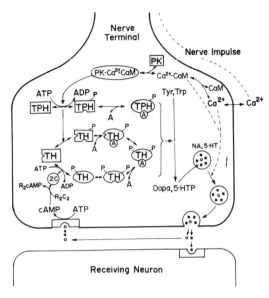

FIGURE 5

Schematic diagram of regulation of biosynthesis of catecholamines and serotonin by protein phosphorylation in the brain. The inactive or less active form of enzymes is expressed as a symbol ▢, and active form of enzymes as ◯. TH = tyrosine 3-monooxygenase; TPH = tryptophan 5-mono-oxygenase; PK = kinase II; R_2C_2 = cAMP dependent protein kinase (R = regulatory subunit; C = catalytic subunit); A = activator protein; CaM = calmodulin; Dopa = 3,4-dihydroxyphenylanaline; NA = noradrenaline; 5-HTP = 5-hydroxytryptophan; 5-HT = serotonin.

mono-oxygenase by cAMP dependent protein kinase did not require the presence of activator protein. Since the effect of kinase II on the activation of tyrosine 3-mono-oxygenase was additive to that of cAMP dependent protein kinase, tyrosine 3-mono-oxygenase was considered to be independently regulated by the Ca^{++}-calmodulin dependent protein kinase (kinase II) system and by the cAMP dependent protein kinase system.

A possible regulatory mechanism of biosynthesis of catecholamines and serotonin by Ca^{++} and/or cAMP in the brain is shown in Fig. 5. The depolarization of noradrenergic or serotonergic nerve terminal results in an influx of Ca^{++} into the nerve terminal, which induces the release of noradrenaline or serotonin from the storage sites into the synaptic space. The released transmitter molecules bind to a specific receptor on the postsynaptic membrane, triggering a series of reactions on the receiving neuron. The Ca^{++}

which has entered the nerve terminals binds to calmodulin and subsequently results in the activation of the protein kinase (kinase II). Tyrosine 3-mono-oxygenase or tryptophan 5-mono-oxygenase are phosphorylated by the activated kinase II, and the phosphorylated mono-oxygenases are then activated by activator proteins. Thus the biosynthesis of catecholamines or serotonin are enhanced within a short time in the terminals of the stimulated neurons. On the other hand, noradrenaline release leads to the stimulation of cAMP synthesis in the presynaptic nerve terminal, resulting in a local increase in the cAMP level. The increased levels of cAMP activated cAMP dependent protein kinase and the activated protein kinase then phosphorylates tyrosine 3-mono-oxygenase, converting it to an activated form. Thus, noradrenaline biosynthesis may be regulated by two distinct protein kinase systems, a Ca^{++}-calmodulin dependent protein kinase (kinase II) system and a cAMP dependent protein kinase system. In contrast to tyrosine 3-mono-oxygenase, tryptophan 5-mono-oxygenase appears to be regulated by only a Ca^{++}-calmodulin dependent protein kinase system, since it was not activated by cAMP dependent protein kinase. The new Ca^{++}-calmodulin dependent protein kinase (kinase II) which is specifically located in brain tissues may be involved not only in the regulation of both noradrenaline and serotonin biosynthesis in nerve terminals but also in other functions in the nervous system, since it shows a broad substrate specificity with respect to endogenous protein substrates in brain tissues.

ACKNOWLEDGEMENTS

The authors would like to thank Mr. Andrew Grenville who kindly read the first draft and offered suggestions for correcting the English phrasing. This work has been supported in part by a Grant-in-Aid for Scientific Research from the Ministry of Education, Science, and Culture of Japan, and by grants from the Naito Foundation Research Grant and the Chiyoda-Seimei Foundation for 1980.

REFERENCES

1. Dabrowska, R. & Hartshorne, D.J. (1978): Biochem. Biophys. Res. Commun. 85:1352-1359.
2. Drummond, G.I. & Bellward, G. (1970): J. Neurochem. 17:475-482.
3. Friedman, P.A., Kappelman, A.H. & Kaufman, S. (1972): J. Biol. Chem. 247:4165-4173.
4. Hamon, M., Bourgoin, S., Hery, F. & Simonnet, G. (1978): Molec. Pharmacol. 14:99-110.
5. Joh, T.H., Park, D.H. & Reis, D.J. (1978): Proc. Nat. Acad. Sci. 75:4744-4748.
6. Kuhn, D.M., O'Callaghan, J.P., Juskevich, J. & Lovenberg, W. (1980): Proc. Nat. Acad. Sci. 77:4688-4691.

7. Kuhn, D.M., Vogel, R.L. & Lovenberg, W. (1978): Biochem. Biophys. Res. Commun. 82:759-766.
8. Lowry, O.H., Rosebrough, N.J., Farr, A.L. & Randall, R. (1951): J. Biol. Chem. 193:236-275.
9. Lysz, T.W. & Sze, P.Y. (1978): J. Neurosci. Res. 3:411-418.
10. Ozawa, E. (1973): J. Neurochem. 20:1487-1488.
11. Perrie, W.T. & Perry, S.V. (1970): Biochem. J. 119:31-38.
12. Raese, D.J., Edelman, A.M., Lazer, M.A. & Barchas, J.D. (1977): In Structure and Function of Monoamine Enzymes, (eds) E. Usdin, N. Weiner & M.B.H. Youdin, Marcel Dekker, New York, pp. 383-400.
13. Rubin, C.S., Erlichman, J. & Rosen, O.M. (1972): J. Biol. Chem. 247:36-44.
14. Takai, Y., Kishimoto, A., Iwasa, Y., Kawahara, Y., Mori, T. & Nishizuka, Y. (1979): J. Biol. Chem. 254:3692-3695.
15. Vulliet, P.R., Langan, T.A. & Weiner, N. (1980): Proc. Nat. Acad. Sci. 77:92-96.
16. Wang, J.H. & Desai, R. (1977): J. Biol. Chem. 252:4175-4184.
17. Weber, K. & Osborn, M. (1969): J. Biol. Chem. 244:4406-4412.
18. Yamamura, H., Nishiyama, K., Shimomura, R. & Nishizuka, Y. (1973): Biochemistry 12:856-862.
19. Yamauchi, T. & Fujisawa, H. (1978): Anal. Biochem. 89:143-150.
20. Yamauchi, T. & Fujisawa, H. (1978): Biochem. Biophys. Res. Commun. 82:514-517.
21. Yamauchi, T. & Fujisawa, H. (1979): Biochem. Biophys. Res. Commun. 90:1172-1178.
22. Yamauchi, T. & Fujisawa, H. (1979): J. Biol. Chem. 254:503-507.
23. Yamauchi, T. & Fujisawa, H. (1979): Arch. Biochem. Biophys. 198:219-226.
24. Yamauchi, T. & Fujisawa, H. (1979): J. Biol. Chem. 254:6408-6413.
25. Yamauchi, T. & Fujisawa, H. (1979): Biochem. Biophys. Res. Commun. 90:28-35.
26. Yamauchi, T. & Fujisawa, H. (1980): Biochemistry Int. 1:98-104.
27. Yamauchi, T. & Fujisawa, H. (1980): FEBS Lett. 116:141-146.
28. Yamauchi, T., Nakata, H. & Fujisawa, H. (1981): J. Biol. Chem. 256:5405-5409.

A HIGHLY ACTIVE CALCIUM ION STIMULATED ENDOGENOUS
PROTEIN KINASE IN MYELIN: KINETIC AND OTHER
CHARACTERISTICS, CALMODULIN DEPENDENCE AND
COMPARISON WITH OTHER BRAIN KINASE(S)

P.V. Sulakhe, E.H. Petrali and B.L. Raney

Department of Physiology, College of Medicine
University of Saskatchewan
Saskatoon S7N OWO Canada

Highly purified myelin preparations from the central and peripheral nervous systems of rat and mouse contained endogenous protein kinases which catalyzed the phosphorylation mainly of large and small myelin basic proteins (LBP and SBP). Ca^{++} in micromolar concentrations markedly stimulated the phosphorylation of LBP and SBP. Triton X-100 further increased the Ca^{++} stimulated phosphorylation of basic proteins; basal phosphorylation (with Mg^{++} and EGTA present) was also increased by Triton X-100 and a modest increase in phosphorylation by cAMP was seen only in the presence of the detergent. Ca^{++} stimulated kinase was enriched in the lighter subfractions, whereas cAMP stimulated phosphorylation was enriched in the heavier subfractions prepared by centrifugation of myelin on discontinuous or continuous sucrose density gradients.

A detailed study of Ca^{++} stimulated kinase in myelin (from rat brain white matter) revealed the following major findings: 1) It is an Mg^{++} dependent enzyme and its nucleotide substrate is $MgATP^{2-}$. Two Mg^{++} binding sites (Site I, high affinity, K_a approximately 0.1 mM; Site II, low affinity, K_a approximately 3 mM) were present. Site I likely represented the binding of Mg^{++} (as $MgATP^{2-}$) to the catalytic site which exhibited a K_m for $MgATP^{2-}$ of about 0.07 mM. Ca^{++} (μM) did not influence the K_a for Mg^{++} or K_m for $MgATP^{2-}$. Ca^{++} (μM) did decrease K_a for Mg^{++} of Site II by three- to five-fold. The results also indicated that Mg^{++} binding to Site II allosterically regulated the catalytic reactivity of both basal and Ca^{++} stimulated kinases. Ca^{++} increased the V_{max} under all conditions except when saturating Mg^{++} (20 mM) and Triton X-100 were present in assay. 2) Divalent cation specific ionophore, A23187, and channel forming ionophore, alamethicin, both increased Ca^{++} stimulated kinase up to 80%; the latter ionophore increased (30%) basal kinase as well. 3) Several lines of evidence were obtained that indicate a role for a calmodulin-like protein, which was present in isolated myelin preparations, in the Ca^{++} stimulation of myelin phosphorylation. 4) interestingly, while Triton X-100 increased Ca^{++} stimulated kinase, the detergent rendered the enzyme insensitive to calmodulin but still sensitive to Ca^{++}. Phenothiazines only at high concentrations (>100 µM) inhibited Ca^{++} stimulated kinase and only when assays were carried out in the absence of the detergent; this was also true for the inhibitory effect of W-7 on myelin Ca^{++} kinase. Basal kinase, on the other hand, was increased at higher concentrations (> 100 µM) of phenothiazines and W-7. 5) Microsomal fraction from rat cerebral cortex gray matter contained Ca^{++}-calmodulin requiring kinase that phosphorylated mainly the peptides of M_r, 65K, 60K and 55K (K = 1000). Triton X-100 and phenothiazines markedly inhibited the Ca^{++} stimulated phosphorylation. Basal kinase in this membrane was increased by Triton X-100, as well as phenothiazines. 6) It is suggested that myelin (as well as gray matter microsomes) contains Ca^{++}-calmodulin requiring kinase, whose activity predominates when assays are carried out in the absence of Triton X-100. On the other hand, when the detergent is present, the Ca^{++} stimulated phosphorylation becomes calmodulin insensitive. The increased Ca^{++} stimulated phosphorylation of basic proteins in the presence of Triton X-100 is likely due to a phospholipid-requiring myelin kinase, which in some ways resembles the brain cytosolic kinase C reported by Nishizuka and associates. Interestingly, for both types of Ca^{++} stimulated myelin kinases, the substrate proteins are basic proteins (SBP and LBP), although the possibility exists that they phosphorylate different amino acids of SBP and LBP.

INTRODUCTION

It has been shown that under in vivo (11,21) and in vitro (1, 11,21) conditions myelin basic proteins are phosphorylated. Subsequently, Miyamoto (9,10,12) established the presence of protein kinases in myelin preparations and studied some of their characteristics. In these and other studies, myelin basic proteins were found to undergo phosphorylation that was catalyzed by endogenous or exogenous protein kinases. We discovered the presence of a Ca^{++} ion stimulated protein kinase in myelin isolated from central and peripheral nervous systems of rat and mouse (15-19, 22-25). Our study further showed that calmodulin-like protein was required for stimulation by Ca^{++} of myelin phosphorylation (23,24); and Endo and Hidaka (3) have recently reported a similar observation. We noted that isolated myelin preparations contained endogenous calmodulin-like protein(s) as well (24); this is also suggested by Endo and Hidaka (3) and Ca^{++} dependent phosphorylation of isolated myelin has now been confirmed by Wuthrich and Steck (30).

In this paper, we described a number of major properites of myelin kinases (basal, cyclic AMP stimulated, and Ca^{++} stimulated). In addition, the distribution of these kinases in myelin subfraction is presented. Further, we compare some of the properties of myelin Ca^{++} kinase with a Ca^{++} sensitive kinase present in rat brain gray matter microsomes as well as with a kinase C, reported by Takai et al. (26) in a cytosolic fraction from rat brain. Some evidence is presented that Triton X-100, which markedly increased the Ca^{++} stimulated phosphorylation of myelin peptides (22), renders myelin Ca^{++} kinase calmodulin insensitive. Thus, these results raise an interesting and intriguing possibility of the presence of two types of Ca^{++} sensitive kinases in myelin, one type that depends on calmodulin and the other that is calmodulin independent. The latter in some ways resembles the kinase C.

METHODS

Isolation of myelin from rat brain white matter. Several procedures were used for isolation of myelin and these have been detailed by Petrali et al. (19), Sulakhe et al. (23,24), and by Waehneldt (27). In addition, myelin subfractions were prepared by centrifugation of myelin on continuous and discontinuous sucrose density gradients as described by Sulakhe et al. (24). A microsomal fraction, termed SN_4 and SN_4-free myelin, both from white matter, were isolated essentially by the method of Waehneldt et al. (28).

Determination of endogenous protein kinases, phosphorylation and electrophoretic separation of phosphoproteins. The details of the phosphorylation assay were provided in earlier publications from our laboratory (19,23,24). Similarly, the separation of solubilized membrane protein by SDS polyacrylamide slab gel electrophoresis by the Laemmli procedure, estimation of protein content of separated polypeptide bands following electrophoresis and autoradiographic determination and quantitation of ^{32}P incorporation into electrophoretically separated polypeptides were all described by us previously (19,24). EGTA extraction of myelin or myelin subfractions as well as determination of calmodulin-like proteins were carried out as previously described (23,24).

RESULTS

Time course of myelin phosphorylation by endogenous kinases. At 30°C, maximal ^{32}P incorporation into myelin was attained between 10 and 15 min incubation. Ca^{++} (μM) increased the rate and the extent of phosphorylation. Addition of Triton X-100 to assay also increased phosphorylation with its stimulatory effect on the Ca^{++} sensitive phosphorylation being greater compared to basal. While Ca^{++} increased the V_{max} of phosphorylation under the standard condition, at saturating Mg^{++} (20 mM) and with Triton X-100 present, Ca^{++} did not increase the V_{max} (see below). Storage of myelin at 4°C for up to three days, which decreased the initial rate of phosphorylation, modestly increased basal but not Ca^{++} stimulated phosphorylation. Interestingly, in freshly prepared myelin preparations, longer incubation (30 min at 30°C) with Ca^{++} (Triton X-100 present) led to a decline in ^{32}P incorporation presumably due to either depletion of the substrate, ATP, or dephosphorylation by phosphatases. Following storage for three days at 4°C, such a decline was not normally observed (19). On the other hand, crude myelin preparations, fresh and stored, showed such decline and further in these preparations, the rate of Ca^{++} stimulated phosphorylation was markedly lower following storage (18) unlike in the case of purified myelin fraction (19). Also, with crude myelin fractions, in the presence of Triton X-100 and subsaturating Mg^{++} (1 mM), Ca^{++} (μM) increased the initial rate but not the maximal level of ^{32}P incorporation. This is also in contrast to the observation made on purified myelin fraction. Not only did Triton X-100 increase basal and Ca^{++} stimulated phosphorylation to different extents but the concentrations of the detergent required for maximal effect were also different (Fig. 1)(see also 18,19).

Effects of temperature and pH of assay. Ca^{++} stimulated phosphorylation was optimal at pH 6-6.5. Increasing the temperature from 4°C to 30°C increased endogenous phosphorylation of myelin when assayed with and without Ca^{++} and with and without Triton X-100. At temperatures higher than 40°C, there was a marked and precipitous decrease in the ^{32}P incorporation under all assay conditions (19).

Nucleotide substrate for myelin kinases. Comparted to ATP, GTP was a poor substrate. For example, GTP supported only about 20% ($-Ca^{++}$) and 5% ($+Ca^{++}$) of phosphorylation of myelin basic proteins compared to that obtained with ATP. Also, $MgATP^{2-}$ and not $CaATP^{2-}$ was the substrate for basal and Ca^{++} kinases. Myelin-like membrane (SN_4) contained cAMP dependent protein kinase whose activity was supported to about 30% by GTP compared to ATP. cGMP dependent protein kinase activity could not be detected in myelin or SN_4, although cGMP was able to stimulate cAMP dependent kinase in SN_4 with the affinity for cGMP being about two orders of magnitude lower than for cAMP (16).

Divalent cation specificity of myelin kinases. Basal kinase was supported maximally by Mg^{++} followed by Mn^{++}, Co^{++}, Zn^{++}, Cu^{++} and Ca^{++} in the order of decreasing potency (18). While nearly

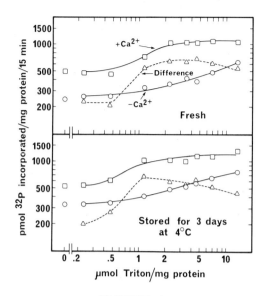

FIGURE 1

Dependence on the concentrations of Triton X-100 of basal and Ca^{++} stimulated (difference between with and without Ca^{++}) phosphorylation. Incubations for phosphorylation at 30°C were carried out for 15 min with and without varying amounts of Triton X-100. Myelin was phosphorylated immediately following its isolation (upper panel) and following storage for three days at 4°C. In this and other figures, when present, $MgCl_2$ was 1 mM, Ca^{++} (free), 22 µM and ATP, 50 µM unless stated otherwise.

20 mM Mg^{++} was needed for maximal activity, lower (2-5 mM) concentrations of Mn^{++} and Co^{++} were necessary. Whereas only micromolar Ca^{++} optimally stimulated basal Mg^{++} dependent kinase, millimolar Ca^{++} was needed in the absence of Mg^{++} and further, the activity reached was 20-fold lower than that attained with 10 mM Mg^{++} (18). Ca^{++} stimulated kinase (with 1 mM Mg^{++}) showed a rigid requirement for Ca^{++} since Mn^{++}, Co^{++}, Ba^{++} or Sr^{++} poorly substituted for Ca^{++}. Also, in the presence of 1 mM Mn^{++} or Co^{++} (but without Mg^{++}), Ca^{++} (µM) either stimulated the activity only modestly (15%) or not at all (18).

Thus, it is clear that myelin Ca^{++} kinase is an Mg^{++} dependent enzyme and its nucleotide substrate is $MgATP^{2-}$. Interestingly, two Mg^{++} binding sites were detected. Site I, which is a high affinity site for $K_{(app)}$ of 90 µM, likely represented the catalytic site since the K_m of about 70 µM for $MgATP^{2-}$ was observed (19). Ca^{++} (µM) did not influence the affinity of Site I toward Mg^{++} or

TABLE 1: Increase by Triton X-100 of basal and Ca^{++} stimulated phosphorylation: dependence on $MgCl_2$ and time of incubation.

$MgCl_2$ (mM)	(pmol ^{32}P incorporated/mg protein)					
	Triton Absent			Triton Present		
	$-Ca^{++}$	$+Ca^{++}$	Dif	$-Ca^{++}$	$+Ca^{++}$	Dif
Experiment 1						
0.1	22	93	71	32	160	128
1	119	460	341	138	632	494
10	308	1046	738	386	1301	915
20	348	871	523	455	903	448
Experiment 2						
0.1	59	290	231	65	299	234
1	220	650	430	246	620	374
10	640	1348	708	823	753	-70
20	785	1294	509	825	750	-75

Isolated myelin was phosphorylated for 2 min at 37°C (experiment 1) and 15 min at 30°C (experiment 2) under standard assay conditions (± Triton and Ca^{++}) except for variation in $MgCl_2$ as indicated. When present, Triton X-100 was 0.025%, and Ca^{++} 50 μM. Separate myelin preparations were used in experiments 1 and 2. Dif = difference.

of the catalytic site towards $MgATP^{2-}$. Site II is a low affinity site with $K_{(app)}$ for Mg^{++} or 3-5 mM and Ca^{++} (μM) decreased $K_{(app)}$ for Mg^{++} by about three- to four-fold. Mg^{++} binding to this site increased the catalytic reactivity without influencing the K_m for $MgATP^{2-}$ of the catalytic site. Thus, Mg^{++} appears to allosterically increase the V_{max} of Ca^{++} kinase (as well as basal kinase) by its interaction with the Site II. However, as shown in Table 1, increases by Triton X-100 in basal and Ca^{++} kinases depended on the concentrations of Mg^{++} in the assay as well as time of incubation. In fact, at saturating Mg^{++} and 15 min incubation, Ca^{++} stimulated phosphorylation (shown as difference) over basal could not be detected when Triton X-100 was present (19).

Affinity towards Ca^{++}. Ca^{++} stimulated phosphorylation of myelin basic proteins was half-maximally stimulated at 6-8 μM Ca^{++} and was maximal at 20 μM Ca^{++}. Higher concentrations of Ca^{++} inhibited the phosphorylation (18,19). While Triton X-100 increased the V_{max}, it did not appear to affect the affinity towards Ca^{++}. This was true when the assay contained subsaturating Mg^{++} (1 mM) and incubations were carried out for 2 min at 30°C.

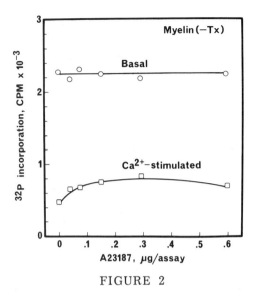

FIGURE 2

Effect of A23187 on basal and Ca^{++} stimulated phosphorylation. Myelin (95 µg) phosphorylation (± Ca^{++}) was carried out with and without varying concentrations of A23187. Triton X-100 (T_X) was absent.

Effects of ionophores. Divalent cation specific ionophore, A23187, had no effect on basal phosphorylation, whereas Ca^{++} stimulated phosphorylation had modestly (up to 80%) increased (Fig. 2). Such an increase, however, could not be seen in the presence of Triton X-100 in assays (not shown in Fig. 2). Alamethicin also modestly (40%) increased Ca^{++} stimulated phosphorylation, although this ionophore increased basal phosphorylation as well (Fig. 3: lower panel). Again, the stimulatory effects of alamethicin could not be detected in the presence of Triton X-100 (not shown in Fig. 3).

Evidence that calmodulin-like protein participates in the Ca^{++} stimulated phosphorylation of myelin. Exposure to EGTA (1 to 5 mM), a powerful Ca^{++} chelator, of isolated myelin followed by its centrifugation selectively decreased the Ca^{++} sensitive phosphorylation by about 60% (23); this was observed whether the fractions were assayed with or without Triton X-100 (Fig. 4: panel A). This decrease was restored by addition of the EGTA extract to assays (Fig. 4: panel B). A similar finding was also obtained when EGTA (2 mM) was added to the homogenizing buffer in which the tissue (white matter) was homogenized and myelin was then subsequently isolated. Exogenous calmodulin also restored the decrease in the Ca^{++} stimulated phosphorylation observed in the EGTA exposed myelin preparations (23). The EGTA extract contained a polypeptide

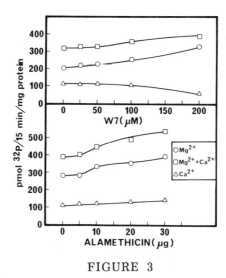

FIGURE 3

Effects of W-7 and alamethicin on myelin phosphorylation. Myelin (20 µg/assay), isolated according to Waehneldt et al. (28), was phosphorylated at 30°C for 15 min (± Ca^{++}). Other additions are as shown.

band that electrophoretically migrated similarly to purified calmodulin (unpublished work). A heat-stable calmodulin binding protein (BP_{70}), but not a heat-labile binding protein (BP_{80}), selectively inhibited (70%) the Ca^{++} stimulated phosphorylation (Fig. 5). Leupeptin, up to 300 µM, did not show any effect on Ca^{++} sensitive phosphorylation and this indicated that a thiol requiring Ca^{++} activated neutral protease was not involved in the Ca^{++} dependent stimulation of myelin phosphorylation. The EGTA extract increased the activity of calmodulin deficient cAMP phosphodiesterase to the same extent, and in a Ca^{++} dependent manner, as purified calmodulin (24). W-7 (5), which inhibits calmodulin sensitive systems, inhibited Ca^{++} dependent myelin phosphorylation (assayed without Triton X-100)(Fig. 3: upper panel). W-7, in a concentration dependent manner, increased basal phosphorylation as well. Trifluoperazine (Fig. 6) and chlorpromazine (16) also showed effects similar to W-7, in that they decreased the Ca^{++} stimulated phosphorylation and increased basal phosphorylation. However, it is important to note that the concentrations of W-7, trifluoperazine and chlorpromazine required for inhibition of the Ca^{++} sensitive phosphorylation are much higher (ten- to twenty-fold) than those required for inhibition of (soluble) Ca^{++}-calmodulin dependent cAMP phosphodiesterases. A cytosolic fraction (S_2) prepared from brain white matter selectively inhibited Ca^{++} stimulated, but not basal, phosphorylation (Fig. 4: panel B). In preliminary experiments

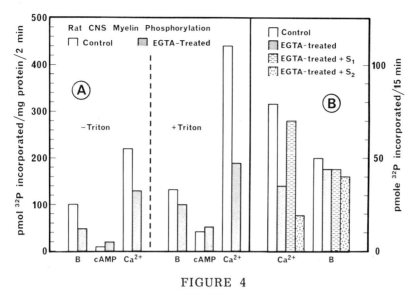

FIGURE 4

Effects of EGTA exposure, EGTA extract (S_1) and cytosolic fraction (S_2) on myelin phosphorylation. Myelin, isolated according to Sulakhe et al. (23), was treated with 2 mM EGTA for 30 min at 4°C, then centrifuged. Treated and untreated myelin (100 µg/assay) was assayed ± Triton X-100 (0.05%) and with and without Ca^{++} and cyclic AMP. Either EGTA extract (S_1) or a cytosolic fraction (S_2, 100,000 x g supernatant) from white matter, was added to the EGTA treated myelin prior to assay. B = basal; cAMP = with 5 µM cyclic AMP added; Ca^{++} = 25 µM Ca^{++} added.

S_2 fraction was found to contain proteins that functionally resembled calmodulin binding proteins, especially the heat-stable (BP_{70}) binding protein (29).

Conditions under which myelin Ca^{++} kinase becomes Ca^{++} insensitive and/or calmodulin insensitive. At saturating Mg^{++} and with Triton X-100 present, Ca^{++} did not increase the V_{max} of phosphorylation of purified myelin (Table 1: experiment 2; see also 19). In the absence of Triton X-100, and even when Mg^{++} was saturating, Ca^{++} did increase the V_{max}. Thus it seems that Triton X-100 has complex interactions. A similar complexity was noted when we attempted to restore the decreased Ca^{++} stimulated phosphorylation of the EGTA exposed myelin by addition of the EGTA extract or exogenous calmodulin. Such restoration was readily and reproducibly achieved when Triton X-100 was absent from the assay. We further observed that Triton X-100 markedly

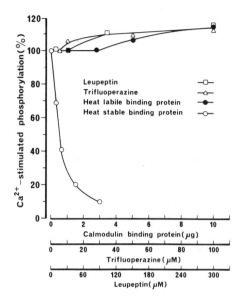

FIGURE 5

Effects of leupeptin, trifluoperazine, and calmodulin binding proteins on Ca^{++} stimulated phosphorylation of myelin. Ca^{++} stimulated phosphorylation was determined (Triton absent) in the absence (set to represent 100%) and presence of varying amounts of the agents shown.

interfered in the calmodulin-Ca^{++} dependent activation of cAMP phosphodiesterase (16). Thus, while a role for calmodulin-like protein in the Ca^{++} stimulation of myelin phosphorylation (assayed without Triton X-100) is suggested from many observations described earlier, the stimulatory effect of Triton X-100 on the Ca^{++} sensitive myelin phosphorylation, when the detergent is added to the assay, could be independent of its effect on Ca^{++}-calmodulin requiring myelin kinase. The detergent is known to interact with (and in fact substitutes for) the membrane phospholipids and thus it is likely that myelin contains a Ca^{++} stimulated, phospholipid requiring kinase, an enzyme which has recently been reported by Nishizuka and associates (26), and others (6). As mentioned above, chlorpromazine and trifluoperazine inhibited myelin Ca^{++} kinase only at high concentrations and at these concentrations, these drugs can interact with the membrane phospholipids.

Presence of phosphatases. As indicated earlier, longer incubation of myelin (particularly with Ca^{++} and Triton X-100 present in assay) decreased the amount of phosphorylation. In the presence of NaF (20 mM), such a decline was essentially prevented, which

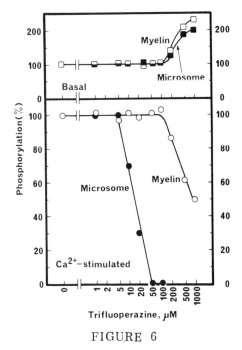

FIGURE 6

Effect of trifluoperazine on basal (upper panel) and Ca^{++} stimulated (lower panel) phosphorylation of myelin and cerebral cortex gray matter microsomes. Note that Triton X-100 was not present in assays. 70 cpm represents one pmol of phosphate incorporated.

indicated the presence of phosphatases (18,19). However, NaF also decreased the rate and extent of phosphorylation (± Ca^{++}), especially when the membrane fraction was stored prior to use (Fig. 7).

Likely mechanisms for stimulatory effect of Triton X-100. As described earlier, Triton X-100 increased the basal and Ca^{++} stimulated phosphorylation of myelin basic proteins; in fact, cAMP stimulated phosphorylation was observed only in the presence of Triton X-100 (19,23,24). Various possibilities could account for the marked stimulating effect of Triton X-100: 1) The detergent increased the accessibility of the substrate ($MgATP^{2-}$), divalent cations, cAMP, etc, to the respective sites on the membrane associated kinases. In other words, the kinases have latency. Similarly, the detergent could also have increased the interaction between the kinase(s) and the substrate proteins (myelin basic proteins). 2) The detergent inhibits phosphatases present in isolated myelin preparations. 3) The detergent interacts with the lipids, especially in the case of myelin, which is a highly lipid-rich

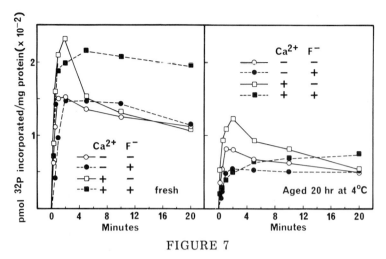

FIGURE 7

Effect of NaF. Myelin enriched fraction (18) was phosphorylated (+ Triton X-100) for the indicated times at 30°C in the absence and presence of Ca^{++} and/or NaF (10 mM).

membrane structure, and in fact could very well displace or substitute for phospholipids of myelin. In the latter case, the detergent could conceivably generate even a greater hydrophobic environment surrounding the enzyme(s) and/or substrate proteins. If this were the case, the properties of the kinases, deduced by determining the phosphorylation of myelin basic proteins, may not be exactly similar to those observed in the absence of the detergent.

While the problems of the accessibility of the substrate and cofactors and thus latency of the kinase(s) require consideration, we obtained some evidence that indicates that these may not explain the marked stimulatory effect of Triton X-100. For example, ionophores such as A23187 and alamethicin, which abolish latency of many membrane enzymes, increased Ca^{++} stimulated phosphorylation only modestly. Also, in preliminary experiments, myelin phosphatases were relatively uninfluenced by the detergent. On the other hand, a number of subtle but interesting differences in the properties of myelin kinase(s) were obtained when the results of the assays carried out with and without detergent were carefully examined. These differences included: 1) the Mg^{++} dependence of the kinases, 2) the ease and degree of restoration of Ca^{++} stimulated phosphorylation following EGTA or EDTA exposure of myelin, and 3) responses to a number of calmodulin inhibitors such as calmodulin binding proteins, phenothiazines, etc (16). Thus it is likely that the detergent could have interacted with myelin lipids and this secondarily resulted in the increased phosphorylation of basic proteins, especially that stimulated by Ca^{++}.

Isolation, polypeptide composition and phosphorylation of myelin subfractions. Myelin isolated from rat brain contains a few specific major polypeptides. Further, the contents of these polypeptides in myelin subfractions has been shown to differ markedly (27). For example, Fig. 8 shows the polypeptide composition of five myelin subfractions prepared by centrifugation of myelin on discontinuous sucrose density gradient. Briefly, the lighter fractions were enriched with respect to proteolipid protein (PLP) and small and large basic proteins (SBP and LBP), whereas Wolfgram polypeptides (WP) and myelin associated glycoprotein (MGP) were enriched in the heavier fractions. Ca^{++} stimulated phosphorylation of basic proteins was enriched in lighter fractions (23). Similar patterns for the distribution of various polypeptides as well as Ca^{++} dependent phosphorylation of basic proteins emerge for myelin subfractions prepared by centrifugation on a continuous sucrose density gradient. Phosphorylation of myelin subfractions obtained by continuous sucrose density gradient centrifugation is shown in Fig. 9. This figure also shows the effect of cAMP on phosphorylation. It is interesting to note that cAMP, which very modestly increased the phosphorylation of basic proteins (LBP and SBP), irrespective of the myelin subfraction assay, did promote phosphorylation of a polypeptide of M_r around 52,000 (see fractions C, D and E). ^{32}P incorporation into this band (R) likely represented the autophosphorylation of the regulatory subunit of the cAMP dependent kinase.

In other studies we decided to remove a microsomal contaminant, termed SN_4, from myelin (27) and test whether this influenced the pattern of phosphorylation, especially in the presence and absence of Ca^{++} and cAMP. In addition, we also removed neuronal membrane contaminants of myelin which were termed fraction SN_1 (27). Note that the polypeptide compositions of purified myelin, SN_4 and SN_1 differ considerably (23). For example, PLP and SBP and LBP represent more than 80% of the total myelin protein. In the case of SN_4 about 50% of the total protein was due to these proteins and, in addition, it contained two higher molecular weight polypeptides, name WP and MGP, which together comprised nearly 20% of the total SN protein. SN_1 fraction, on the other hand, contained less than 10% of its total protein due to PLP and SBP and LBP. Instead it contained considerably higher amounts of polypeptides with M_r greater than 45,000. From the phosphorylation of SN_1, SN_4 and myelin we noted an enrichment of the regulatory subunit of cAMP kinase in SN_4 (23). We also observed that cAMP stimulation of phosphorylation of basic protein was only evident in SN_4 and SN_1, but not in highly purified myelin. Ca^{++} stimulated phosphorylation of basic protein was higher in myelin compared to SN_4 or SN_1 (16). Thus it is evident that cAMP dependent protein kinase is primarily present in SN_4, whereas highly purified myelin is almost devoid of this enzyme. The reverse may be true for the Ca^{++} stimulated kinase which is present primarily in myelin and the presence of this enzyme in SN_4 could likely be due to a small amount of

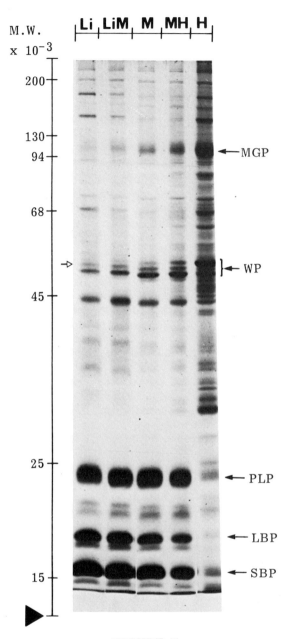

FIGURE 8

Polypeptide profiles of myelin subfractions prepared by discontinuous sucrose density gradient centrifugation. Myelin fractions were prepared as described by Sulakhe et al. (24). MGP = myelin associated glycoprotein; WP = Wolfgram peptides; PLP = proteolipid protein; LBP = large basic protein; SBP = small basic protein.

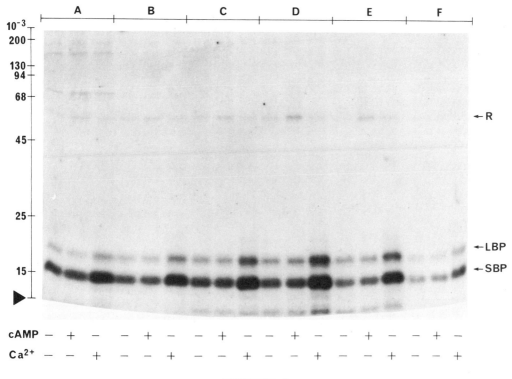

FIGURE 9

Effects of Ca^{++} and cyclic AMP on phosphorylation of myelin subfractions. Fractions were isolated by continuous sucrose density gradient centrifugation (24). Phosphorylation assay mixture contained Triton X-100 (0.05%). The figure shows autoradiogram following electrophoretic separation of phosphorylated peptides. R = regulatory subunit of cAMP dependent protein kinase.

contaminating myelin. A much improved resolution between myelin and SN_4 will be required to fully substantiate such a view.

Comparison between myelin (white matter) and microsomes (gray matter). Myelin and microsomes possessed endogenous kinases including Ca^{++} stimulated kinase. In the former, phosphorylation of basic proteins (M_r, 18,000 and 16,000) and in the latter, of polypeptides with M_r 65,000, 60,000 and 55,000, was markedly increased in the presence of Ca^{++}. Further, for either membrane we obtained evidence supporting the role of calmodulin in the Ca^{++} stimulated phosphorylation of the respective polypeptides provided that phosphorylation was carried out in the absence of Triton

TABLE 2: Ca^{++} stimulated phosphorylation of microsomal and myelin polypeptides: effects of Triton X-100.

	Triton X-100	Phosphorylation (arbitrary units)		
		Basal	$+ Ca^{++}$	Ca^{++}-stimulated
Microsome				
65K	−	15	55	40
	+	22 (147)	44	22 (55)
55K	−	10	34	24
	+	18 (180)	21	3 (12.5)
Myelin				
18K	−	9	14	5
	+	17 (188)	27	10 (200)
16K	−	18	28	10
	+	26 (144)	69	43 (430)

Following phosphorylation, solubilized membrane proteins were separated by slab gel electrophoresis and ^{32}P incorporation into individual polypeptide bands detected by autoradiography and estimated by scanning on X-ray film. Number in parentheses represents ratio (x 100) of phosphorylation with and without Triton X-100.

X-100 (16). Because of the differences in lipid compositions, as well as in the structure between myelin (multilamellar compact sheath) and microsomes (typical membrane vesicle), we decided to investigate the effects on Ca^{++} stimulated phosphorylation of calmodulin inhibitors, protease inhibitors as well as nonionic detergents. A number of interesting differences were noted: 1) Triton X-100 stimulated Ca^{++} kinase of myelin but inhibited microsomal kinase; basal kinase was increased by Triton X-100 in either membrane with the increment of the myelin enzyme being greater (Table 2). 2) Leupeptin, which had no effect on basal kinase in either membrane, inhibited Ca^{++} kinase of microsomes but not of myelin (16). 3) Trifluoperazine inhibited microsomal Ca^{++} kinase at lower concentration and only at higher concentration (> 100 µM) its inhibitory effect was noted on the myelin enzyme (Fig. 6). Basal kinases in either membrane were increased at higher concentrations (>100 µM) of chlorpromazine. 4) Heat-labile calmodulin binding protein (BP_{80}) readily inhibited Ca^{++} kinase of microsomes but not of myelin (16). On the other hand, heat-stable calmodulin binding protein (BP_{70}) inhibited Ca^{++} kinase in either membrane (see Fig. 5). Thus, it seems likely that membrane structure and (lipid) compositions influence the effects of calmodulin inhibitors.

Interestingly, chlorpromazine effect on microsomal Ca^{++} kinase resembled the effect on protein kinase C (13) and this raises an interesting possibility of the presence of kinase C in microsomal membrane. Leupeptin inhibition of microsomal Ca^{++} kinase indicated a likely role for a thiol requiring neutral protease as well. Calmodulin binding protein(s) results suggested a role for calmodulin in microsomal Ca^{++} kinase. Thus, it is becoming evident that in both myelin and microsomes two separate Ca^{++} kinases are present. Without Triton X-100 in assay, Ca^{++} stimulated phosphorylation of the respective peptides is catalyzed by a calmodulin requiring enzyme. On the other hand, with the detergent present, Ca^{++} stimulated phosphorylation is catalyzed by phospholipid requiring kinase that resembles kinase C (8) in many of its properties.

DISCUSSION

It is evident from our study of myelin kinases that this membrane contains at least three separate protein kinases, namely Ca^{++} dependent protein kinase, cAMP dependent protein kinase and Ca^{++} and cyclic nucleotides independent protein kinase (basal kinase). A number of pertinent findings dealing with myelin Ca^{++} kinase deserve comment at this stage. Although we have obtained some evidence that calmodulin-like protein is involved, our data to date do not exclude the possibility that myelin additionally contains calmodulin independent Ca^{++} stimulated protein kinase. For example, the meager inhibition of myelin kinase by phenothiazines, which is evident only at rather high concentrations of various phenothiazines tested, would raise the possibility that either myelin Ca^{++} kinase is similar to protein kinase C in its behavior (6,13), or the calmodulin-like protein is deeply buried within the membrane such that only at high concentration of phenothiazine it is accessible for interaction. Since at such high concentrations phenothiazines likely interact with membrane lipids (13), it is difficult then to decide whether the inhibition of myelin Ca^{++} kinase by trifluoperazine and other phenothiazines is due to binding of these agents to calmodulin-like proteins or binding to phospholipids. Further, we found that, especially when Mg^{++} concentrations are saturating and Triton X-100 was present, Ca^{++} did not increase the V_{max}, and also exogenous calmodulin now became ineffective. This observation again suggests two possibilities. Similar to the findings obtained with other calmodulin requiring enzymes (e.g. Ca^{++}-ATPase from erythrocytes; see 14) which can become calmodulin independent following altered hydrophobicity, we have created a similar situation in the presence of Triton X-100. If this were true, then one may still presume a role for calmodulin in myelin Ca^{++} kinase. On the other hand, it could be that Triton X-100 has displaced the phospholipids and this has rendered the enzyme calmodulin insensitive. This view will then support the possible presence of calmodulin independent protein kinase in myelin.

We have observed that at very low concentrations, Triton X-100 interferes markedly in the stimulatory effect of calmodulin on calmodulin dependent cAMP phosphodiesterase. In fact, we noted that at Triton X-100 concentration of 0.01%, there was >90% inhibition of Ca^{++} calmodulin dependent stimulation of the phosphodiesterase (16). The stimulatory effect of Triton X-100 on myelin phosphorylation becomes evident only when the detergent concentration is greater than 0.01% and in fact, as a routine, the phosphorylation assay for Triton stimulated activity has the detergent present at 0.05%. We also observed that phenothiazines such as trifluoperazine did not inhibit myelin Ca^{++} kinase in the presence of Triton X-100; the inhibitory effect of W-7 or the stimulatory effects of ionophores were also detected only in the absence of Triton X-100. In rat gray matter microsomes, Ca^{++}-calmodulin dependent phosphorylation was markedly decreased in the presence of Triton X-100. Thus, it is very likely that in the presence of Triton X-100, calmodulin-like proteins are either removed from myelin and/or inactivated by the detergent binding to these proteins. Further, the detergent could bind to these proteins such that drug (e.g. chlorpromazine) binding sites on these are blocked resulting in the lack of inhibitory effect of these drugs (20). Thus, the stimulatory effect of Triton X-100 cannot be readily explained as due to greater functional interaction between the myelin kinase and calmodulin-like proteins. On the other hand, the detergent could interact with myelin phospholipids to create an altered phospholipid environment either surrounding the kinase and myelin basic proteins (portions of which are hydrophobically anchored in myelin; see 7) or both. Myelin basic protein reportedly interacts preferentially with acidic lipids such as phosphatidylserine and phosphatidylinositol (2). It may be that Triton X-100 removed the myelin basic protein from the membrane by solubilization, a condition which, according to Gwarsha et al. (4), leads to increased lamellar separation and thus increased exposure of phospholipids. Thus, if phospholipid requiring Ca^{++} stimulated kinase were to be present in myelin, addition of Triton X-100 appears to favor the conditions for its detection and at the same time, the detergent is blocking the activity of calmodulin requiring myelin kinase. The latter of course then appears responsible for Ca^{++} sensitive phosphorylation of myelin assayed without Triton X-100. For this, we have gathered strong, albeit indirect, evidence that supports the role for calmodulin in myelin Ca^{++} kinase: 1) For example, EGTA extraction selectively decreases Ca^{++} dependent phosphorylation and this decrease can be almost fully restored by adding the EGTA extract or exogenous calmodulin. 2) Additionally, EGTA extract stimulated calmodulin deficient cAMP phosphodiesterase in a Ca^{++} dependent manner to the same extent as purified calmodulin. 3) EGTA extract contained a polypeptide band that migrated identically to purified calmodulin. 4) The heat-stable calmodulin binding protein inhibited myelin Ca^{++} kinase. 5) The affinity towards Ca^{++} of myelin kinase is in the same range reported for calmodulin. 6) Since Ca^{++} stimulates phosphorylation without Triton

X-100, and since this is increased only moderately in the presence of the Ca^{++} ionophore, A23187, or channel forming ionophore, alamethicin, it seems that Ca^{++} binding sites on myelin associated calmodulin are accessible in isolated myelin. Thus, it may be that when we determine Ca^{++} dependent phosphorylation of myelin it represents a composite activity of these two Ca^{++} sensitive kinases. This raises an interesting possibility that in the same membrane such as myelin two different Ca^{++} responsive protein kinases can be present. Moreover, for both kinases, the substrate proteins in myelin are identical - namely, basic proteins. Currently, we are examining whether or not the same serines and/or threonines of myelin basic proteins are phosphorylated by two Ca^{++} kinases in myelin.

There is little doubt from our findings that cAMP dependent protein kinase, if present in highly purified myelin, contributes very little to the total myelin kinase activity and our generous estimates will indicate this to be no more than 10% of the total myelin kinase activity (18,19). Further, we believe that cAMP kinase is primarily that of an "early" myelin (SN_4)(16,24). Interestingly though, it is the cAMP kinase that attracted considerable attention in the earlier years from Miyamoto, Appel and Carnegie. We have further observed that cGMP dependent protein kinase is either absent in myelin or if it is present it must indeed be in exceedingly low amounts. While it is difficult to exclude the possibility that cAMP kinase under *in vivo* situation phosphorylates myelin basic proteins, our *in vitro* data so far do not support such a possibility. It has not as yet been possible for us to comment regarding what basal kinase of myelin is. There are some findings that may suggest a basal kinase being an enzyme completely separate from Ca^{++} kinase (23, 24) and yet its distribution and some of its properties might indicate it to be a part of myelin Ca^{++} kinase complex (18). It is obvious that purification of various myelin kinases should help to resolve this and many other important questions about myelin protein kinases.

ACKNOWLEDGEMENTS

This study was supported by a grant from the Medical Research Council of Canada. EHP is a Postdoctoral Fellow of the Muscular Dystrophy Association of Canada. We acknowledge the able assistance of Ms. B. Thiessen, Mr. P. Dessens, Ms. E. Davis and Mr. D. Harley in some of the experiments described here. We are grateful to Professor H. Hidaka for the gift of W-7.

REFERENCES

1. Carnegie, P.R., Dunkley, P.R., Kemp, B.E. & Murray, A.W. (1973): Nature (Lond.) 249:147-150.
2. Demel, R.A., London, Y., Geurts van Kessel, W.S.M., Vossenberg, F.G.A. & van Deenen, L.L.M. (1973): Biochim. Biophys. Acta 311:507-519.
3. Endo, T. & Hidaka, H. (1980): Biochim. Biophys. Acta 97: 553-558.
4. Gwarsha, K., Rumsby, M.G. & Little, C. (1980): Biochem. Soc. Trans. 8:600-601.
5. Hidaka, H., Yamaki, T., Totsuka, T. & Asano, M. (1979): Molec. Pharmacol. 15:49-59.
6. Kuo, J.F., Andersson, R.G.G., Wise, B.C., Mackerlova, L., Salomonsson, I., Brackett, N.L., Katoh, N., Shoji, M. & Wrenn, R.W. (1980): Proc. Nat. Acad. Sci. 77:7039-7043.
7. London, Y., Demel, R.A., Geurts van Kessel, W.S.M., Vossenberg, F.G.A. & van Deenen, L.L.M. (1973): Biochim. Biophys. Acta 311:520-530.
8. Minakuchi, R., Takai, Y., Yu, B. & Nishizuka, Y. (1981): J. Biochem. 89:1651-1654.
9. Miyamoto, E. (1975): J. Neurochem. 24:503-512.
10. Miyamoto, E. (1976): J. Neurochem. 26:573-577.
11. Miyamoto, E. & Kakiuchi, S. (1974): J. Biol. Chem. 249:2769-2777.
12. Miyamoto, E., Miyazaki, K. & Hirose, R. (1978): J. Neurochem. 31:269-275.
13. Mori, T., Takai, Y., Minakuchi, R., Yu, B. & Nishizuka, Y. (1980): J. Biol. Chem. 255:8378-8380.
14. Penniston, J.T., Graf, E., Niggli, V., Verma, A.J. & Carafoli, E. (1980): In Calcium Binding Proteins: Structure and Function, Vol. 14. (eds) F.L. Siegel, E. Carafoli, R.H. Kretsinger, D.H. MacLennan & R.H. Wasserman, Elsevier/North-Holland, Amsterdam, pp. 23-30.
15. Petrali, E.H. & Sulakhe, P.V. (1979): Canad. J. Physiol. Pharmacol. 57:1200-1204.
16. Petrali, E.H. & Sulakhe, P.V. (1981): Prog. Brain Res. (in press)
17. Petrali, E.H., Thiessen, B.J. & Sulakhe, P.V. (1978): Proc. 23rd Ann. Psych. Res. Meet., pp. 33-34.
18. Petrali, E.H., Thiessen, B.J. & Sulakhe, P.V. (1979): Int. J. Biochem. 11:21-36.
19. Petrali, E.H., Thiessen, B.J. & Sulakhe, P.V. (1980): Arch. Biochem. Biophys. 205:520-535.
20. Sharma, R.K. (1981): Fed. Proc. 40:1739.
21. Steck, A.J. & Appel, S.H. (1974): J. Biol. Chem. 249:5416-5420.
22. Sulakhe, P.V., Petrali, E.H. & Thiessen, B.J. (1978): Proc. Soc. Neurosci. 4:249.
23. Sulakhe, P.V., Petrali, E.H., Thiessen, B.J. & Davis, E.R. (1980): Biochem. J. 186:469-473.

24. Sulakhe, P.V., Petrali, E.H., Davis, E.R. & Thiessen, B.J. (1980): Biochemistry 19:5363-5371.
25. Sulakhe, P.V., Petrali, E.H. & Harley, D.L. (1980): In Calcium Binding Proteins: Structure and Function, Vol. 14. (eds) F.L. Siegel, E. Carafoli, R.H. Kretsinger, D.H. MacLennan & R.H. Wasserman, Elsevier/North-Holland, Amsterdam, pp. 239-240.
26. Takai, Y., Kishimoto, A., Iwasa, Y., Kawahara, Y., Mori, T. & Nishizuka, Y. (1979): J. Biochem. 86:575-578.
27. Waehneldt, T.V.(1978): Brain Res. Bull. 3:37-44.
28. Waehneldt, T.V., Matthieu, J.-M. & Neuhoff, V. (1977): Brain Res. 138:29-43.
29. Wang, J.H., Sharma, R.K. & Tam, S.W. (1980): In Calcium and Cell Function, Vol. 1. (ed) W.Y. Cheung, Academic Press, New York, pp. 305-328.
30. Wuthrich, C. & Steck, A. (1981): Biochim. Biophys. Acta 640:195-206.

INTERACTION OF CALMODULIN WITH cAMP DEPENDENT PROTEIN KINASE IN BRAIN

C.B. Klee, M.H. Krinks, D.R. Hathaway* and
D.A. Flockhart**

Laboratory of Biochemistry, National Cancer Institute
and *Cardiology Branch, National Heart, Lung and
Blood Institute, National Institutes of Health,
Bethesda, Maryland 20205 USA, and
**Howard Hughes Medical Institute, Department of
Physiology, Vanderbilt University Medical School,
Nashville, Tennessee 37232, USA

INTRODUCTION

In most eukaryotic cells, cellular responses to external stimuli are mediated by each of the two second messengers Ca^{++} and cAMP (for review see 3). The actions of each of the two intracellular signals are modulated in a similar fashion by an intracellular receptor protein, the regulatory subunit of cAMP dependent protein kinase in the case of cAMP (22) and calmodulin, or in some other tissues more specific intracellular Ca^{++} binding proteins, in the case of Ca^{++} (23). Although these actions are usually mediated by protein kinases, Ca^{++} can also act directly on its target proteins by means of interaction of the protein with the Ca^{++}-calmodulin complex (5, 21, 25, 32, 33).

Because Ca^{++} and cAMP occur within the same cell, the importance of interactions between the two second messengers during cell activation, originally pointed out by Rasmussen (27), is worthy of investigation. Calmodulin and Ca^{++} can lower cellular levels of cAMP, by stimulation of the ubiquitous Ca^{++} dependent cyclic nucleotide phosphodiesterase discovered by Kakiuchi and Yamazaki (19). In some specific tissues Ca^{++}-calmodulin can raise cAMP levels by stimulation of adenylate cyclase (4,6). Conversely cAMP has been reported to stimulate Ca^{++} efflux from internal stores in some tissues and to stimulate Ca^{++} uptake in other tissues (3).

These actions of the two second messengers may also be modulated by their concerted action on the same substrate (13,18,24,30, 34). It has also been observed that cAMP dependent phosphorylation of an enzyme can modify its response to the Ca^{++} signal. In skeletal muscle, cAMP dependent phosphorylation of phosphorylase kinase lowers the Ca^{++} concentration needed for the activation of this enzyme (7) and increases the V_{max} of the Ca^{++} dependent activity (7,16). On the other hand, phosphorylation of myosin kinase in smooth muscle decreases the ability of the enzyme to respond to Ca^{++} and calmodulin (1,9).

In brain, we have now obtained evidence that the two receptor proteins of Ca^{++} and cAMP interact with each other (15) and that this interaction modifies the response of the cAMP dependent protein kinase to its own signal, cAMP. This interaction is the subject of the present communication.

Interaction of brain cAMP dependent protein kinase with calmodulin. In brain extracts a fraction (10-40%) of the cAMP dependent protein kinase can be adsorbed on calmodulin sepharose in the presence of Ca^{++} and is specifically released by addition of the chelating agent EGTA to the eluting buffer (15). As shown in Table 1, cAMP phosphodiesterase is also associated with cAMP dependent protein kinase during the early purification steps but can be partially resolved by sepharose 6B gel filtration and subsequent DEAE-sephacel chromatography. Upon affinity chromatography on calmodulin-sepharose (Fig. 1A) under non-saturating conditions, as previously described (21), phosphodiesterase is completely adsorbed on the column, whereas only 50% of the kinase is bound to the column. Both activities are eluted with an EGTA containing buffer. On the other hand, most of the protein phosphatase activity is not significantly retained by the calmodulin-sepharose column (Table 1).

As shown in Fig. 1B, a large number of polypeptides are detected in the EGTA eluate of the non-saturated calmodulin-sepharose column. Proteins with a relatively low affinity for the calmodulin-Ca^{++} complex are eluted first. Two high M_r polypeptides (235,000 and 230,000) which have been described in postsynaptic densities (8) and which may be involved in the regulation of glutamate receptors (2) are eluted in the early fractions. These are membrane associated actin binding proteins with a low affinity for calmodulin (12). Proteins with high affinity for the calmodulin Ca^{++} complex such as cAMP phosphodiesterase and calcineurin are eluted later. Like calcineurin A (M_r 61,000) and calcineurin B (M_r 15,000), the two polypeptides with M_r of 55,000 and 40,000 corresponding to the R and C subunits of protein kinase are eluted late and appear to represent major components of the class of calmodulin binding proteins in brain.

TABLE 1: Purification of the calmodulin interacting component of brain cAMP dependent protein kinase.

	Protein Kinase		Phosphodiesterase*	Protein Phosphatase	
Crude extract	12	(3)	796	2.8	$\times 10^{-3}$
Ammonium sulfate (30-60%)	5.3	(3)	460	0.5	$\times 10^{-3}$
Sepharose 6B	11.2	(10)	486	0.6	$\times 10^{-3}$
DEAE-sephacel	11.6	(10)	145	0.6	$\times 10^{-3}$
Calmodulin-sepharose bound	4.0	(12)	140	0.01	$\times 10^{-3}$
Not bound	8.0		2	0.3	$\times 10^{-3}$

Purification procedures as described in Ref. 15, using 600 g bovine brain. The numbers in parentheses indicate the fold stimulation by cAMP. Phosphodiesterase and protein phosphatase activities contaminating the protein kinase fractions indicated above.
* Total units (μmol/min).

The major histone kinase that interacts with the calmodulin-sepharose eluted prior to cAMP phosphodiesterase (M_r 120,000) (29), calcineurin (M_r 80,000)(21), and the residual protein phosphatase upon gel filtration on ultragel ACA 22 in agreement with the known larger M_r (180,000) of cAMP dependent protein kinase (26,28). It activity was stimulated ten- to twelve-fold by addition of 20 μM cAMP and the specific activity of the activated enzyme was 1-1.5 units/mg of protein. Analysis of its subunit composition by sodium dodecyl sulfate (SDS) gel electrophoresis revealed the presence of two major polypeptides with M_r of 55,000 and 40,000 similar to those of the Type II cAMP dependent protein kinase, the predominant form of cAMP dependent kinase in brain extracts (10) along with the two subunits of calcineurin (15). The 55,000 M_r subunit was identified as the regulatory subunit of protein kinase on the basis of its labelling with the affinity reagent, 8-azido-cAMP (31; and J. Haiech, unpublished observation), and its specific autophosphorylation in the presence of $^{32}P\gamma ATP$ (14,17). As described for the Type II cyclic AMP dependent protein kinase of brain (28), the electrophoretic mobility of the phosphorylated polypeptide was similar to that of the non-phosphorylated protein. In contrast, as was reported previously (17), we observed a decreased electrophoretic mobility of heart Type II regulatory subunit after reconstitution with its catalytic subunit and autophosphorylation performed under similar conditions.

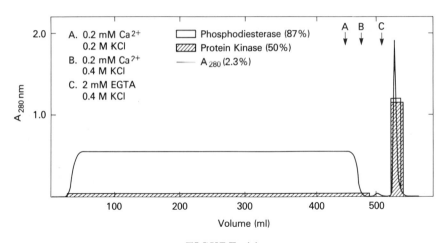

FIGURE 1A

Chromatography of bovine brain proteins on calmodulin-sepharose. The DEAE-sephacel fraction (Table 1) was purified on a 1.5 x 10 cm column of calmodulin-sepharose as previously described (21). The arrows indicate the washes with Ca^{++} containing loading buffer (A) and 0.4 M NaCl buffer (B) and EGTA containing buffer (C).

The regulatory subunit of cAMP dependent protein kinase interacts with calmodulin. The protein kinase activity bound to calmodulin sepharose in the presence of Ca^{++} can be dissociated from the calmodulin sepharose by addition of cAMP in the buffer solutions. The dissociated enzyme no longer requires cAMP for activity and is composed of a single polypeptide of 40,000 M_r, corresponding to the catalytic component of the enzyme. The regulatory subunit, together with calcineurin, remains bound to calmodulin under these eluting conditions and can only be dissociated by removal of Ca^{++} (15). A weak interaction of calmodulin with the catalytic subunit has also been detected by cross-linking experiments but it is not strongly dependent on Ca^{++}. A direct interaction of the regulatory subunit with calmodulin has not yet been demonstrated since the calcineurin present in the preparation could participate in the formation of the complex. Brain cAMP dependent protein kinase sediments in gradients of glycerol with the anomalously high sedimentation coefficient of 8S, compatible with that of a calcineurin-protein kinase complex. Furthermore, despite the low sedimentation coefficient of calcineurin (4.5S), a fraction of calcineurin is consistently found to be associated with the protein kinase peak (Fig. 2). When similar experiments were performed in the presence of excess calmodulin (10^{-8}-10^{-7} M) the sedimentation coefficient of the kinase was increased to 9S (Fig. 3). A similar behavior was observed for the enzyme after autophosphorylation monitored by activity or ^{32}P

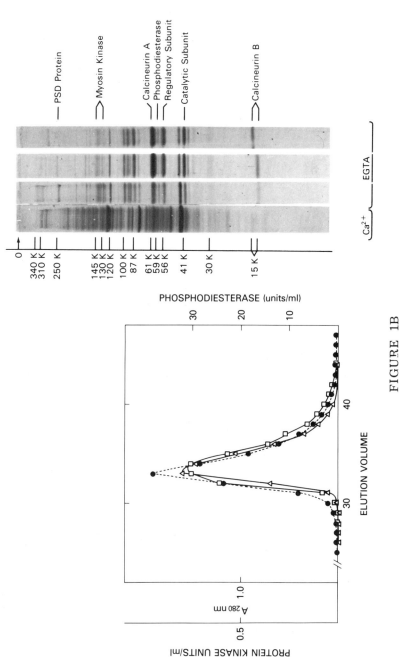

FIGURE 1B

Elution profile of the proteins eluted with buffer C. Protein kinase (□); cAMP phosphodiesterase (△); A280 nm (●——●). The SDS gel electrophoretic pattern of the proteins not adsorbed on the column (Ca^{++}) and in tubes 31, 34 and 38 (EGTA) is shown on the left of the figure. The estimated M_r of the polypeptides is indicated on the right of the gel pattern.

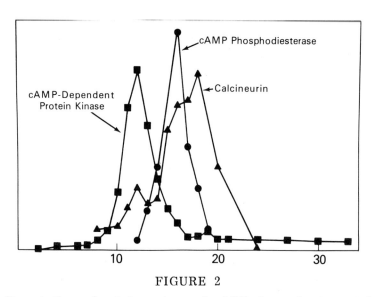

FIGURE 2

Co-sedimentation of calcineurin and cAMP dependent protein kinase in a gradient of glycerol. The centrifugation was as described in Ref. 15. Calcineurin levels were determined by densitometirc analysis of the SDS gell patterns of the fractions.

TABLE 2: Effect of calmodulin on cAMP dependent protein kinase from bovine brain.

	Stimulation	Activity*	
	-Fold	+cAMP	-cAMP
No addition	3.0	37.5 (3.7)	12.3 (1.1)
Calmodulin (2×10^{-6} M)	5.1	45.8 (2.3)	8.8 (1.3)

* Activity = nmol Pi incorporated/min/ml enz
The standard deviation from the means is indicated in parentheses.

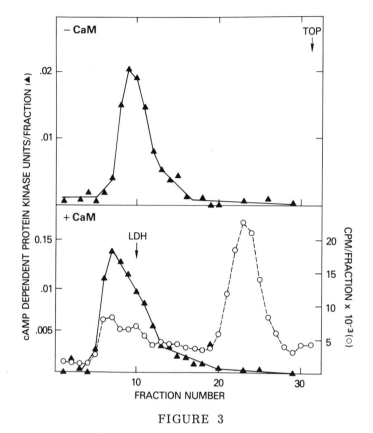

FIGURE 3

Glycerol gradient centrifugation of cAMP dependent protein kinase in the presence and absence of calmodulin (From Ref. 15).

label strictly associated with the R subunit. Cross-linking experiments also confirmed the identical behavior of the phosphorylated and non-phosphorylated enzymes. Addition of calmodulin and Ca^{++}, but not calmodulin and EGTA, markedly increased the complexity of the gel patterns observed after cross-linking of both the phosphorylated and non-phosphorylated regulatory subunits with the other components of the system.

The interaction of cAMP dependent protein kinase with calmodulin is specific for the brain enzyme. Attempts to detect a complex formation between calmodulin and reconstituted heart Type II cyclic AMP dependent protein kinase by glycerol gradient centrifugation in the presence of calmodulin, or after cross-linking experiments with dimethyl suberimidate (11) were unsuccessful. Addition of exogenous calcineurin at concentrations similar to those observed in the brain preparation did not promote interaction, suggesting that the regulatory subunit of the brain enzyme which is immunologically different (28) and behaves differently after phosphorylation (28), is also different in its ability to interact with calmodulin.

Effect of calmodulin and Ca^{++} on the enzymatic activity of cAMP dependent protein kinase. Calmodulin and Ca^{++} have a small, probably insignificant effect on the activity of the protein kinase measured in the presence of cAMP. In the absence of cyclic nucleotide, the basal activity is reproducibly inhibited by calmodulin in the presence of Ca^{++} (Table 2), but not in the presence of EGTA (data not shown). These experiments suggest that the interaction of the R and C subunits is stronger in the presence of the Ca^{++} binding protein. This conclusion is strengthened by the increased stability of the RC complex as detected by cross-linking experiments and by the lower affinity of the enzyme for cAMP in the presence of calmodulin and Ca^{++}.

DISCUSSION

The molecular mechanism of the interaction of bovine brain cAMP dependent kinase with the calmodulin-Ca^{++} complex is not yet fully understood. The large S value of the brain enzyme (15) as opposed to that of the heart enzyme or of the fraction of the brain enzyme which does not interact with calmodulin, suggests that calcineurin, which co-purifies with the protein kinase, may participate in this interaction. The lack of effect of calcineurin on the ability of the heart Type II cAMP dependent protein kinase to interact with calmodulin cannot be taken as evidence that calcineurin is not needed for interaction since the kinases isolated from the two tissues differ chemically and immunologically (28). The precise nature of these differences is not known, but since the regulatory subunit of protein kinase contains at least two phosphorylation sites, it is possible that they differ in the extent of phosphorylation. Although phosphorylation of the brain kinase does not seem to affect its interaction with calmodulin, definitive conclusions cannot be drawn without prior determination of the number and nature of the phosphorylated sites. A dependence of the interaction on some reversible chemical modification would explain the variability in the amount of protein kinase which interacts with calmodulin in different brain extracts (15). It is equally possible that, because of the heterogeneous nature of brain tissue, the heterogeneity of protein kinase is a consequence of its multicellular origin.

Although the effects of cAMP and Ca^{++} in nervous tissue are usually synergistic, the apparent negative control exerted by Ca^{++} and calmodulin on the regulation of protein kinase by cAMP may in fact insure a tighter coupling of the kinase and cAMP levels. As indicated by our in vitro activity measurements, the holoenzyme in the absence of Ca^{++} and calmodulin is already significantly activated and can be further activated only two- to three-fold by cAMP. These results may also lead to reservations about the use of the activity ratios as a means of determining cAMP levels. Our data indicate that the activity of cAMP dependent protein kinase may be regulated by mechanisms other than changes in cAMP levels and that this regulation may be different in different tissues. Clearly, the inter-relationships between Ca^{++} and cAMP provide a network of coupled reactions that controls much of cellular function.

REFERENCES

1. Adelstein, R.S., Conti, M.A., Hathaway, D.R. & Klee, C.B. (1978): J. Biol. Chem. 253:8347-8350.
2. Bandry, M., Bundman, M.C., Smith, E.K. & Lynch, G.S. (1981): Science 212:937-938.
3. Berridge, M.J. (1975): Adv. Cyclic Nucleotide Res. 6:1-98.
4. Brostrom, C.O., Huang, Y.C., Breckenridge, B.McL. & Wolff, D.J. (1975): Proc. Nat. Acad. Sci. 72:64-68.
5. Cheung, W.Y. (1980): Science 207:19-27.
6. Cheung, W.Y., Bradham, L.S., Lynch, T.J., Lin, Y.M. & Tallant, E.A. (1975): Biochem. Biophys. Res. Commun. 66:1055-1062.
7. Cohen, P. (1980): Eur. J. Biochem. 111:563-574.
8. Cohen, R.S., Blomberg, F., Berzins, K. & Siekevitz, P. (1977): J. Cell. Biol. 74:181-203.
9. Conti, M.A. & Adelstein, R.S. (1981): J. Biol. Chem. 256:3178-3181.
10. Corbin, J.D., Keely, S.L. & Park, C.R. (1975): J. Biol. Chem. 250:218-225.
11. Davies, G.E. & Stark, G.R. (1970): Proc. Nat. Acad. Sci. 66:651-656.
12. Davies, P.J.A. & Klee, C.B. (1981): Biochem. Int. 3:203-212.
13. DePaoli-Roach, A.A., Roach, P.J. & Larner, J. (1979): J. Biol. Chem. 254:12062-12068.
14. Erlichman, J., Rosenfeld, R. & Rosen, O.M. (1974): J. Biol. Chem. 249:5000-5003.
15. Hathaway, D.R., Adelstein, R.S. & Klee, C.B.(1981): J. Biol. Chem. 256:8183-8189.
16. Heilmeyer, L.M.G. Jr., Groschel, S.U., Jahnke, U., Kilimanan, M.W., Kohse, K.P. & Varsanyi, M. (1980): Adv. Enz. Reg. 18:121-144.
17. Hofman, F., Beavo, J.A., Bechtel, P.J. & Krebs, E.G. (1975): J. Biol. Chem. 250:7795-7801.
18. Huttner, W.B. & Greengard, P. (1979): Proc. Nat. Acad. Sci. 76:5402-5406.

19. Kakiuchi, S. & Yamazaki, R. (1970): Biochem. Biophys. Res. Commun. 41:1104-1110.
20. Klee, C.B., Crouch, T.H. & Richman, P.G. (1980): Ann. Rev. Biochem. 49:489-515.
21. Klee, C.B. & Krinks, M.H. (1978): Biochemistry 17:120-126.
22. Krebs, E.G. (1972): Curr. Top. Cell. Reg. 5:99-133.
23. Kretsinger, R.H. (1979): Adv. Cyclic Nucleotide Res. 11:1-26.
24. LePeuch, C.J., Haiech, J. & Demaille, J. (1979): Biochemistry 18:5150-5157.
25. Means, A.R. & Dedman, J.R. (1980): Nature 285:73-77.
26. Nimmo, H.G. & Cohen, P. (1977): Adv. Cyclic Nucleotide Res. 8:145-266.
27. Rasmussen, H. (1970): Science 170:404-412.
28. Rubin, C.S., Rangel-Aldeo, R., Sarkar, D., Erlichman, J. & Fleischer, N. (1979): J. Biol. Chem. 254:3797-3805.
29. Sharma, R.K., Wang, T.H., Wirch, E. & Wang, J.H. (1980): J. Biol. Chem. 255:5916-5923.
30. Srivastava, A.K., Waisman, D.M., Brostrom, C.O. & Soderling, T.R. (1979): J. Biol. Chem. 254:583-586.
31. Walter, V., Uno, I., Liu, A.Y.C. & Greengard, P. (1977): J. Biol. Chem. 252:6588-6590.
32. Wang, J.H. & Waisman, D.M. (1979): Curr. Top. Cell. Reg. 15:47-107.
33. Wolff, D.J. & Brostrom, C.O. (1979): Adv. Cyclic Nucleotide Res. 11:28-88.
34. Yamazaki, T. & Fujisawa, H. (1980): Biochem. Int. 1:98-104.

REGULATION OF CONTRACTILE PROTEINS IN SMOOTH MUSCLE AND PLATELETS BY CALMODULIN AND CYCLIC AMP

R.S. Adelstein, P. de Lanerolle, J.R. Sellers, M.D. Pato and M.A. Conti

Laboratory of Molecular Cardiology, National Heart, Lung and Blood Institute, National Institutes of Health, Bethesda, Maryland 20205, USA

INTRODUCTION

The two major contractile proteins actin and myosin are now known to exist throughout nature (for reviews see 1,25). Of particular interest is the regulation of the interaction between actin and myosin in vertebrate smooth muscle and non-muscle cells. The form of regulation to be discussed in this paper involves the reversible phosphorylation of the myosin molecule (see Fig. 1) which appears to play a major role in regulating contractile activity in smooth muscle and non-muscle cells such as platelets and macrophages.

There are at least three major regulatory systems that govern the interaction of actin with myosin for which a detailed mechanism has been worked out. Each of these systems requires calcium to initiate the contractile process, but the site of calcium binding differs in all three (Fig. 2).

In the case of skeletal and cardiac muscle, Ca^{++} binds to troponin C and relieves the inhibition imposed on the actin-activated MgATPase activity by the troponin-tropomyosin complex (15). Recently Chalovich and Eisenberg (7) examined the kinetics of this inhibition in vitro and found that it is not the binding of actin to myosin that is interfered with, but a specific kinetic step, possibly involving the release of phosphate. In this scheme troponin-tropomyosin is thought to prevent the rotation of the globular portion of the myosin molecule from the 90°C (ATP) state to the 45°C (ADP) state (7). It is of interest that the protein,

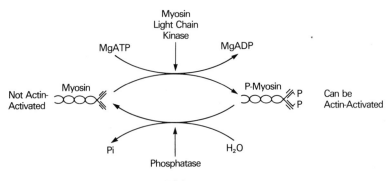

FIGURE 1

Stick diagram of the myosin molecule illustrating the reversible phosphorylation of the regulatory light chains of myosin and the effect on the actin-activated MgATPase activity of myosin isolated from smooth muscle and non-muscle cells.

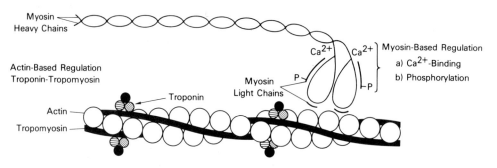

FIGURE 2

Diagrammatic representation of the three major regulatory systems (see text). Each of these regulatory systems plays a dominant role in a different type of muscle (from Ref. 1).

troponin C, to which Ca^{++} binds, shares a number of structural similarities to the calcium binding protein, calmodulin (22,24). As outlined below, this latter protein plays an important role in regulating contractile proteins in smooth muscle and non-muscle cells.

A second regulatory mechanism has been described by Szent-Gyorgyi and his colleagues for molluscan muscles (8,9). In this system Ca^{++} binds directly to myosin, probably at a site shared by the myosin heavy chains and a pair of regulatory light chains (Fig. 2). The binding of Ca^{++} acts to relieve an inhibition conferred by the regulatory light chains on the actin activated Mg-ATPase activity of molluscan myosin. This form of regulation differs from that described for skeletal and cardiac muscle in that it does not require a separate set of proteins, such as troponin, for the regulation of the actin activated MgATPase activity (Fig. 2).

A third mechanism for regulating contractile activity is the mechanism which requires phosphorylation of myosin, comprises the main subject of this paper. This mechanism appears to be a dominant mode of regulation in smooth muscle and non-muscle cells. (However, see below for a discussion on the leiotonin system, which is outlined in detail by Ebashi and co-workers elsewhere in This Volume). The phosphorylation system is similar to that described for the mollusc in that the interaction of actin and myosin appears to be inhibited by a pair of 20,000 dalton light chains on myosin (see Fig. 2). This inhibition is relieved by the covalent attachment of a phosphate group to a particular serine residue located on the 20,000 dalton myosin light chain (21). The phosphate donor is ATP and the transfer of phosphate to myosin is catalyzed by the substrate-specific enzyme, myosin light chain kinase. This enzyme is strictly dependent on Ca^{++}-calmodulin for activity and it is the binding of Ca^{++} to calmodulin that appears to initiate contractile activity in these cells. The reaction is reversible in that phosphatases catalyze the dephosphorylation of myosin and thereby restore it to a form that cannot undergo actin activation (Fig. 1).

Having described three different mechanisms for regulating the interaction of actin and myosin it is important to stress the following: 1) There are other important regulatory mechanisms which are presently under investigation, such as the leiotonin regulatory system that appears to be present in smooth muscle (15). 2) All muscle and non-muscle cells probably contain more than one mechanism for regulating their contractile activity. Some of these may only be active under special circumstances. For example, skeletal muscle contains the necessary enzymes and substrate (the 18,500 dalton light chain of myosin) for a reversible phosphorylating system involving myosin. However, myosin phosphorylation appears to only <u>modulate</u> skeletal muscle contraction and does not play the same <u>dominant</u> role in these cells that it does in smooth muscle and non-muscle cells. Phosphorylation of skeletal muscle myosin appears to result in potentiation of peak twitch tension (26) and/or a decrease in ATP utilization (11,12) in skeletal muscle. Phosphorylation is not required for in vitro actin activation of the MgATPase activity of skeletal or cardiac muscle myosin (4,27).

In this paper we shall present information of two different reversible phosphorylation reactions that regulate contractile activity in smooth muscle and non-muscle cells. The first of these concerns the reversible phosphorylation of myosin isolated from smooth muscle and platelets. We shall present evidence that this phosphorylation is capable of regulating the actin-activated MgATPase activity of myosin isolated from these cells. The second reaction involves the phosphorylation of the enzyme, myosin kinase, by cAMP dependent protein kinase. This phosphorylation may play a role in decreasing contractile activity in smooth muscle and non-muscle cells by reducing the activity of myosin kinase. As we shall see, the calcium binding protein, calmodulin, plays an important role in both phosphorylation reactions.

Phosphorylation regulates the actin activated MgATPase activity of smooth muscle and platelet myosin. In order to study the role of myosin light chain phosphorylation in regulating the actin activated MgATPase activity of myosin the following approach was undertaken: The individual protein components of the phosphorylating system were purified to apparent homogeneity. These included smooth muscle and platelet myosin, the chymotryptic fragment of smooth muscle myosin (heavy meromyosin or HMM)(33), smooth muscle myosin kinase (2), smooth muscle phosphatase I (29), calmodulin (23) and actin. A Coomassie blue-stained gel of these proteins following SDS-polyacrylamide gel electrophoresis is shown in Fig. 3.

The smooth muscle myosin fragment, HMM, was utilized in these experiments to see if the role of phosphorylation in regulating the MgATPase activity could be separated from its role in filament formation. Reports by Suzuki et al. (34) and Scholey et al (32) showed that phosphorylation of myosin isolated from smooth muscle and non-muscle cells plays an important part in stabilizing myosin filaments formed in vitro. This raised the possibility that myosin filament formation might be a prerequisite for actin activation of the MgATPase activity of myosin. In order to examine this possibility we used the soluble fragment, turkey gizzard smooth muscle HMM, which cannot form filaments. This fragment, which retains a phosphorylatable 20,000 dalton light chain, can undergo reversible phosphorylation (33).

The experiment was conducted as follows: Myosin (from turkey gizzard smooth muscle or human platelets) was isolated in the unphosphorylated state (see Fig. 4, gel U) and assayed for actin activation of its MgATPase activity. It was then phosphorylated

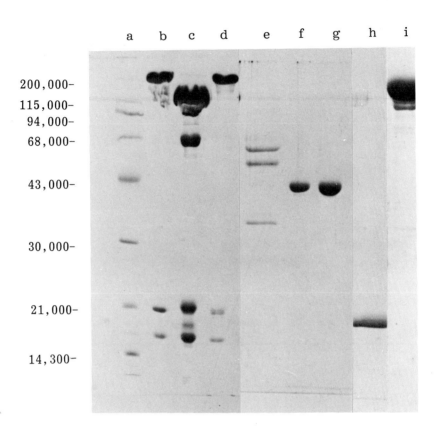

FIGURE 3

1% SDS-12.5% polyacrylamide slab gel electrophoresis of the purified proteins used in studying the effects of reversibly phosphorylating myosin and HMM: a) molecular weight standards (BioRad), 200,000, myosin heavy chains; 115,000, β-galactosidase; 94,000, phosphorylase b; 68,000, bovine serum albumin; 43,000 ovalbumin; 30,000 carbonic anhydrase; 21,000, soybean trypsin inhibitor; 14,300, lysozyme. b) turkey gizzard smooth muscle myosin. c) turkey gizzard heavy meromyosin. d) human platelet myosin. e) smooth muscle phosphatase I. f) rabbit skeletal muscle actin, g) turkey gizzard smooth muscle actin. h) porcine brain calmodulin. i) turkey gizzard myosin kinase (from Ref. 33).

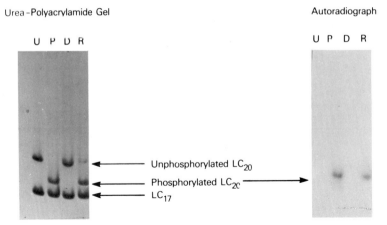

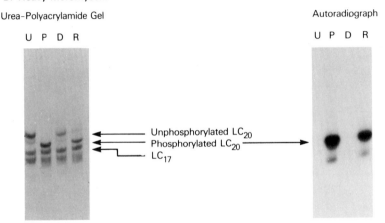

LC$_{20}$ — 20,000-dalton myosin light chain
LC$_{17}$ — 17,000-dalton myosin light chain
U — Unphosphorylated myosin
P — Phosphorylated myosin
D — Dephosphorylated myosin
R — Rephosphorylated myosin

FIGURE 4

Reversible phosphorylation of turkey gizzard smooth muscle myosin and heavy meromyosin. A. the left panel shows a urea-polyacrylamide gel of unphosphorylated (U), phosphorylated (P), dephosphorylated (D) and rephosphorylated (R) turkey gizzard myosin. In this gel system (29) the unphosphorylated LC$_{20}$ migrates more slowly than the phosphorylated LC$_{20}$. The right panel shows an autoradiograph of the gel on the left. B. The left panel shows a

FIGURE 4 legend (continued)
urea-polyacrylamide gel of unphosphorylated (U), phosphorylated (P), dephosphorylated (D), and rephosphorylated (R) turkey gizzard heavy meromyosin. Note the presence of extra bands due to partial proteolysis. The right panel shows an autoradiograph of the gel on the left. Note that the majority of the radioactivity is confined to the LC_{20} band. (LC_{20}, 20,000 dalton light chain. LC_{17}, 17,000 dalton light chain of myosin)(From Ref. 33).

---

by addition of Ca^{++} calmodulin, myosin kinase, and ATP, and reassayed (Fig. 4, gel P). Next the kinase was inhibited by addition of EGTA to chelate the Ca^{++} (under which conditions calmodulin can no longer bind to myosin kinase) and smooth muscle phosphatase I (in the dissociated form, see below) was added to dephosphorylate the myosin (gel D). The myosin was then rephosphorylated following inhibition of the phosphatase by addition of 2 mM ATP and the same myosin was rephosphorylated by adding a small amount of kinase in the presence of Ca^{++} (Gel R). The MgATPase activity was assayed under complete dephosphorylation and rephosphorylation of the myosin. Figure 4 shows how the state of myosin and HMM phosphorylation was monitored using urea-polyacrylamide gels. The extent of incorporation was also quantitated by using $[\gamma\text{-}32P]ATP$ of known specific activity (33). Table 1 shows the extent of myosin phosphorylation and the actin-activated MgATPase activity for each step. Wherease the MgATPase activity of the unphosphorylated and dephosphorylated forms of myosin could not be activated by actin to any significant extent, the phosphorylated and rephosphorylated forms of myosin could be activated.

The results using smooth muscle HMM are particularly impressive. The results confirm that the effect of phosphorylation on myosin MgATPase activity can be distinguished from any effect that phosphorylation may have on filament formation. In other words, the actin activated MgATPase activity appears to be dependent on the state of myosin phosphorylation and not on the ability of myosin to form ilaments.

Having reconstituted a reversible phosphorylating system using purified, well-characterized components, we are now in a position to see whether the addition of other purified proteins from smooth muscle might play a role in modulating this regulatory system. Below we discuss some of the important properties of myosin kinase and smooth muscle phosphatase I, two of the components of this system.

Myosin light chain kinase. The enzyme catalyzing phosphorylation of myosin has been purified from a number of cells including smooth muscle (2,13), platelets (17), brain (18), cardiac (36) and skeletal muscle (5,38). All of these enzymes share one important property: they are completely inactive in the absence of Ca^{++}-calmodulin. The mechanism of activation, suggested in a report on the skeletal muscle myosin kinase (5), as well as for other calmodulin dependent enzymes, is that Ca^{++} first binds to three or four sites on calmodulin and that Ca_3-calmodulin or Ca_4-calmodulin binds to and activates myosin kinase (for a review, see 37).

A second important property of the myosin kinases isolated from smooth muscle and non-muscle cells is that they are substrates for cAMP dependent protein kinase (3,10,19). As illustrated in Fig. 5, myosin kinase incorporates either one or two moles of phosphate, depending on whether calmodulin is bound to the enzyme. When calmodulin is bound, myosin kinase incorporates only one mole of phosphate with no effect on the activity of the enzyme. Extensive tryptic digestion of the denatured myosin kinase previously labelled using $[\gamma-^{32}P]ATP$ shows that the label is incorporated into a single small peptide. On the other hand, when myosin kinase is phosphorylated in the absence of bound calmodulin, phosphate is incorporated into two different sites, with a marked decrease (15- to 20-fold) in the ability of myosin kinase to bind calmodulin (10). As we shall see in the next section, this effect on calmodulin binding is reversible when the phosphorylated enzyme is dephosphorylated.

To date, two different myosin kinases have been shown to undergo phosphorylation catalyzed by the catalytic subunit of cAMP dependent protein kinase. One of these is the smooth muscle enzyme isolated from turkey gizzard (10), the other is the enzyme isolated from human platelets (19). In both cases phosphorylation results in a decrease in myosin kinase activity at a given concentration of calmodulin. The implication of these findings, to be elaborated on in the section below, is that a rise in cAMP in smooth muscle and non-muscle cells might be expected to be associated with a decrease in contractile activity. (The cardiac kinase has been reported to undergo phosphorylation, but the effect of this phosphorylation is uncertain [36]).

The sites that are phosphorylated in myosin kinase can be removed from the native enzyme by very brief proteolysis with trypsin (Fig. 6). The myosin kinase that remains after such a digestion retains the calmodulin binding site and thus remains dependent on Ca^{++}-calmodulin for activity. Further digestion of the myosin kinase with trypsin results in a myosin kinase that is no longer dependent on Ca^{++}-calmodulin for activity (Table 2; see also ref. 35). This two-step proteolysis of the smooth muscle kinase with trypsin is summarized diagrammatically in Fig. 6.

TABLE 1

	Phosphate Incorporation (mol P_i/mol myosin)	MgATPase (nmol P_i released/min/mg)	
		+ Actin	- Actin
A. Smooth muscle myosin			
unphosphorylated	0	4	2
phosphorylated	1.9	51	2
dephosphorylated	0.1	5	
rephosphorylated	2.0	46	
B. Smooth muscle HMM			
unphosphorylated	0	10	2
phosphorylated	1.9	357	2
dephosphorylated	0.1	20	
rephosphorylated	2.1	371	
C. Platelet myosin			
unphosphorylated	0	5	
phosphorylated	1.9	89	
dephosphorylated	0.1	5	

(From Ref. 33).

TABLE 2: Ca^{++} dependence of smooth muscle myosin kinase after tryptic digestion.

Time of Digestion	Specific Activity (μmol P_i transferred/mg kinase/min)	
	+ Ca^{++}	- Ca^{++}
0	2.6	-
1 min	0.7	0.3
30 min	0.9	0.9

Conditions for digestions: 0°C, 1:100 (mg trypsin:mg myosin kinase)

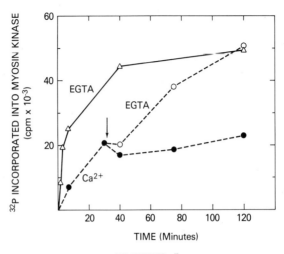

FIGURE 5

Time course of phosphorylation of myosin kinase by the catalytic subunit of cAMP dependent protein kinase in the presence of bound Ca^{++}-calmodulin (●—●) of EGTA and calmodulin (△—△). At 30 min (arrow) EGTA was added to an aliquot (○—○) of the assay initiated in the presence of Ca^{++} calmodulin 50,000 cpm = approximately 2 mol P_i/mol myosin kinase (From Ref. 10).

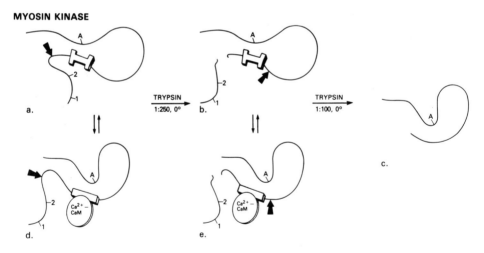

FIGURE 6

Diagram showing that brief tryptic digestion first results in removal of the two sites of phosphorylation (b. and e.) and then of the calmodulin binding site from myosin kinase. The enzyme c., requires neither Ca^{++} nor calmodulin for activity.

TABLE 3: Inhibition of myosin kinase activity in extracts of turkey gizzard smooth muscle and human platelets.

	pmol ^{32}P Incorporated into Smooth Muscle Myosin Light Chains/min
Gizzard	
non-retarded IgG	76
affinity purified antibody	9
Platelet	
non-retarded IgG	132
affinity purified antibody	68

(From Ref. 16).

Affinity-purified antibodies to myosin kinase. Myosin kinases isolated from turkey gizzards and human platelets appear to share a number of important structural and functional properties. In addition to requiring Ca^{++}-calmodulin for activity and acting as a substrate for cAMP dependent protein kinase, these proteins share antigenic determinants. This was shown by preparing antibodies to turkey gizzard myosin kinase which inhibited the activity of the human platelet myosin kinase (see Table 3)(16).

This experiment was carried out as follows: The antibodies prepared to the turkey gizzard myosin kinase were purified on a column of turkey gizzard myosin kinase bound to sepharose 4B. The non-retarded antibodies (those that failed to bind to this column) were used for control experiments. Those antibodies that did bind were eluted from the column with 6 M guanidine·HCl and were utilized as a source of affinity-purified antibodies (14). As shown in Table 3, the activity of myosin kinase in extracts from turkey gizzards and human platelets could be inhibited by the affinity-purified antibodies. The non-retarded IgG fraction did not inhibit kinase activity in either extract. As expected, the affinity purified antibodies also inhibited the activity of purified human platelet myosin kinase (data not shown).

The affinity-purified antibodies were also used in immunofluorescence studies (in collaboration with J. Feramisco and K. Burridge of Cold Spring Harbor Laboratory; see 14) to localize myosin kinase in non-muscle cells. Figure 7 shows that the fluorescently labelled antibodies to myosin kinase can be localized to the stress fibers, structures which are known to contain the contractile proteins, in human fibroblasts. The studies suggest the widespread distribution in vertebrate non-muscle cells of an enzyme similar to the myosin kinase found in smooth muscle cells.

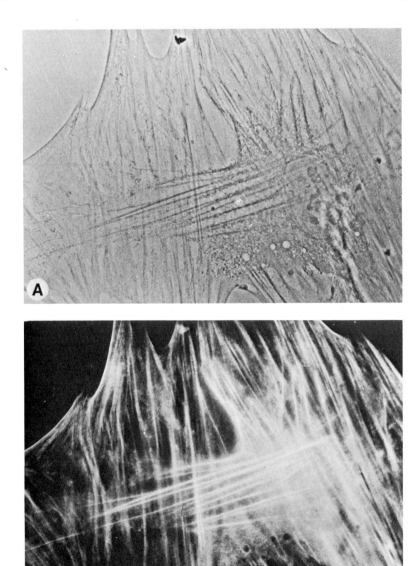

FIGURE 7

Phase and fluorescence micrographs of cells stained with the affinity-purified rabbit anti-myosin light chain kinase antibodies. The phase-contrast micrograph for a human fibroblast cell is shown in A, while the fluorescence micrograph of the same cell is shown in B. Panel C is a fluorescence micrograph of a gerbil fibroma cell. Panel D is a fluorescence micrograph of a chick embryo fibroblast. The primary antibodies were visualized using FITC-labelled goat

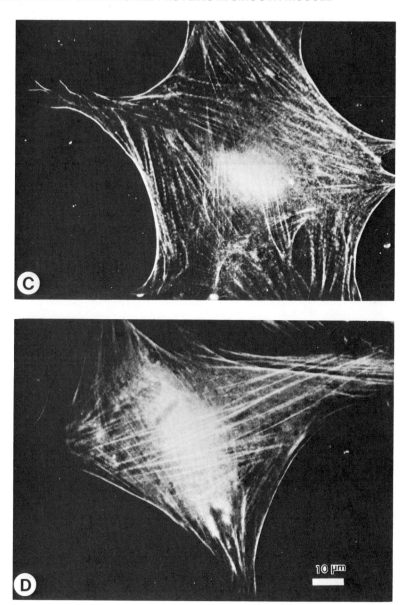

anti-rabbit IgG antibody in these experiments. Note the prominent staining of the stress fibers in every case. Cells incubated with pre-immune IgG did not show stained stress fibers but contained some bright fibrillar staining around the nucleus. This perinuclear staining has been shown to be due to a spontaneously occurring antibody to the protein vimentin found in the pre-immune IgG (from Refs. 14,16).

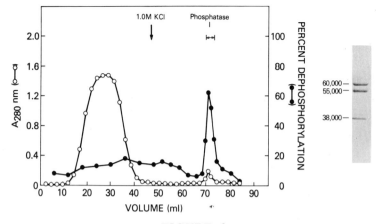

FIGURE 8

Affinity chromatography of smooth muscle phosphatase I on thiophosphorylated myosin light chain-sepharose. The column (1.5 x 11.5 cm) was equilibrated with 20 mM KCl, 20 mM Tris·HCl (pH 7.4), 2 mM EGTA, 5 mM EDTA, 1 mM dithiothreitol, at 4°C at 35 ml/hr. After loading the sample, the column was washed with about 1.5 column vol of the equilibrating buffer. Elution of the phosphatase was carried out with a buffer containing 1 M KCl. Fractions of 1.2 ml were collected and assayed for phosphatase activity (●). The absorbance at 280 nm was measured (○). SDS-polyacrylamide gel of the purified phosphatase is shown on the right. SMP-I = smooth muscle phosphatase I (From Ref. 29).

Smooth Muscle Phosphatases

If cAMP dependent phosphorylation of myosin kinase is to be of physiological significance, the effect of phosphorylation should be readily reversible. In order to examine this we have isolated a number of different phosphatases from smooth muscle (29,30) and examined their activity using phosphorylated myosin kinase as a substrate. Three different phosphatases were identified following sephacryl S-300 chromatography. Two of these phosphatases were purified to apparent homogeneity using affinity chromatography. For this purpose the 20,000 dalton light of smooth muscle myosin was thiophosphorylated using ATPγS and the light chain was covalently bound to sepharose 4B. Figure 8 shows the elution pattern of smooth muscle phosphatase I from this affinity column. It also shows an SDS-polyacrylamide gel of the purified enzyme. The enzyme is composed of three different subunits (M_r = 60,000, 55,000 and 38,000) which are present in a molar ratio of 1:1:1. The holoenzyme has a molecular weight of 165,000 as determined by sedimentation equilibrium under non-denaturing conditions. This

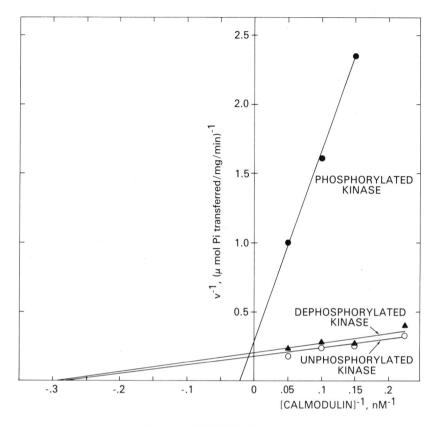

FIGURE 9

Double reciprocal plot of the activation curves obtained by addition of increasing concentrations of calmodulin to aliquots of unphosphorylated (○——○), phosphorylated (●——●) and dephosphorylated (▲——▲) smooth muscle myosin kinase. Myosin kinase was phosphorylated by incubation with the catalytic subunit of cAMP dependent protein kinase and [γ-^{32}P]ATP. It was dephosphorylated by incubation with smooth muscle phosphatase I. Note the marked effect of phosphorylation on the K_{app} for calmodulin (for details see Ref. 10).

phosphatase is similar in structure and activity to phosphatases previously isolated from a number of other sources. The holoenzyme can be dissociated by freezing in 2-mercaptoethanol. The catalytic subunit (M_r = 38,000 daltons) can then be separated from the two other subunits by gel filtration through sephadex G-200 (30).

The catalytic subunit differs from the holoenzyme with respect to at least one important substrate. The catalytic subunit dephosphorylates smooth muscle myosin but the holoenzyme does not. Indeed, it was the catalytic subunit that dephosphorylated myosin in the reconstituted, reversible phosphorylating system that we described above. Both holoenzyme and the catalytic subunit are active in dephosphorylating the isolated 20,000 dalton light chain of myosin.

Reversible phosphorylation of myosin kinase. The holoenzyme of smooth muscle phosphatase I catalyzed the dephosphorylation of myosin light chain kinase. We have used it to see if dephosphorylation of myosin kinase will restore the tight affinity of the enzyme for calmodulin. Figure 9 shows the results of such an experiment. This figure is a double reciprocal plot of calmodulin concentration and myosin kinase activity. It shows that both the original, unphosphorylated myosin kinase as well as the dephosphorylated enzyme bind calmodulin about 15- to 20-fold more tightly than the enzyme that has been phosphorylated in the absence of bound calmodulin.

Significance of myosin kinase phosphorylation. The reversible phosphorylation of myosin kinases isolated from smooth muscle and human platelets suggests a mechanism by which cAMP might modulate contractile activity in these cells. A rise in intracellular cAMP would result in activation of cAMP dependent protein kinase activity which catalyzes the phosphorylation of myosin light chain kinase. Phosphorylation would result in a decrease in myosin kinase activity provided that calmodulin was not bound to myosin kinase, so that two moles of phosphate could be incorporated.

This decrease in myosin kinase activity can be translated into a decrease in contractile activity if we assume that the state of phosphorylation of myosin is determined by an equilibrium between myosin kinase and phosphatase activities. Thus a reduction in myosin kinase activity due to cAMP-dependent phosphorylation of myosin kinase would favor the dephosphorylation of myosin and result in a decrease in contractile activity.

One important problem that remains to be solved is how the various phosphatases isolated from smooth muscle are regulated. To date we have isolated two different phosphatases capable of dephosphorylating smooth muscle myosin. As outlined above, one of these enzymes (smooth muscle phosphatase I) is only active after it has undergone dissociation, forming a free catalytic subunit. We are presently investigating mechanisms for regulating this dissociation. The second enzyme, smooth muscle phosphatase III, appears to be more specific than I, but has not yet been purified to homogeneity.

SUMMARY

Contractile activity in both smooth muscle and vertebrate non-muscle cells appears to be regulated by reversible phosphorylation of myosin. Whereas phosphorylated myosin can interact with actin to hydrolyze MgATP, myosin that is not phosphorylated cannot catalyze this hydrolysis.

The existence of this regulatory system does not preclude an important role for other forms of regulation in smooth muscle and non-muscle cells. For example, the leiotonin system outlined by Ebashi and co-workers (this volume), as well as the direct binding of Ca^{++} to phosphorylated smooth muscle myosin (6,31) may also play a role in regulating actin-myosin interaction.

The important role that calmodulin plays in regulating contractile activity appears to be related to its role in activating myosin kinase activity (20,28). Thus a rise in the intracellular Ca^{++} concentration results in the binding of Ca^{++} to calmodulin and the subsequent activation of myosin kinase activity by the Ca^{++}-calmodulin complex.

cAMP can modulate contractile activity by decreasing the activity of myosin kinase. This decrease in activity results from the decreased affinity of phosphorylated myosin kinase for calmodulin. This possible mechanism for how cAMP-mediated phosphorylation might decrease contractile activity does not preclude other mechanisms which might also involve cAMP. For example, a cAMP dependent phosphorylation might also be involved in lowering the intracellular concentration of Ca^{++}. Future experiments with purified proteins and intact cells should help to establish the relative importance of each of these mechanisms.

ACKNOWLEDGEMENTS

The authors wish to acknowledge the technical assistance of C.R. Eaton, W. Anderson Jr. and J.M. Miles. They are indebted to E. Murray for expert editorial assistance.

REFERENCES

1. Adelstein, R.S. & Eisenberg, E. (1980): Ann. Rev. Biochem. 49:921-956.
2. Adelstein, R.S. & Klee, C.B. (1981): J. Biol. Chem. 256:7501-7509.
3. Adelstein, R.S., Conti, M.A., Hathaway, D.R. & Klee, C.B. (1978): J. Biol. Chem. 253:8347-8350.
4. Bahn, A., Malhotra, A., Scheur, J., Conti, M.A. & Adelstein, R.S. (1981): J. Biol. Chem. 256:7741-7743.
5. Blumenthal, D.K. & Stull, J.T. (1981): Biochemistry 19:5608-5614.

6. Chacko, S., Conti, M.A. & Adelstein, R.S. (1977): Proc. Nat. Acad. Sci. 74:129-133.
7. Chalovich, J.M., Chock, P.B. & Eisenberg, E. (1981): J. Biol. Chem. 256;575-578.
8. Chantler, P.D. & Szent-Gyorgyi, A.G. (1980): J. Molec. Biol. 138:473-492.
9. Chantler, P.D., Sellers, J.R. & Szent-Gyorgyi, A.G. (1981): Biochemistry 20:210-216.
10. Conti, M.A. & Adelstein, R.S. (1981): J. Biol. Chem. 256: 3178-3181.
11. Cooke, R., Franks, K., Ritz-Gold, C.J., Toste, T., Blumenthal, D.K. & Stull, J.T. (1981): Biophys. J. 33:235a.
12. Crow, M.T.&Kushmerick, M.J. (1981):J.Biol.Chem. 257:2121-2124.
13. Dabrowska, R., Aromatorio, D., Sherry, J.M.F. & Hartshorne, D.J. (1977): Biochem. Biophys. Res. Commun. 78:1263-1272.
14. de Lanerolle, P., Adelstein, R.S., Feramisco, J.R. & Burridge, K. (1981): Proc. Nat. Acad. Sci. 78:4738-4742.
15. Ebashi, S. (1980): Proc. Roy. Soc. Lond. B. 207:259-286.
16. Feramisco, J.R., Burridge, K., de Lanerolle, P. & Adelstein, R.S. (1981): Proc. Cold Spring Harbor Conf. on Cell Proliferation 8:855-868.
17. Hathaway, D.R. & Adelstein, R.S. (1979): Proc. Nat. Acad. Sci. 76:1653-1657.
18. Hathaway, D.R., Adelstein, R.S. & Klee, C.B. (1981): J. Biol. Chem. 256:8183-8189.
19. Hathaway, D.R., Eaton, C.R. & Adelstein, R.S. (1981): Nature 291:252-254.
20. Hidaka, H., Naka, M. & Yamaki, T. (1979): Biochem. Biophys. Res. Commun. 90:694-699.
21. Jakes, R., Northrup, F. & Kendrick-Jones, J. (1976): FEBS Lett. 70:229-234.
22. Kakiuchi, S., Yamazaki, R., Teshima, Y., Uenishi, K., Yasuda, S., Kashiba, A., Sobue, K., Ohshima, M. & Nakajima, T. (1978): Adv. Cyclic Nucleotide Res. 9:253-263.
23. Klee, C.B. (1977): Biochemistry 16:1017-1024.
24. Klee, C.B., Crouch, T.H. & Richman, P.G. (1980): Ann. Rev. Biochem. 49:489-515.
25. Korn, E.D. (1978): Proc. Nat. Acad. Sci. 75:588-599.
26. Manning, D.R. & Stull, J.T. (1979): Biochem. Biophys. Res. Commun. 90:164-170.
27. Morgan, M., Perry, S.V. & Ottaway, J. (1976): Biochem. J. 157:687-697.
28. Nishikawa, M., Tanaka, T. & Hidaka, H. (1980): Nature 287: 863-865.
29. Pato, M.D. & Adelstein, R.S. (1980): J. Biol. Chem. 255: 6535-6538.
30. Pato, M.D. & Adelstein, R.S. (1981): Biophys. J. 33:278a.
31. Rosenfeld, A. & Chacko, S. (1982):Proc. Nat. Acad. Sci. 79: 292-296.
32. Scholey, J.M., Taylor, K.A. & Kendrick-Jones, J. (1980): Nature 287:233-235.

33. Sellers, J.R., Pato, M.D. & Adelstein, R.S. (1981): J. Biol. Chem. 256:9274-9278.
34. Suzuki, H., Onishi, H., Takahashi, K. & Watanabe, S. (1978): J. Biochem. 84:1529-1542.
35. Tanaka, T., Naka, M. & Hidaka, H. (1980): Bichem. Biophys. Res. Commun. 92:313-318.
36. Wolf, H. & Hofmann, F. (1980): Proc. Nat. Acad. Sci. 77:5852-5855.
37. Wolff, D.J. & Brostrom, C.O. (1979): Adv. Cyclic Nucleotide Res. 11:27-88.
38. Yagi, K., Yazawa, M., Kakiuchi, S., Ohshima, M. & Uenishi, K. (1978): J. Biol. Chem. 253:1338-1340.

TWO TRANSMEMBRANE CONTROL MECHANISMS FOR PROTEIN PHOSPHORYLATION IN BIDIRECTIONAL REGULATION OF CELL FUNCTIONS

Y. Takai, K. Kaibuchi, T. Matsubara, K. Sano, B. Yu and Y. Nishizuka

Department of Biochemistry, Kobe University School of Medicine, Kobe 650, Japan, and
Department of Cell Biology, National Institute for Basic Biology, Okazaki 444, Japan

Two Transmembrane Control Mechanisms

There appear to be two major receptor mechanisms for the control of cellular activities by various extracellular stimulants. Goldberg and his associates (9) have introduced the terms "mono" and "bidirectional" for describing the main features of these two mechanisms. In monodirectional systems, the control is exercised by a simple on-off principle. When a stimulant is present cells are switched on, and when such stimulant is removed they become quiescent. In bidirectional systems, cellular activation and recovery therefrom are mediated by two opposing stimulants. It is proposed that in both systems cellular activities may be regulated by interactions between two intracellular second messengers, cyclic AMP (cAMP) and cyclic GMP (cGMP). In monodirectional systems both nucleotides appear to function in concert, whereas in bidirectional systems they antagonize each other. This antagonism between these nucleotides have been incorporated sometime into the Yin-Yang hypothesis. Later, Berridge (5) has indicated that Ca^{++} may be a more important second messenger than cGMP. In general, in monodirectional systems, two stimulants act in parallel with each other, and both Ca^{++} and cAMP may serve as positive messengers for the respective stimulants; whereas, in bidirectional systems, Ca^{++} normally serves as a positive messenger for stimulants that activate cells, and the action of antagonistic stimulants may be mediated by cAMP.

Another line of evidence indicates that a variety of extracellular stimulants induce phosphatidylinositol (PI) turnover in their target cell membranes. This phospholipid turnover was first described in 1955 by Hokin and Hokin (19) in acetylcholine sensitive tissues, and subsequently shown by many investigators in a large number of tissues that are stimulated by various extracellular stimulants. These stimulants include neurotransmitters, peptide hormones, mitogenic substances, oncogenic viruses and many other biologically active substances (17,34,35). PI turnover has been interpreted as a closed circle of PI breakdown and resynthesis via diacylglycerol (DG) and phosphatidic acid (PA)(34). This phospholipid breakdown is initiated by the action of phospholipase C which produces unsaturated DG and cyclic inositol phosphate as primary products. Michell (34) proposed that extracellular stimulants which increase the cytosolic Ca^{++} concentration always induce PI turnover, and very often increase cGMP, as exemplified in Table 1. It is noted that these stimulants never increase cAMP and that, inversely, cAMP elevating stimulants do not induce this phospholipid turnover. Thus, a variety of extracellular stimulants may be divided into two groups; one group of those related to Ca^{++}, cGMP as well as to PI turnover, and the other of those related to cAMP. In this report the former group of stimulants will be referred to tentatively as an α-stimulant, and the latter group as a β-stimulant.

It has been generally accepted that a wide variety of regulatory effects of cAMP and cGMP may be mediated through the actions of cAMP dependent protein kinase (A-kinase) and cGMP dependent protein kinase (G-kinase), respectively (30,49). Although the factors mediating the intracellular regulatory properties of Ca^{++} are far less understood, this divalent cation seems to play its diverse roles through the actions of some specific proteins and enzymes such as troponin (8), calmodulin (6,23), gelsolin (51), actinogelin (36) and Ca^{++} dependent proteases (11,33). Michell (35) has proposed a possible role of PI turnover in the regulation of calcium gates, but this hypothesis has not yet been substantiated. We have recently found that DG derived from PI turnover may serve as a second messenger for the selective activation of a new species of cyclic nucleotide independent multifunctional protein kinase (C-kinase)(47). The proposed mechanism of activation of C-kinase by α-stimulant is shown schematically in Fig. 1. The enzyme is normally inactive, but is activated by DG in the presence of micromolar concentrations of Ca^{++} and membrane phospholipid, particularly phosphatidylserine (PS). Kinetically, DG sharply increases the affinity of enzyme for Ca^{++} as well as for phospholipid, and thereby activates C-kinase without net increase of Ca^{++} concentrations. During this activation the enzyme attaches to membranes, and the process is reversible. The enzyme shows a molecular weight of about 77,000 and is composed of a single polypeptide with at least two functionally different domains; hydrophobic membrane binding and hydrophilic catalytic domains. The

TABLE 1: Effects of extracellular stimulants on Ca^{++}, PI turnover, and cyclic nucleotides.

Stimulant	Target Tissue	Ca^{++}	PI Turnover	cGMP	cAMP
Adrenalin (α)	Various	↑	↑	↑	→or ↓
Acetyl-choline (m)	Various	↑	↑	↑	→or ↓
Histamine (H1)	Smooth muscle, Brain	↑	↑	↑	→
Angiotensin	Liver	↑	↑	?	→
Vasopressin	Liver	↑	↑	?	→
Thrombin	Platelets	↑	↑	↑	→
FMet-Leu-Phe	Leukocytes	↑	↑	↑	→
Plant lectin	Lymphocytes	↑	↑	↑	→or ↓
Glucagon	Liver, Fat cells	→	→	→	↑
Adrenalin (β)	Various	→ or ↓	→	→	↑
Histamine (H2)	Smooth muscle, Brain	→	→	→	↑

enzyme is able to phosphorylate many proteins, probably those located in or just below membranes, and plays pleiotropic functions. C-Kinase does not require cAMP, cGMP nor calmodulin, and is clearly distinguishable from other well defined protein kinases such as A- and G-kinases, glycogen phosphorylase kinase, and myosin light chain kinase. The precise properties of this unique protein kinase have been described elsewhere (21,22,29, 31,37-40,45-47).

Arachidonic acid peroxide and prostaglandin endoperoxide, both of which are derived from DG upon the action of DG lipase, may activate guanylate cyclase to produce cGMP (10,18,39). Thus, the transmembrane control mechanisms by a variety of extracellular stimulants may be outlined as given in Fig. 2. At least three pathways are distinguished in the action of α-stimulant, whereas only cAMP A-kinase system may be clarified in the action of β-stimulant. The subsequent description will be concerned primarily with the relationship between these two major receptor functions in bidirectional systems, with the special emphasis on the mechanism of inhibition by β-stimulant of the cellular activation which is induced by α-stimulant.

Inhibition of PI Turnover by cAMP

In bidirectional systems cellular activation by α-stimulant is often inhibited by cAMP as mentioned above. Two possible mechanisms for such inhibition have thus far been proposed: 1) A-kinase

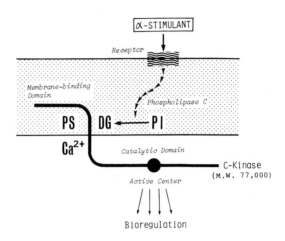

FIGURE 1

Proposed mechanism of transmembrane control system for activation of C-kinase.

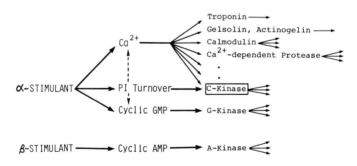

FIGURE 2

Transmembrane control mechanisms of α- and β-stimulants.

phosphorylates a microsomal protein having a molecular weight of about 22,000 resulting in the activation of microsomal Ca^{++} activated ATPase followed by decrease in the cytosolic Ca^{++} concentration (5,24); and 2) A-Kinase phosphorylates Ca^{++} activated myosin light chain kinase resulting in the inhibition of its catalytic activity (2). The latter enzyme is known to phosphorylate specifically myosin light chain and thereby activate myosin ATPase (1). In addition to these proposed mechanisms we wish to describe here that cAMP strongly inhibits PI turnover, and thereby counteracts the activation of C-kinase.

For these studies the following three bidirectional systems were employed; thrombin-induced serotonin release in human platelets, phytohemagglutinin (PHA)-induced DNA synthesis in human peripheral lymphocytes, and fMet-Leu-Phe (fMLP)-induced release of N-acetylglucosaminidase in rat peritoneal polymorphonuclear leukocytes. It is well known that these cellular activations are all inhibited by prostaglandin E1 (PGE1), which elevates cAMP in these cells (5,13,43,50). As shown in Table 2, PGE1 inhibited simultaneously both cellular activation and PI turnover which were induced by the respective α-stimulants. Such effects of PGE1 were reproduced by dibutyryl cAMP (dbcAMP). Butyrate itself showed no effect. PGE1 or dbcAMP alone did not affect cellular activation nor PI turnover in resting cells. Both cellular activation and PI turnover were inhibited roughly in parallel manners by increasing concentrations of PGE1 and dbcAMP. Kennerly et al. (27) recently described the inhibition of PI turnover by PGE1, as well as histamine release in mast cells which are stimulated by concanavalin A plus PS. Although such an inhibitory action by cAMP has been studied in a limited number of tissues, it is possible to propose that cAMP-elevating β-stimulant antagonizes the cellular activation induced by α-stimulant, at least in part, by inhibiting PI turnover.

Counteraction of C-Kinase Activation by cAMP

PI turnover has usually been assayed in most tissues by measuring the incorporation of ^{32}Pi into PI or PA. The initial reaction of PI turnover, that is, PI hydrolysis to produce DG and inositol cyclic phosphate, can be directly measured for several tissues such as platelets (4,41), thyroid gland (20) and mast cells (28), which are stimulated by thrombin, thyrotropin and compound 48/80, respectively. Among these tissues, thrombin-induced PI turnover in platelets were most extensively investigated. When platelets are stimulated by thrombin, several proteins, particularly those having molecular weights of about 40,000 (40K) and 20,000 (20K) are heavily phosphorylated (7,14,32). It is generally accepted that the phosphorylation of these proteins may be directly related to the release reaction of serotonin. 20K protein is identified as myosin light chain, and the protein kinase responsible for this protein phosphorylation is identified as myosin light chain

TABLE 2: Inhibition of PI turnover by cAMP in three typical bidirectional systems.

Experiment	Addition	Cellular Activity	PI Turnover	cAMP
# 1		(cpm)	(cpm)	(pmol/1x10^8 cells)
	None	640	200	6.4
	Thrombin	4,870	2,240	6.3
	Thrombin + PGE1	650	650	–
	Thrombin + DbcAMP	630	250	–
	PGE1	610	170	18.8
	DbcAMP	630	200	–
# 2		(cpm)	(cpm)	(pmol/1x10^7 cells)
	None	420	1,090	2.2
	PHA	1,710	2,730	2.0
	PHA + PGE1	560	1,250	–
	PHA + DbcAMP	670	1,450	–
	PGE1	510	790	22.0
	DbcAMP	430	820	–
# 3		(percent)	(cpm)	(pmol/1x10^7 cells)
	None	16.5*	160	1.7
	FMLP	47.2*	420	1.9
	FMLP + PGE1	22.4*	200	–
	FMLP + DbcAMP	19.4*	180	–
	PGE1	22.1*	180	3.5
	DbcAMP	22.7*	190	–

* These numbers indicate percentages of the enzymatic activity released into the medium with the total activity as 100.

In Experiment #1, human platelets (6 x 10^8 cells/ml) which were pre-labelled with either [^{3}H]arachidonic acid or [^{14}C]serotonin, were incubated with either 1 x 10^{-5} M PGE1 or 1 x 10^{-4} M dbcAMP for 5 min at 37°C, and then stimulated by 0.3 unit/ml of thrombin for 15 sec. PI turnover was assayed by measuring the formation of radioactive DG. Serotonin release was assayed by measuring the radioactive serotonin which was released into the medium. Cyclic AMP was measured by radioimmunoassay after incubation for 30 sec at 37°C with either PGE1 or thrombin. Detailed procedures for these assays were described previously (26,44). In Experiment #2, human peripheral lymphocytes (1 x 10^6 cells/ml), which were pre-labelled with ^{32}Pi, were stimulated by 10 μg/ml of PHA for 30 min at 37°C in the presence and absence of 1 x 10^{-5} M PGE1 or

TABLE 2 (continued)

1 x 10^{-3} M dbcAMP as indicated. PI turnover was assayed by measuring the incorporation of ^{32}Pi into PI. Another suspension of non-radioactive lymphocytes was stimulated by PHA for 72 hr in the presence and absence of 1 x 10^{-5} M PGE1 or 1 x 10^{-3} M dbcAMP and then incubated with [^{3}H]thymidine for 2 hr at 37°C. DNA synthesis was assayed by measuring the incorporation of radioactive thymidine into DNA. Cyclic AMP was measured after incubation for 10 min at 37°C with either PGE1 or PHA. In Experiment #3, rat peritoneal polymorphonuclear leukocytes (2 x 10^{7} cells/ml), which were prelabelled with ^{32}Pi, were preincubated with 1 x 10^{-5} M cytochalasin B for 10 min at 37°C, incubated with either 1 x 10^{-4} M PGE1 or 1 x 10^{-3} M dbcAMP for 10 min, and then stimulated by 1 x 10^{-8} M fMLP for 5 min. PI turnover was assayed by measuring the incorporation of ^{32}Pi into PI. Another suspension of non-radioactive leukocytes was pretreated and stimulated by fMLP for 10 min under similar conditions, and N-acetylglucosaminidase released into the medium was assayed. Cyclic AMP was measured after incubation for 5 min at 37°C with either PGE1 or fMLP. Detailed procedures and assay conditions employed for these experiments will be described elsewhere.

---

kinase (7). However, a role of 40K protein as well as a protein kinase responsible for its phosphorylation has not yet been defined. A series of analyses in this laboratory has led us to the conclusion that PI turnover may serve as a trigger for 40K protein phosphorylation, and that C-kinase is directly involved in this reaction (25, 26, 44).

Table 3 shows that upon stimulation of platelets by thrombin PI rapidly disappears with the concomitant formation of DG. Although this DG appeared to be derived from PI by receptor-linked hydrolysis, the amount of DG produced was far less than that of the PI that had disappeared. It is likely that this difference may be due to rapid conversion of DG back to PI by way of PA, and also to further degradation of DG to monoacylglycerol and free fatty acid by the action of DG lipase. This DG lipase has recently been found in platelets (3). In any case, the thrombin induced formation of DG was always accompanied with 40K protein phosphorylation and serotonin release, and these reactions are parallel with one another. It is also noted that these reactions are simultaneously inhibited to the same extent by the addition of PGE1 and dbcAMP. Apparently both agents show practically no effect on resting platelets. Such inhibitory effects of PGE1 and dbcAMP on platelet activation are more quantitatively given in Fig. 3. DG formation, 40K protein phosphorylation and serotonin release were

TABLE 3: Inhibition by cAMP of thrombin induced PI hydrolysis, DG formation, 40K protein phosphorylation and serotonin release.

Addition	PI Hydrolysis* (Δ cpm)	DG Formation (cpm)	40K Protein Phosphorylation** (ΔO.D. at 430 nm)	Serotonin Release (cpm)
None	0	200	0	640
Thrombin	15,640	2,240	0.30	4,870
Thrombin + PGE1	4,700	650	0.02	650
Thrombin + DbcAMP	150	250	0	630
PGE1	100	170	0	610
DbcAMP	180	200	0	630

Human platelets (6×10^8 cells/ml), which were prelabelled with either [^{3}H]arachidonic acid, ^{32}Pi or [^{14}C]serotonin, were incubated for 5 min at 37°C with either 1×10^{-5} M PGE1 or 1×10^{-4} M dbcAMP, and then stimulated by 0.3 unit/ml of thrombin for 15 sec at 37°C. PI hydrolysis, DG formation, 40K protein phosphorylation and serotonin release were assayed as described previously (26, 44).

* The basal level of radioactive PI in resting platelets was 39,700 cpm. The numbers given in this table were obtained by subtracting the radioactivity of PI in each experiment from this basal value.
** The background of O.D. at 430 nm for this autoradiograph was 0.19 and this value was subtracted from each experimental value.

progressively inhibited in parallel manners by increasing concentrations of these agents. cAMP was increased inversely by the addition of PGE1. Under these conditions PGE1 and dbcAMP stimulated the phosphorylation of several distinct proteins having molecular weights of about 240,000 (240K), 50,000 (50K), 24,000 (24K) and 22,000 (22K) proteins. 240K protein has been identified as an actin-binding protein, presumably filamin (48). 50K protein has not yet been identified. 24K and 22K proteins are probably related to the activation of microsomal Ca^{++} ATPase as mentioned above (15). The phosphorylation of these four proteins was enhanced by increasing concentration of PGE1 and dbcAMP, and was inversely proportional to the formation of DG; Fig. 4 shows such inverse relationships between 50K protein phosphorylation and DG formation. The results, as briefly described above, seem to indicate that PGE1 blocks PI turnover at the level of DG formation, and thereby counteracts the activation of C-kinase. Since all

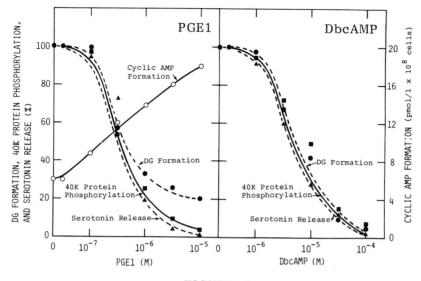

FIGURE 3

Effects of PGE1 and dbcAMP on DG formation, 40K protein phosphorylation and serotonin release. Radioactive human platelets were incubated with various concentrations of either PGE1 or dbcAMP as indicated, and then stimulated by thrombin under the conditions similar to those described in the legend to Table 3. DG formation, 40K protein phosphorylation and serotonin release were assayed as described previously (26,44). Cyclic AMP was measured after incubation for 30 sec at 37°C with various concentrations of PGE1 as indicated.

effects of PGE1 could be reproduced by dbcAMP, it seems most likely that PGE1 plays its role through the activation of A-kinase. Presumably, the phosphorylation of some proteins such as 50K protein may be intimately related to this inhibitory process.

PI Turnover and Role of cGMP

Since the first proposal of Yin-Yang theory by Goldberg and his associates (9), much attention has been paid to cGMP. However, the definitive role of this cyclic nucleotide in biological regulation has not yet been fully understood. It still remains puzzling that G-kinase shows very similar, if not identical, catalytic properties to those of A-kinase as far as tested in in vitro systems, as first reported from this laboratory (12). Recently Schultz et al. (42) proposed that cGMP may be a feedback inhibitor rather than a positive messenger of muscarinic cholinergic stimulants for smooth muscle contraction. This assumption was based on the observations that cGMP elevating agents such as sodium

TABLE 4: Inhibition by cGMP of thrombin induced DG formation, 40K protein phosphorylation and serotonin release.

Addition	DG Formation (cpm)	40K Protein Phosphorylation* (ΔO.D. at 430 nm)	Serotonin Release (cpm)	cGMP (pmol/1x10^8 cells)
None	210	0	700	0.4
Thrombin	1,160	0.40	6,050	0.62
Thrombin + SNP	720	0.19	3,600	-
Thrombin + 8bcGMP	560	0.17	2,040	-
SNP	230	0	650	1.5
8bcGMP	210	0	720	-

Radioactive human platelets were incubated for 10 min at 37°C with 1×10^{-4} M SNP or 1×10^{-3} M 8bcGMP, and then stimulated by thrombin for 15 sec at 37°C under conditions similar to those described in the legend to Table 3. DG formation, 40K protein phosphorylation and serotonin release were assayed as described previously (26,44). Cyclic GMP was measured after incubation for 10 min at 37°C with SNP or thrombin.
* The background of O.D. at 430 nm for this autoradiograph was 0.19 and this value was subtracted from each experimental value.

nitroprusside (SNP) and 8-bromo-cGMP (8bcGMP) cause muscle relaxation, although muscarinic cholinergic stimulants enhance cGMP. More recently, Haslam et al. (16) proposed that, although cGMP is increased in human platelets stimulated by thrombin and collagen, this cyclic nucleotide probably inhibits platelet activation, since SNP is known as a potent platelet inhibitor and markedly increases cGMP. An immediate possibility may arise then as to whether SNP and 8bcGMP inhibit PI turnover in an analogous manner to cAMP.

The experiment shown in Table 4 was designed to show that both SNP and 8bcGMP in fact inhibited the thrombin induced DG formation as well as 40K protein phosphorylation and serotonin release. Under these conditions SNP increased cGMP. These agents did not affect resting platelets. The concentration of SNP or 8bcGMP needed for the inhibition of DG formation was nearly the same as the concentration needed for the inhibition of 40K protein phosphorylation as well as that of serotonin release. It

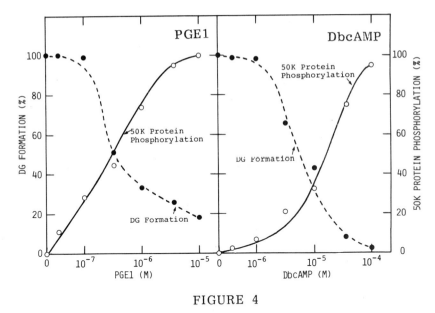

FIGURE 4

50K Protein phosphorylation and inhibition of DG formation. 50K Protein phosphorylation and DG formation were assayed as described in the legends to Table 3 and Fig. 3, except that various concentrations of either PGE1 or dbcAMP were added as indicated.

was also noted that these agents stimulated the phosphorylation of a distinct protein having a molecular weight of about 50,000, and the reaction was again inversely proportional to the inhibition of DG formation. This protein was probably identical with 50K protein that is phosphorylated by A-kinase upon the addition of PGE1 or dbcAMP. As mentioned above, cAMP stimulated the phosphorylation of several proteins such as 240K, 24K and 22K proteins in addition to 50K protein. However, both SNP and 8bcGMP stimulated the phosphorylation of 50K protein more selectively. Available evidence suggests that cGMP inhibits PI turnover, presumably through activation of G-kinase. Thus, in the receptor-linked cascade reactions cGMP may be involved in an intracellular short circuit leading to the feedback inhibition of PI turnover, and again counteracts C-kinase activation as shown in Fig. 5.

SUMMARY AND DISCUSSION

Two major receptor functions appear to be responsible for activation and inhibition of many biological systems. In such bidirectional control systems a wide variety of cellular activities may be regulated by two opposing extracellular stimulants. In most tissues

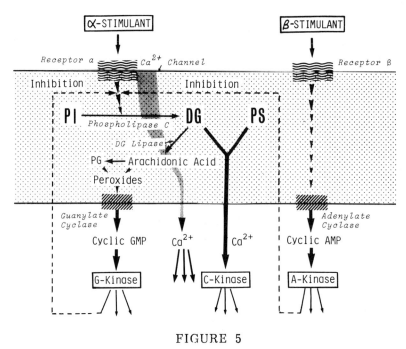

FIGURE 5

Two major receptor functions in bidirectional systems.

it seems likely that cAMP plays roles in the inhibitory processes which may be induced by a group of stimulants that is designated as β-stimulant in this report. In contrast, for the cellular activation by another group of stimulants, namely α-stimulant, both Ca^{++} and DG play indispensable roles. This DG may be derived from PI turnover which is provoked by α-stimulant in its target cell membranes. Ca^{++} is not necessarily increased during C-kinase activation since DG dramatically increases the affinity of C-kinase for Ca^{++} to the micromolar range. The plausible evidence strongly indicates that PI turnover is an integral part of this receptor function. However, the precise relationship between Ca^{++} influx or movement and PI turnover remains to be explored.

C-Kinase activated in this way reveals substrate specificity distinctly different from A-kinase, and may regulate several proteins presumably related to cellular activation. A series of experiments using human platelets and other bidirectional control systems suggests that this receptor-linked transmembrane control mechanism for protein phosphorylation does operate in physiological processes. It is emphasized here that cAMP, as well as cGMP, blocks such receptor-linked hydrolysis of PI, and thereby counteracts the selective activation of C-kinase. Although the mechanism of such blockade is not clear at this time, it is suggestive that this blockade of PI hydrolysis may be mediated through the phosphorylation of specific proteins, such as 50K protein in platelets, by A-kinase as well as by G-kinase. It is attractive to propose, therefore, that both cAMP and cGMP serve as negative messengers which are involved in the extracellular and intracellular circuits, respectively, leading to the feedback control of cellular activation processes in bidirectional systems.

ACKNOWLEDGEMENTS

T. Matsubara is on leave from the Department of Orthopedic Surgery, Kobe University School of Medicine, Kobe 650, Japan. B. Yu is on leave from the Department of Chemistry, Chinese Medical College, Shinyan, China. The authors are grateful to Mrs. S. Nishiyama and Miss K. Yamasaki for their skillful secretarial assistance. This investigation was supported in part by research grants from the Scientific Research Fund of the Ministry of Education, Science, and Culture, Japan (1979-1981), the Intractable Diseases Division, Public Health Bureau, the Ministry of Health and Welfare, Japan (1979-1981), a Grant-in-Aid of New Drug Development from the Ministry of Health and Welfare, Japan (1979-1981), the Yamanouchi Foundation for Research on Metabolic Disorders (1977-1981), and the Foundation for the Promotion of Research on Medicinal Resources, Japan (1977-1981).

REFERENCES

1. Adelstein, R.S. & Conti, M.A. (1975): Nature 256:597-598.
2. Adelstein, R.S., Conti, M.A., Hathaway, D.R. & Klee, C.B. (1978): J. Biol. Chem. 253:8347-8350.
3. Bell, R.L., Kennerly, D.A., Stanford, N. & Majerus, P.W. (1979): Proc. Nat. Acad. Sci. 76:3238-3241.
4. Bell, R.L. & Majerus, P.W. (1980): J. Biol. Chem. 255:1790-1792.
5. Berridge, M.J. (1975): Adv. Cyclic Nucleotide Res. 6:1-98.
6. Cheung, W.Y. (1970): Biochem. Biophys. Res. Commun. 38:533-538.
7. Daniel, J.L., Holmsen, H. & Adelstein, R.S. (1977): Thrombos. Haemostas. 38:984-989.
8. Ebashi, S. & Kodama, K. (1964): J. Biochem. 58:107-108.
9. Goldberg, N.D., Haddox, M.K., Dunham, E., Lopez, C. & Hadden, J.W. (1974): In Control of Proliferation in Animal Cells. (eds) B. Clarkson & R. Baserga, Cold Springs Harbor Laboratory, pp. 609-625.
10. Graff, G., Stephenson, J.H., Glass, D.B., Haddox, M.K. & Goldberg, N.D. (1978): J. Biol. Chem. 253:7662-7676.

11. Guroff, G. (1964): J. Biol. Chem. 239:149-155.
12. Hashimoto, E., Takeda, M., Nishizuka, Y., Hamana, K. & Iwai, K. (1976): J. Biol. Chem. 251:6287-6293.
13. Haslam, R.J., Davidson, M.M.L., Davies, T., Lynham, J.A. & McClenaghan, M.D. (1978): Adv. Cyclic Nucleotide Res. 9: 533-552.
14. Haslam, R.J. & Lynham, J.A. (1977): Biochem. Biophys. Res. Commun. 77:714-722.
15. Haslam, R.J., Lynham, J.A. & Fox, J.E.B. (1979): Biochem. J. 178:397-406.
16. Haslam, R.J., Salam, S.E., Fox, J.E.B., Lynham, J.A. & Davidson, M.M.L. (1980): In Cellular Response Mechanisms and Their Biological Significance. (eds) A. Rotman, F.A. Meyer, C. Gilter & A. Silberberg, John Wiley & Sons, New York, pp. 213-231.
17. Hawthorne, J.N. & White, D.A. (1975): Vitamin Hormones 33: 529-573.
18. Hidaka, H. & Asano, T. (1977): Proc. Nat. Acad. Sci. 74: 3657-3661.
19. Hokin, L.E. & Hokin, M.R. (1955): Biochim. Biophys. Acta 18:102-110.
20. Igarashi, Y. & Kondo, Y. (1980): Biochem. Biophys. Res. Commun. 97:759-765.
21. Iwasa, Y., Takai, Y., Kikkawa, U. & Nishizuka, Y. (1980): Biochem. Biophys. Res. Commun. 96:180-187.
22. Kaibuchi, K., Takai, Y. & Nishizuka, Y. (1981): J. Biol. Chem. 256:7146-7149.
23. Kakiuchi, S. & Yamazaki, R. (1970): Biochem. Biophys. Res. Commun. 41:1104-1110.
24. Katz, A.M., Tada, M. & Kirchberger, M.A. (1975): Adv. Cyclic Nucleotide Res. 5:453-472.
25. Kawahara, Y., Takai, Y., Minakuchi, R., Sano, K. & Nishizuka, Y. (1980): J. Biochem. 88:913-916.
26. Kawahara, Y., Takai, Y., Minakuchi, R., Sano, K. & Nishizuka, Y. (1980): Biochem. Biophys. Res. Commun. 97:309-317.
27. Kennerly, D.A., Secosan, C.J., Parker, C.W. & Sullivan, T.J. (1979): J. Immunol. 123:1519-1524.
28. Kennerly, D.A., Sullivan, T.J., Sylwester, P. & Parker, C.W. (1979): J. Exp. Med. 150:1039-1044.
29. Kishimoto, A., Takai, Y., Mori, T., Kikkawa, U. & Nishizuka, Y. (1980): J. Biol. Chem. 255:2273-2276.
30. Kuo, J.F. & Greengard, P. (1970): J. Biol. Chem. 245: 2493-2498.
31. Ku, Y., Kishimoto, A., Takai, Y., Ogawa, Y., Kimura, S. & Nishizuka, Y. (1981): J. Immunol. 127:1375-1379.
32. Lyons, R.M., Stanford, N. & Majerus, P.W. (1975): J. Clin. Invest. 56:924-936.
33. Meyer, W.L., Fischer, E.H. & Krebs, E.G. (1964): Biochemistry 3:1033-1039.
34. Michell, R.H. (1975): Biochim. Biophys. Acta 415:81-147.

35. Michell, R.H. (1979): Trends Biochem. Sci. 4:128-131.
36. Mimura, N. & Asano, A. (1979): Nature 282:44-48.
37. Minakuchi, R., Takai, Y., Yu, B. & Nishizuka, Y. (1981): J. Biochem. 89:1651-1654.
38. Mori, T., Takai, Y., Minakuchi, R., Yu, B. & Nishizuka, Y. (1980): J. Biol. Chem. 255:8378-8380.
39. Nishizuka, Y. & Takai, Y. (1981): In Cold Spring Harbor Conf. on Cell Proliferation 8:237-249.
40. Ogawa, Y., Takai, Y., Kawahara, Y., Kimura, S. & Nishizuka, Y. (1981): J. Immunol. 127:1369-1374.
41. Rittenhouse-Simmons, S. (1979): J. Clin. Invest. 63:580-587.
42. Schultz, K.D., Schultz, K. & Schultz, G. (1977): Nature 265:750-751.
43. Smith, J.W., Steiner, A.L. & Parker, C.W. (1971): J. Clin. Invest. 50:442-448.
44. Takai, Y., Kaibuchi, K., Matsubara, T. & Nishizuka, Y. (1981): Biochem. Biophys. Res. Commun. 101:61-67.
45. Takai, Y., Kishimoto, A., Iwasa, Y., Kawahara, Y., Mori, T. & Nishizuka, Y. (1979): J. Biochem. 86:575-578.
46. Takai, Y., Kishimoto, A., Iwasa, Y., Kawahara, Y., Mori, T. & Nishizuka, Y. (1979): J. Biol. Chem. 254:3692-3695.
47. Takai, Y., Kishimoto, A., Kikkawa, U., Mori, T. & Nishizuka, Y. (1979): Biochem. Biophys. Res. Commun. 91:1218-1224.
48. Wallach, D., Davies, P.J.A. & Pastan, I. (1978): J. Biol. Chem. 253:3328-3335.
49. Walsh, D.A., Perkins, J.P. & Krebs, E.G. (1968): J. Biol. Chem. 243:3763-3765.
50. Weismann, G., Dukor, P. & Zurier, R.B. (1971). Nature New Biol. 231:131-135.
51. Yin, H.L. & Stossel, T.P. (1979): Nature 281:583-586.

CALCIUM REGULATION IN AMOEBOID MOVEMENT

D.L. Taylor and M. Fechheimer

Cell and Developmental Biology, Harvard University
Cambridge, Massachusetts 02138, USA

Calcium regulation of cytoplasmic structure and contractility in amoeboid cells was first demonstrated in single cell models of C. carolinensis (48). Evidence has mounted in subsequent years for the role of calcium in regulating the motile activities of other types of amoeboid cells, including macrophages (60), Acanthamoeba (34), D. discoideum (15,49) and mammalian cells (27,32). The present paper aims to discuss calcium regulation of amoeboid movement in relation to four fundamental questions: 1) Can fluctuations in the free [Ca^{++}] regulate amoeboid movement? 2) What is the relationship between free [Ca^{++}] and pH? 3) Do secondary messengers such as calcium and protons change primarily the distribution or activity of contractile proteins? 4) What are the cellular targets of elevated free [Ca^{++}]?

We address the above questions with the results obtained primarily from our investigations on the giant amoeba, C. carolinensis, the cellular slime mold amoeba, D. discoideum, and the murine peritoneal macrophage.

Can fluctuations in the free [Ca^{++}] regulate amoeboid movement? Results from single cell models, cytoplasmic extracts and single cells support the hypothesis that fluctuations in the free [Ca^{++}] can regulate amoeboid movement. The original observation of calcium regulation of contractility and cytoplasmic streaming was made on cytoplasm isolated from single specimens of C. carolinensis (48). Single cells were incubated in solutions designed to simulate the intracellular environment. One of these buffers maintained the free [Ca^{++}] well below about 10^{-8} M. Cytoplasm isolated in this buffer was non-motile, viscoelastic (gelled), and optically isotropic.

Optically anisotropic fibrils, shown to contain actin filaments, were induced by straining the viscoelastic cytoplasm up to the fracture point (Fig. 1). When the cytoplasmic fibrils were released, the birefringence did not return to the original value, but showed a positive deformation (Fig. 2). These physical properties are consistent with the presence of an initially random orientation of crosslinked polymers (actin filaments). The positive deformation indicated that the oriented filaments remained partially aligned even after the strain was removed. The ability to induce strain birefringence over an extended length of stretch indicated that either the polymers were very flexible, and/or the cross-linking agents were flexible. Addition of a contraction solution (the same buffer containing about 10^{-6} M free [Ca^{++}]) to the oriented fibrils caused the fibrils to shorten to more than 1/20 of the original length while generating force.

The free [Ca^{++}] determined the type and rate of motility exhibited by the isolated cytoplasm. The effect of the free [Ca^{++}] on the average rate of contraction of a population of oriented fibrils was determined. Surprisingly, the rate of contraction showed a linear increase as the free [Ca^{++}] was varied from 10^{-6} to 10^{-3} M. This was in contrast to the known properties of striated muscle which exhibit a plateau at about 10^{-5} M (48).

The effect of the free [Ca^{++}] on the optically isotropic (unstrained) droplets of cytoplasm was also determined. Free [Ca^{++}] above about 10^{-6} M caused rapid isodiametric contractions of the cytoplasm. However, free [Ca^{++}] around the threshold level (about 5.0×10^{-7} M to 10^{-6} M) caused streaming within the cytoplasmic droplets, and extension of pseudopodium-like structures (Fig. 1). Free [Ca^{++}] around the threshold level permitted streaming of cytoplasm in the droplets. The rate of streaming was also sensitive to the free [Ca^{++}] with higher free [Ca^{++}] causing faster rates of streaming up to the point where isodiametric contractions dominated the motility (48).

A more detailed analysis of the structural and contractile properties of cytoplasm was performed using extracts from both A. proteus and D. discoideum (15,49,53). Extracts prepared from D. discoideum in a buffer containing a low free [Ca^{++}] formed a non-motile, viscoelastic (gelled) and optically isotropic mass when warmed to room temperature (15,49)(Fig. 3: upper). This gelled extract was stable and did not contract for several hours at room temperature. Application of a contraction solution containing a free [Ca^{++}] about 10^{-6}M initiated contractions and cytoplasmic streaming reminiscent of single cell models. The method of addition of the contraction solution determined the behavior of the extract. If the free [Ca^{++}] was raised much above the threshold level before warming, then the extent of gelation was decreased and the warmed extracts exhibited

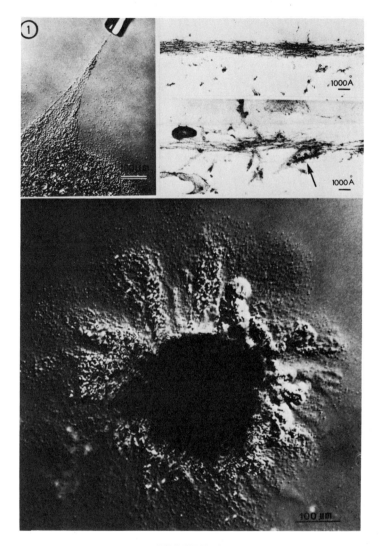

FIGURE 1

Cytoplasm isolated from single cells of C. carolinensis. (upper left) Cytoplasm was non-motile and viscoelastic when isolated in buffers containing free [Ca^{++}] equal to about 10^{-8} M. Fibrils were induced by applying tension with a micropipette. (upper right) The cytoplasmic fibrils contained both actin and myosin filaments (see arrow). (lower figure) Cytoplasm from single cells exhibited cytoplasmic streaming and amoeboid-like movements when isolated in buffers containing about 5.0×10^{-7} M to 1.0×10^{-6} M free calcium (48).

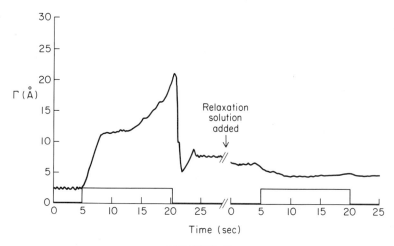

FIGURE 2

Photoelectrically measured changes in retardation due to birefringence upon stretching isolated cytoplasm in buffers containing about 10^{-8} M free [Ca^{++}]. Cytoplasm was stretched by drawing it into a micropipette under constant negative pressure. The increased birefringence fell off sharply when the deforming force was removed, but some birefringence remained at 25 sec (positive deformation). (Γ) retardation (48).

a superprecipitation without forming a contracted mass. In contrast, if the extract was permitted to gel at the subthreshold free [Ca^{++}], and a contraction solution was subsequently layered on top of the gelled cytoplasm, then contraction was initiated at the interface (49). Force-producing contractions apparently required the presence of at least a partially structured gel to transmit the tension. In the absence of any gel only superprecipitation was possible. Therefore, some relationship existed between the state of gel formation and contraction.

A dual effect of calcium on gelation and contraction was demonstrated by removing myosin from an extract of D. discoideum amoebae. This myosin-free preparation exhibited gelation at low free [Ca^{++}], and solation when the free [Ca^{++}] was raised to the threshold level (15). Calcium-sensitive contraction was observed following addition of purified D. discoideum myosin to the preparation. These experiments demonstrated conclusively that both gelation and contraction were sensitive to changes in free [Ca^{++}]. In addition, it was shown that the fraction remaining soluble after sedimentation of the contracted mass of cytoplasm could form a gel when warmed to room temperature. Therefore, neither actin nor gelation activity was completely retained in the mass of filaments which were readily sedimentable following contraction.

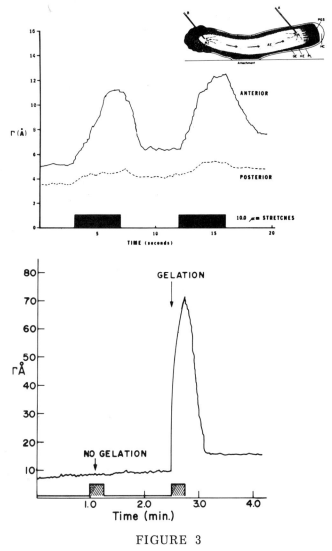

FIGURE 3

Strain birefringence induced in cell extracts and in cytoplasm of living cells. (lower figure) Strain birefringence assay for gelation. The phase retardation (Γ) was monitored vs time with the application of 10 μm stretches by a micropipette inserted into the extract (shaded blocks). Strain birefringence was induced only when the extract gelled (15). (upper figure) Line drawing representing a strip chart recording of strain birefringence induced in anterior endoplasm at the fountain zone (——), and the recruitment zone in the tail endoplasm (----). Phase retardation (Γ) is plotted vs time when a 10 μm stretch was applied to a micropipette inserted into the endoplasm (46).

Studies in vivo

The effect of calcium on the structure and contractility of cytoplasm has also been characterized in living cells. The anterior endoplasm just behind the fountain zone exhibited strain birefringence when tension was applied with a microneedle inserted into C. carolinensis (46). In contrast, very little strain birefringence was induced in the tail endoplasm. Therefore, the endoplasm at the tips of advancing pseudopods had the consistency of a weak gel, while the endoplasm in the tail was more "solated" (a much weaker gel)(Fig. 3: lower). In contrast, the ectoplasm has been shown to be a relatively rigid gel (see 50 for a review).

Microinjection of our standard contraction solution (free [Ca^{++}] about 10^{-6} M) into the tail ectoplasm of amoebae (C. carolinensis) caused the ectoplasm to visibly shorten. The endoplasm then streamed forward at more than twice the original rate of streaming. Repeated injection of contraction solution caused the cells to become monopodial, and the ectoplasm lost its characteristic phase dense appearance. Microinjection of a large volume of contraction solution caused the cytoplasm to contract into a central mass leaving the plasma membrane intact at the cell periphery (46).

A direct demonstration of fluctuations in the free [Ca^{++}] in motile cells is an absolute requirement for identifying calcium as a physiologically relevant secondary messenger in amoeboid movement. We approached this problem by using aequorin luminescence (3,6) as an indicator of the free [Ca^{++}] in single C. carolinensis amoebae (47). Cells injected with aequorin exhibited normal patterns and rates of movement for several hours. The luminescence was measured from whole cells using either a photon-counting system attached to a microscope (6,7), or a four-stage image intensifier (37). Luminescence from motile cells was characterized by a continuous luminescence that varied between 5 and 10 times the dark current of the photomultiplier, and by intermittent pulses of luminescence of variable amplitude (Fig. 4: upper). The use of an image intensifier allowed us to localize the two types of luminescent signals. The continuous luminescence was localized in the tails of actively motile cells (Fig. 4: lower), and the spontaneous pulses occurred primarily over the anterior regions of cells. The intermittent pulses were sometimes correlated with extending pseudopods. Calibration of the luminescence in relation to the free [Ca^{++}] indicated that the free [Ca^{++}] increased to between $1.0-5.0 \times 10^{-7}$ M during normal amoeboid movement (Fig. 5).

The source of calcium has not yet been defined. However, several observations have implicated both external and internal sources. The elevated free [Ca^{++}] was sustained over an extended area in the tail during movement and was not affected immediately by decreasing the extracellular free [Ca^{++}]. Similarly, cytoplasmic

FIGURE 4

Aequorin luminescence in C. carolinensis (upper) Aequorin luminescence measured from an entire cell moving normally. These cells exhibit both continuous luminescence and intermittent pulses of luminescence. The shutter between the cell and the photon counting system was closed at the beginning of the record. (lower) Image of C. carolinensis loaded with aequorin and viewed with an image intensifier. The cell is outlined with dashes. Continuous luminescence can be detected in the tails of moving cells (47).

streaming out of the tail region did not cease immediately after removal of extracellular calcium (47). A cellular source of calcium was therefore implicated. This source could have been the glycocalyx on the cell surface, the cytoplasmic surface of the plasmalemma (16), or calcium-sequestering vesicles (12,36). In contrast, the transient elevated free [Ca^{++}] in the anterior region of cells was rapidly abolished by lowering the extracellular free [Ca^{++}](47). In addition, pseudopods stopped extending during the same time period after lowering the external [Ca^{++}], suggesting an external source of calcium for pseudopod extension.

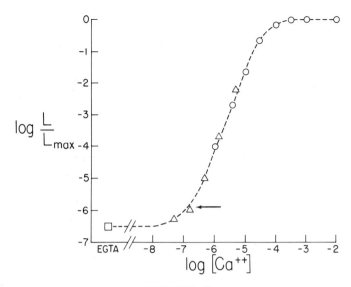

FIGURE 5

Calibration curve relating the intensity of aequorin luminescence to free [Ca^{++}] under conditions believed to be appropriate to amoeba cytoplasm. Curves were determined in vitro in solutions containing 75 mM KCl, 1.0 mM $MgCl_2$, 2.5 mM Pipes buffer, pH 7.0 at 22°C. Three rounded, non-motile cells exhibited an average log fractional luminescence of -5.98 (arrow). (O) $CaCl_2$ dilutions. (△) Ca-EGTA buffers (0.5 mM EGTA). (□) 0.5 mM EGTA with no added Ca^{++}. The arrow indicates the [Ca^{++}] measured in resting (non-motile) cells (47).

What is the relationship between pCa and pH in amoeboid cells? The free [Ca^{++}] and pH are interrelated parameters. For example, the pH of a solution has a dramatic effect on the K'app of calcium binding to EGTA (4,35). Therefore, the free [Ca^{++}] must be defined in a Ca^{++}/EGTA buffer relative to the pH. The lower the pH the higher the free [Ca^{++}]. Furthermore, titration of Ca^{++}/EGTA buffers with calcium causes a drop in pH. This drop in pH is caused by the release of protons when EGTA binds calcium. This basic principle also applies to solutions containing calcium binding proteins. Most data indicate that protons can compete with divalent cations for binding sites (19). For example, Ca^{++} binding to rabbit muscle troponin C is affected by pH changes (38). Therefore the pH in cells, cell extracts and reconstituted models must be defined carefully.

Changes in cytoplasmic pH have also been implicated as a direct secondary messenger (19,23). For example, it has been demonstrated that both gelation and contraction of cytoplasmic extracts can be regulated by changing the pH, while maintaining a subthreshold free [Ca^{++}] (15,49). Increasing the pH above about pH 6.8 induces the solation of the gel and contraction. Therefore, small increases in pH (0.2 to 0.4 units) can simulate the effects of about 10^{-6} M free [Ca^{++}](49). It is important to note that small increases in pH which induce solation would tend to decrease the free [Ca^{++}]. Thus, the effect on cytoplasmic consistency of a change in pH could not be due to an indirect effect of pH on the free [Ca^{++}]. In addition, an increase in the free [Ca^{++}] was not detectable by use of aequorin during contraction induced by an increase of the pH in an extract of D. discoideum (15).

The potential regulatory role of pH was also demonstrated in a partially reconstituted cytoskeletal and contractile system. A myosin-free preparation derived from the contracted pellet of D. discoideum exhibited maximal gelation at pH 6.8. Both lower and higher pH inhibited gelation (25). pH regulated contraction was observed following the addition of purified myosin back into the model system. Contraction was not observed at pH 6.8 where the gel was stable. The rate of contraction increased as the pH was raised to pH 7.6. Lowering the pH below 6.8 inhibited both gelation and contraction (25). A direct effect of pH on contractility has also been demonstrated by microinjecting a calcium-free buffer (relaxation solution) at pH 7.5 into the tails of amoebae (46). The elevated pH induced contractions were similar to those described after injection of calcium.

We have attempted to measure the cytoplasmic pH and changes in pH in specific regions of single, motile cells. Microelectrodes are not suited for determining the spatial distribution of pH in motile cells, and weak acids and pH sensitive indicator dyes are valuable only for obtaining an average cellular pH (average of organelles and cytoplasm). We therefore developed an optical technique to determine cytoplasmic pH by measuring the fluorescence of fluorescein-thiocarbamyl (FTC)-ovalbumin after microinjection into individual cells. The method is based on the fact that the fluorescence excitation spectrum of fluorescein is pH sensitive. The FTC-ovalbumin conjugate is retained in the cytoplasm and does not permeate organelles (24).

The average cytoplasmic pH measured in C. carolinensis amoebae was 6.75 (SD ± 0.3)(Fig. 6). The average pH in different cells ranged from 6.3 to 7.4. Comparison of tails and advancing pseudopod tips revealed no significant differences in cytoplasmic pH at the level of spatial (measuring aperture of 50 μm) and temporal (1.3 sec) resolution available in this study. Further investigations will employ a detection system with better spatial and temporal resolution (Tanasugarn and Taylor, unpublished observation).

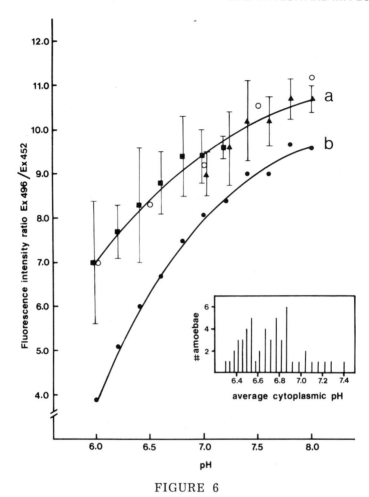

FIGURE 6

Standard curves of fluorescence intensity ratio ($Ex_{496}:Ex_{452}$, $EM_{520-560}$) vs pH for FTC-ovalbumin in situ (a) and in solution (b), as measured with the microfluorometer. The standard curve in situ utilized cells equilibrated in weak acids and based (■,▲) or injected with strong buffers (○). INSET: a bar graph of the distribution of average cytoplasmic pH among 52 FTC-ovalbumin injected amoebae using the in situ calibration curve (a) (24).

The average cytoplasmic pH of 6.75 is the pH at which regulation of gelation and contraction by calcium is most readily demonstrated in cell extracts and reconstituted model systems in vitro (15,25,49). It is possible that changes in pH could at least "fine tune" the regulation by calcium (24).

Thus, experimentally induced changes in the concentration of either calcium or protons appear to modulate the structure and contractility of cytoplasm both in vitro and in intact cells. In order to assess the relative importance of these potential secondary messengers in regulation of amoeboid movement, it is necessary to make additional measurements of the concentration of these ions in cells, and to correlate changes in concentration with cellular activities. In addition, it is necessary to determine whether the concentration of protons and of calcium ions in cells vary independently, or whether some degree of coordinate regulation may occur.

Do secondary messengers such as calcium and protons change primarily the distribution or activity of contractile proteins? When an amoeboid cell is stimulated to move, at least two underlying mechanisms could be involved. The stimulation could induce primarily a redistribution of the major cytoskeletal proteins to the site of pseudopod extension (21,42). Alternatively, the stimulation could induce primarily a change in the activity of an otherwise uniformly distributed cytoskeletal and contractile machinery (47,54). This change in activity could be due to changes in assembly, interaction, or chemical modification of the contractile proteins. The two mechanisms are not mutually exclusive. We shall attempt to assess the evidence for each postulate which has been derived from studies of cell extracts, reconstituted systems and whole cells.

Any theory of the role of calcium in amoeboid locomotion must be consistent with data concerning the distribution and organization of the major cytoskeletal and contractile proteins in locomoting cells. Immunofluorescence microscopy has been widely used in defining these parameters. Images obtained by this technique must be interpreted with caution, since it is difficult to assess possible effects of fixation, extraction, the availability of antigenic determinants, and the pathlength or accessible volume in different regions of the cells. In addition, only static images may be obtained by this technique (56,58). Although the method has certain limitations which have been outlined above, immunofluorescence microscopy has been used to obtain a great deal of information concerning the distribution of contractile proteins in cells (20).

The technique of fluorescent analog cytochemistry, formerly called molecular cytochemistry, was used to determine the distribution and organization of 5-iodoacetamidofluorescein labelled actin (AF-actin) in living C. carolinensis during amoeboid movement (54, 58). The fluorescent images of actin were compared to the fluorescent images of ovalbumin labelled with rhodamine. Therefore, the effects of pathlength and accessible volume were evaluated in the same cells. The ectoplasm in the tails of actively motile cells exhibited distinct fluorescent bundles containing AF-actin (Fig. 7). Furthermore, the plasmagel sheets, which "peel" off of the cytoplasm-membrane interface at the tips of advancing pseudopods also

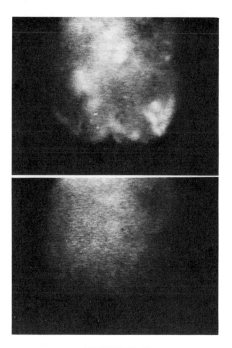

FIGURE 7

Fluorescent analog cytochemistry of actin in C. carolinensis. AF-actin was injected together with rhodamine labelled ovalbumin into C. carolinensis. The tails exhibited distinct actin fluorescence in the form of fibrils (upper figure), while only uniform fluorescence of ovalbumin was observed (lower figure)(54).

contained fluorescent AF-actin bundles (Fig. 8)(54). Otherwise, actin appeared to be rather uniformly distributed in actively motile cells. Recently it has been demonstrated that actively motile Hela cells exhibited a relatively homogeneous distribution of actin and myosin (26). Thus there is evidence that the distribution of both actin and myosin is rather uniform in motile cells. These results suggest that extensive changes in the concentration of actin is not required for amoeboid movement. The fluorescent bundles of actin observed in the tail cortex and in the plasmagel sheet of amoebae may reflect local increases in the actin concentration which result from contractions induced by local stimulation by secondary messengers such as Ca^{++} or protons.

More recently, the distribution of AF-actin in murine peritoneal macrophages has been studied by fluorescent analog cytochemistry (Amato and Taylor, unpublished observations). Macrophages exhibit a rather homogeneous distribution of actin. A small increase in the local concentration of AF-actin is detectable in the pseudopods

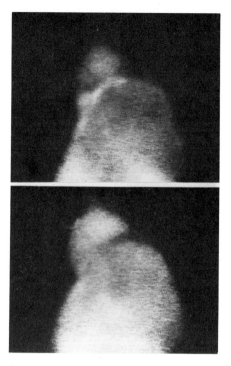

FIGURE 8

Fluorescent analog cytochemistry of actin in C. carolinensis. Plasmagel sheets in extending pseudopods contained actin fibrils (arrow, upper figure). In contrast, rhodamine labelled ovalbumin remained uniformly distributed (54).

engulfing opsonized particles. However, extensive redistribution of actin is not observed in macrophages engaged in locomotion or phagocytosis.

The distribution of other contractile/cytoskeletal proteins has been studied by the immunofluorescence technique. A dramatic polarization and an apparent extensive redistribution of myosin and actin binding protein in macrophages engaged in phagocytosis has been observed (5,42). These results support the hypothesis that transcellular redistribution of contractile proteins, manifested by large fluctuations in concentration, does occur during amoeboid movement.

It is not possible to decide on the basis of these results whether some cytoskeletal proteins such as myosin and actin binding protein do redistribute extensively during movement, while others such as actin do not, since the techniques used to determine the distribution of these proteins were quite different. By application of fluorescent analog cytochemistry to other components of the contractile apparatus, it may be possible to determine whether the differences in distribution of cytoskeletal proteins described above are significant.

Cortical wound healing has been described in a variety of cells (50). This process of resealing a damaged region of the membrane-cortex has served as a valuable model of contractions. We have shown that C. carolinensis amoebae exhibited a rise in the free [Ca^{++}] and a decrease in the cytoplasmic pH during the wound healing process (24,47). Fluorescent analog cytochemistry of AF-actin demonstrated that actin increased in concentration only transiently during the wound healing process (54)(Fig. 9). Damage to the cell cortex initiated a contraction which reached an extent not observed during normal amoeboid movement. The increase in the local actin concentration was probably the result of contraction and was transient in time. The rise in free [Ca^{++}] and the decrease in pH lasted only a few seconds and the local increase in the actin concentration persisted for only a few minutes. Thus, large transcellular redistribution of contractile/cytoskeletal proteins did not appear to be characteristic of cells exhibiting normal amoeboid locomotion.

Studies of cell extracts and reconstituted systems support the concept that local activation of an initially uniformly distributed cytoskeletal-contractile system might direct cell movement. Cell extracts were allowed to gel either in a quartz capillary (46), or in a microscope observation chamber (49,53). One end of the gelled extract was then induced to contract by the local addition of the contraction solution. Cytoplasmic streaming and contraction of fibrils was initiated only at the site of calcium addition. Contractions were defined as the shortening and thickening of fibrils, and the development of strain birefringence. During contraction, birefringent fibrils developed at the site of calcium addition and extended toward the distal end of the preparation which initially remained non-motile, viscoelastic and isotropic. A wave of contraction forming a small contracted mass was observed to proceed with time from the point of addition of calcium to the opposite end of the preparation. It is important to note that no net extension of cytoplasm occurred from the distal end of the capillary. Under these conditions no equivalent of pseudopod extension occurred.

Similar experiments have been performed on myosin-depleted fractions prepared from the extract of D. discoideum (15,25). Local addition of calcium to such a myosin-depleted fraction induced a wave of solation from the site of addition of calcium towards the opposite end (Taylor and Hellewell, unpublished observations). These results suggested that the initial response to a contraction stimulus was a solation at the site of stimulation (51).

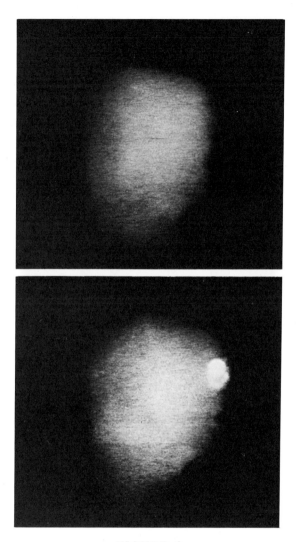

FIGURE 9

Damage response to insertion of a microneedle into the tail cortex of C. carolinensis preloaded with AF-actin and rhodamine labelled ovalbumin. AF-actin fluorescence became distinct as the cortical wound-healing contraction was initiated (lower figure), whereas no distinct ovalbumin fluorescence was observed in the region of the wound (upper figure)(54).

Stendahl and Stossel (43) studied reconstituted mixtures of actin binding protein, actin, muscle myosin and macrophage gelsolin placed in capillaries. A net displacement of the protein complex away from the site of application of calcium was also observed, but it was suggested that the displacement was due to the increased efficiency of contraction in the more cross-linked (gelled) domains at the distal ends of the capillaries. They reasoned that solation of the gel at the site of calcium addition would decrease the efficiency of contraction at that site. A displacement of the lattice network in the capillary is consistent with the proposed redistribution of actin binding protein and myosin at the leading edge of migrating cells (43). This interpretation implies that maximal force generation is occurring at the end opposite to that defined in our experiments (46,51).

The solation-contraction coupling hypothesis can explain the capillary and observation chamber experiments quite simply. Application of calcium at one end of the capillary induces a wave of solation and contraction. As the gel begins to solate its rigidity decreases maximally at the site of calcium addition, permitting extensive sliding of actin and myosin filaments. It is important to note that actin cross-links remaining in the partially solated region are required to transmit tension generated by the cyclic interaction of actin and myosin, and to maintain the integrity of the contracting mass. In this regard, we agree with the concept that actin cross-linking factors enhance the efficiency of the transmission of contractile force (43,51). In addition, the active interactions of actin and myosin in the contracting region may also serve to generate and to transmit contractile force. The gelled region distal to the site of contraction may serve as the anchor for the contractile force. Thus, the contracting mass pulls itself towards the non-motile gelled region. The solation-contraction is propagated away from the site of stimulation as the calcium diffuses into the capillary. In contrast to the proposal of Stendahl and Stossel (43), we suggested that solation and contraction are coupled, and that both occur at the site of stimulation (15,25,51).

A corollary to the hypothesis that calcium induces a local change in activity of a relatively homogeneous contractile apparatus is that changes in the concentration of calcium in cells must also be local. This proposal is supported by the observation that exposure of cell extracts to a high (about 10^{-6} M), uniform concentration of calcium results in superprecipitation without streaming (49). Moreover, local increases in the free [Ca^{++}] have been observed directly in C. carolinensis amoebae (47). Thus we propose that calcium may induce local changes in the activity of a rather uniformly distributed cytoskeletal-contractile apparatus, and that localized and transient increases in concentration of contractile proteins may result from contractile events.

What are the cellular targets of elevated free [Ca^{++}]? Understanding the regulatory action of calcium on amoeboid movement requires the study of mechanisms for controlling the free [Ca^{++}] in the cytoplasm, and the study of possible targets of calcium action. Calcium might exert a direct effect on actin (17,29), or an indirect effect on actin by affecting factors which regulate the length (8, 22,61) and/or cross-linking of actin filaments (25,32,52). Similarly, calcium could underlie either direct or indirect mechanisms of myosin linked regulation. Myosin isolated from some non-muscle cells appears to be regulated by a calcium and calmodulin-dependent phosphorylation (1). Phosphorylation of the 20,000 dalton light chain appears to affect both the actin activated ATPase activity of the myosin, and the competence of the myosin for filament formation (39,45). Calcium-independent phosphorylation of the heavy chain modulates actin-activated ATPase activity and filament formation of myosin isolated from D. discoideum amoebae (40). A direct effect of calcium on myosin from non-muscle cells has not been observed (28). Calcium could also exert a pleiotropic effect on cells as a consequence of its binding to calmodulin, since this protein appears to activate a variety of cellular enzyme systems (31). This list is by no means all inclusive and other possible sites of calcium action may be proposed as components of the contractile apparatus. A number of cytoskeletal and contractile proteins have been identified in the amoeboid stage of D. discoideum, and the activity of many of these proteins has been shown to be sensitive to about 10^{-6} M calcium in vitro (Table 1).

It has become evident that actin plays a dual role as part of the cytoskeleton and the contractile apparatus (15,34,44,49). Unfortunately, most of the methods for analyzing actin have poor time resolution (i.e. sedimentation, viscosity and flow birefringence). We decided that a more sensitive assay of actin structure and function was needed. Therefore, a fluorescence energy transfer technique was developed to study actin assembly-disassembly, and actin subunit exchange (52,55,59). This approach will permit us to investigate the effects of calcium, pH and other environmental conditions on the structure and dynamics of actin. In addition, these techniques will be useful in characterizing the interaction of specific factors with actin.

We initiated our studies on the actin accessory proteins responsible for both structural and contractile properties of cytoplasm by isolating the contracted pellets from extracts of D. discoideum (25). In this original study two fractions were isolated from the contracted pellet which formed highly viscous complexes when mixed with rabbit muscle actin. The viscosity was maximal at free [Ca^{++}] less than or equal to about 10^{-8} M, while the gel solated when the free [Ca^{++}] was raised to about 0.5 to 1.0 x 10^{-6} M. The fractions contained prominent polypeptides with apparent molecular weights

TABLE 1: Cytoskeletal and contractile proteins in D. discoideum.

Protein	Apparent Function	Reference
1. Actin	Cytoskeleton, contraction	57
2. Myosin	Contraction	10
3. Calmodulin	Unknown	9
4. 120 K Factor	Gelation of actin	14
5. 95 K Factor	Ca^{++} sensitive; actin cross-linking	25, 52
6. 30 K Factor	Ca^{++} sensitive; actin cross-linking	25, 52
7. 40 K Factor	Ca^{++} sensitive; actin shortening	40
8. 47 K plus Factor	Actin filament capping	52

of: (I) 43,000 (actin) and 95,000 daltons; and (II) 43,000 (actin), 30,000, and 18,000 daltons (25).

The 95,000 dalton polypeptide has been purified (52; and Fechheimer, Brier and Taylor, unpublished observations). Low shear viscosity assays with rabbit skeletal muscle actin (41) indicated that the viscosity increased as the concentration of the 95,000 dalton protein was increased if the free [Ca^{++}] was maintained at about 10^{-8} M. The increase in viscosity was not observed in the presence of about 10^{-6} M free [Ca^{++}] (Fig. 10). Mixtures of 0.8 mg/ml actin plus 95,000 dalton protein above about 50 µg/ml exhibited strain birefringence when tension was applied with a microneedle. Preliminary analysis by sedimentation and electron microscopy suggested that the 95,000 dalton protein does not dramatically restrict the length of actin filaments in the presence of calcium. Furthermore, low angle shadowing demonstrated that this protein is a rod shaped molecule, 40 nm in length (Fig. 10: insert). This is similar to the structure of α-actinin (33).

Preliminary calcium binding studies have been performed on the purified 95,000 dalton protein (Virgin, Hellewell and Taylor, unpublished observations). The rate of dialysis flow technique (13, 18) was used to show that the 95,000 dalton protein binds calcium with high affinity in the micromolar range (Fig. 11). The precise determination of the equilibrium constant has not yet been completed.

We have also purified the 30,000 dalton calcium sensitive actin binding protein from the contracted pellet (Fig. 12)(52; and Brier and Taylor, unpublished observations). Peptide mapping indicated that this factor is not a proteolytic fragment of the 95,000 dalton factor (Luna, unpublished observation). Mixtures of this protein with rabbit muscle actin also exhibited high viscosity if the free [Ca^{++}] was about 10^{-8} M. The 30,000 dalton factor was more active on a weight basis than the 95,000 dalton protein. Interestingly, the viscosity of mixtures of the 30,000 dalton factor with actin was

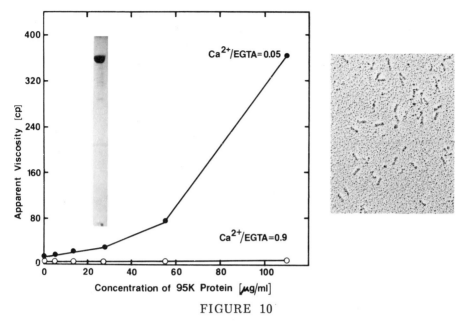

FIGURE 10

95,000 dalton calcium sensitive actin binding protein from D. discoideum. The purified 95,000 dalton factor induces an increase in viscosity measured at a low shear rate when mixed at increasing concentrations with 0.8 mg/ml actin. Assay conditions and methods for the measurement of viscosity were as previously described (25). The increase in viscosity is observed if the free [Ca^{++}] is lower than about 1.0×10^{-7} M. Raising the free [Ca^{++}] above about $1.0-5.0 \times 10^{-7}$ M reversibly inhibits the formation of a highly viscous complex. The 95,000 dalton factor is a rod-shaped molecule about 40 nm in length as judged by shadowing techniques (52).

higher if measured at 0°C as compared to 28°C. Formation of viscous complexes in mixtures of actin and 30,000 dalton factor was not observed in the presence of about 10^{-6} M [Ca^{++}] (Fig. 12).

We have initiated investigations on two separate calcium sensitive factors which can regulate the structure of actin filament networks. However, there are at least two other high affinity calcium binding proteins in D. discoideum. Calmodulin (9) has not been characterized as to function in D. discoideum, but it could exhibit several functions, including regulation of myosin phosphorylation. In addition, Spudich and colleagues (40) have identified a 40,000 dalton factor having properties reminiscent of gelsolin and fragmin (22,61). This factor appears to regulate the length of actin filaments in a calcium dependent manner. Therefore, the number of possible targets for calcium in the cytoskeletal and contractile apparatus is

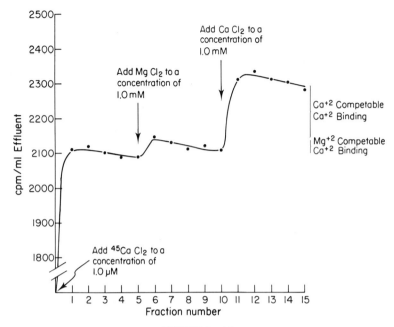

FIGURE 11

$^{45}Ca^{++}$ binding to 95K polypeptide. Calcium binding of the 95,000 dalton factor as determined by the rate of dialysis flow cell technique. This factor exhibits high affinity, magnesium non-competable binding of calcium (Virgin, Hellewell and Taylor, unpublished observation).

large. More complete characterization of these and other unidentified calcium binding proteins must be completed before defining the total role of calcium in amoeboid movement.

SUMMARY

Preliminary answers are available for the four questions posed in this manuscript: 1) Fluctuations in the free [Ca^{++}] can regulate cytoplasmic consistency and contractility and thus amoeboid movement. Evidence has been obtained from studies of single cell models, cell extracts, reconstituted model systems, purified proteins, and living cells. An important fact is that the amplitude of the fluctuation of free [Ca^{++}] is very small during amoeboid movement. In contrast to other stimulus-response coupling processes which are initiated by increases in the free [Ca^{++}], amoeboid movement appears to be regulated at a submicromolar free [Ca^{++}] (see also 11). Therefore it is necessary to determine the local concentration and affinity constants of the potential calcium regulatory proteins in

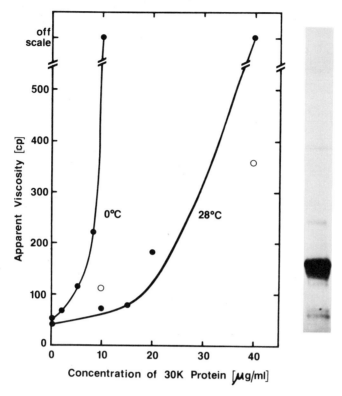

FIGURE 12

30,000 dalton calcium sensitive actin-binding protein from D. discoideum. This factor induces an increase in low shear viscosity when mixed at increasing concentrations with 0.8 mg/ml actin (methods as described in Fig. 10). The increase in viscosity is observed if the free [Ca^{++}] is lower than about 1.0×10^{-7} M (●). Raising the free [Ca^{++}] above about 1.0–5.0×10^{-7} M (○) reversibly inhibits the formation of a highly viscous complex. This factor is more active at 4°C than at 28°C (52).

order to evaluate their participation in a mechanism of regulation of amoeboid movement by calcium. 2) At the present time no decision can be made concerning the relative importance of pH and pCa in regulation. Both parameters have been shown to regulate cytoplasmic structure and contractility. In addition, a change in the free [Ca^{++}] could induce a change in pH, and vice versa. Calcium and pH may also act in concert to modulate cellular activity. An average cytoplasmic pH of 6.75 measured in C. carolinensis is consistent with the pH optimum for calcium regulation identified in cell extracts and partially reconstituted systems. 3) Secondary messengers appear to activate the cytoskeletal and contractile

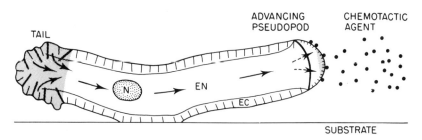

FIGURE 13

Summary of the distribution of actin fibrils and threshold free [Ca^{++}] in motile C. carolinensis. Transient increase in the free [Ca^{++}] occur at the tips of advancing pseudopods (shaded region), possibly due to the binding of chemotactic agents. The plasmagel sheet (dark line near the tip of the advancing pseudopod) contains actin. Continuous elevated free [Ca^{++}] occur in the tails of motile cells (shaded region). Distinct AF-actin fluorescent fibrils are localized in the same region of the tail (see dark lines in the tail cortex). The arrows indicate the direction of streaming. N = nucleus; EN = endoplasm; EC = ectoplasm. (See text for details of the working hypothesis.)

apparatus locally. The initial and primary effect of the local stimulation may be to change the activity of the cytoskeletal and contractile machinery. Following stimulation, there could be a local and transient increase in the concentration of actin at the site(s) of contraction. Contractions do not usually result in extensive transcellular variation in the concentration of actin. More extensive redistributions of other cytoskeletal-contractile proteins cannot be excluded at this time. Application of fluorescent analog cytochemistry with other cytoskeletal and/or contractile proteins should help to solve this question. 4) There is a growing list of factors which both bind calcium with high affinity and exert a calcium sensitive effect on some of the cytoskeletal and contractile proteins. New fluorescence assays should aid in the molecular dissection of the interactions of possible secondary messengers with actin, myosin and accessory proteins of the contractile apparatus. We have previously proposed a working hypothesis of amoeboid movement including the possible interactions of secondary messengers and contractile/cytoskeletal proteins in the process (47,51)(Fig. 13).

Our working hypothesis of amoeboid movement can explain many of the experimental facts described over many years. The local rise in secondary messenger(s) (i.e. calcium and/or pH) in the tail causes solation-contraction as described in studies on cell extracts. Tension generated by the contractions causes the formation of actin containing fibrils which actively shorten to produce force (Fig. 13). The solation of the cortex which accompanies the contraction in a "self destruct" mechanism forms the new endoplasm. The endoplasm is more solated than the cortex and is pushed forward by the force of the contraction (30,46). The free [Ca^{++}] is lowered by an unknown calcium sequestering system and the endoplasm is gradually returned to the conditions favoring gelation. Thus, anterior endoplasm is actually a weak gel (51). The endoplasm becomes ectoplasm whenever maximal gelation occurs, and the major site of gel formation is the anterior ectoplasm/endoplasm interface.

Chemotactic agents could bind to the surface of the cells and cause localized pseudopod extension by inducing transient and localized changes in the concentrations of secondary messengers (i.e. free Ca^{++}, Fig. 13). For example, localized increases in free [Ca^{++}] would solate the gel and, if myosin were present, induce a local contraction. The plasmagel sheet which "peels off" the cytoplasmic surface of the plasmalemma at the tips of advancing pseudopods is probably formed during the transient rise in free [Ca^{++}]. Removal of part of the cortex from the membrane makes this site weak, thus permitting cytoplasmic streaming in that direction.

The evidence is strong for contractions in the tails of locomoting amoebae. However, it is also possible that force producing contractions also occur in the endoplasm at the tips of advancing pseudopods (2). This latter question will require more extensive investigations.

REFERENCES

1. Adelstein, R.S. & Eisenberg, E. (1980): Ann. Rev. Biochem. 49:921-956.
2. Allen, R.D. & Allen, N.S. (1978): Ann. Rev. Biophys. Bioengin. 7:469-495.
3. Allen, D.G. & Blinks, J.R. (1979): In Detection and Measurement of Free Calcium in Cells. (eds) C.C. Ashley & A.K. Campbell, Elsevier/North-Holland, Amsterdam, pp. 155-174.
4. Amos, W.B., Routledge, L.M., Weis-Fogh, T. & Yew, F.F. (1976): Symp. Soc. Exp. Biol. 30:273-299.
5. Berlin, R.D. & Oliver, J.M. (1978): J. Cell Biol. 77:789-804.
6. Blinks, J.R. (1978): Photochem. Photobiol. 27:423-432.
7. Blinks, J.R., Mattingly, P.H., Jewell, B.R., Van Leeuwen, M., Harrer, G.C. & Allen, D.G. (1979): Meth. Enzymol. 57:292-328.
8. Bretscher, A. & Weber, K. (1980): Cell 20:839-847.
9. Clark, M., Bazari, W.L. & Kayman, S.C. (1980): J. Bacteriol. 141:397-400.
10. Clarke, M. & Spudich, J.A. (1974): J. Molec. Biol. 86:209-222.
11. Cobbold, P.H. (1980): Nature 285:441-446.
12. Coleman, J.R., Nilsson, J.R., Warner, R.R. & Batt, P. (1973): Exp. Cell Res. 76:31-40.

13. Colowick, S.P. & Womack, F.C. (1969): J. Biol. Chem. 244: 774-777.
14. Condeelis, J.S. (1981): Neurosci. Res. Prog. Bull. 19:83-99.
15. Condeelis, J.S. & Taylor, D.L. (1977): J. Cell Biol. 74:901-927.
16. Cramer, E.B. & Gallin, J.I. (1979): J. Cell Biol. 82:369-379.
17. Dancker, P. & Low, I. (1977): Biochim. Biophys. Acta 484: 169-176.
18. Feldmann, K. (1978): Anal. Biochem. 88:225-235.
19. Gillies, R.J. (1981): In The Transformed Cell. (eds) I. Cameron & T. Pool, Academic Press, New York, pp. 347-395.
20. Groschel-Stewart, U. (1980): Int. Rev. Cytol. 65:193-254.
21. Hartwig, J.H., Davies, W.A. & Stossel, T.P. (1977): J. Cell Biol. 75:956-967.
22. Hasegawa, T., Takahashi, S., Hayashi, H. & Hatano, S. (1980): Biochemistry 19:2667-2683.
23. Heiple, J. (1981): Ph.D. Thesis, Harvard University, Cambridge.
24. Heiple, J.M. & Taylor, D.L. (1980): J. Cell Biol. 86:885-890.
25. Hellewell, S.B. & Taylor, D.L. (1979): J. Cell Biol. 83:633-648.
26. Herman, I.M., Crisona, N.J. & Pollard, T.D. (1981): J. Cell Biol. 90:84-91.
27. Izzard, C.S. & Izzard, S.L. (1975): J. Cell Sci. 18:241-256.
28. Korn, E.D. (1978): Proc. Nat. Acad. Sci. 75:588-599.
29. Maruyama, K. (1981): Biochim. Biophys. Acta 667:139-142.
30. Mast, S.O. (1926): J. Morph. Physiol. 41:347-425.
31. Means, A.R. & Dedman, J.R. (1980): Nature 285:73-77.
32. Mimura, N. & Asano, A. (1979): Nature 282:44-48.
33. Podlubnaya, Z.A., Tskhovrebova, L.A., Zaalishivili, M. & Stefanenko, G.A. (1975): J. Molec. Biol. 92:357-359.
34. Pollard, T.D. (1976): J. Cell Biol. 68:579-601.
35. Portzehl, H., Caldwell, P.C. & Ruegg, J.C. (1964): Biochim. Biophys. Acta 79:581-591.
36. Reinhold, M. & Stockem, W. (1972): Cytobiologie 6:182-194.
37. Reynolds, G.T. (1968): Adv. Opt. Electron Microsc. 2:1-40.
38. Robertson, S.P., Johnson, J.D. & Potter, J.D. (1978): Biophys. J. 21:16a.
39. Scholey, J.M., Taylor, K.A. & Kendrick-Jones, J. (1980): Nature 287:233-235.
40. Spudich, J.A., Kuczmarski, E.R., Pardee, J.D., Simpson, P.A., Yamamoto, K. & Stryer, L. (1981): In Cold Spring Harbor Symposium on Quantitative Biology on Organization of the Cytoplasm. (in press).
41. Spudich, J.A. & Watt, S. (1971): J. Biol. Chem. 246:4866-4871.
42. Stendahl, O.I., Hartwig, J.H., Brotschi, E.A. & Stossel, T.P. (1980): J. Cell Biol. 84:215-224.
43. Stendahl, O.I. & Stossel, T.P. (1980): Biochem. Biophys. Res. Commun. 92:675-681.
44. Stossel, T.P. & Hartwig, J.H. (1976): J. Cell Biol. 68:602-619.
45. Suzuki, H., Onishi, H., Takahashi, K. & Watanabe, S. (1978): J. Biochem. 84:1529-1542.

46. Taylor, D.L. (1977): Exp. Cell Res. 105:413-426.
47. Taylor, D.L., Blinks, J.R. & Reynolds, G. (1980): J. Cell Biol. 86:599-607.
48. Taylor, D.L., Condeelis, J.S., Moore, P.L. & Allen, R.D. (1973): J. Cell Biol. 59:378-394.
49. Taylor, D.L., Condeelis, J.S. & Rhodes, J.A. (1977): In Proceedings of the Conference on Cell Shape and Surface Architecture. Alan Liss, Inc., New York, pp. 581-603.
50. Taylor, D.L. & Condeelis, J.S. (1979): Int. Rev. Cytol. 56: 57-144.
51. Taylor, D.L., Hellewell, S.B., Virgin, H.W. & Heiple, J.M. (1979): In Cell Motility: Molecules and Organization (eds) S. Hatano, H. Ishikawa & H. Sato, University of Tokyo Press, Tokyo, pp. 363-377.
52. Taylor, D.L., Heiple, J., Wang, Y.-L., Luna, E.J., Tanasugarn, L., Brier, J., Swanson, J., Fechheimer, M., Amato, P., Rockwell, M. & Daley, G. (1981): In Cold Spring Harbor Symposium on Quantitative Biology on Organization of the Cytoplasm. (in press).
53. Taylor, D.L., Rhodes, J.A. & Hammond, S.A. (1976): J. Cell Biol. 70:123-143.
54. Taylor, D.L., Wang, Y.-L. & Heiple, J.M. (1980): J. Cell Biol. 86:590-598.
55. Taylor, D.L., Reidler, J., Spudich, J.A. & Stryer, L. (1981): J. Cell Biol. 89:362-367.
56. Taylor, D.L. & Wang, Y.-L. (1980): Nature 284:405-410.
57. Uyemura, D.G., Brown, S.S. & Spudich, J.A. (1978): J. Biol. Chem. 253:9088-9096.
58. Wang, Y.-L., Heiple, J.M. & Taylor, D.L. (1981): In Handbook for Cell Motility. (ed) L. Wilson, Academic Press, New York, (in press).
59. Wang, Y.-L. & Taylor, D.L. (1981): Cell 27:429-436.
60. Yin, H.L. & Stossel, T.P. (1979): Nature 281:583-586.
61. Yin, H.L., Zaner, K.S. & Stossel, T.P. (1980): J. Biol. Chem. 255:9494-9500.

ACTINOGELIN: A CALCIUM SENSITIVE REGULATOR OF MICROFILAMENT SYSTEM

A. Asano, N. Mimura and P.F. Kuo*

Institute for Protein Research, Osaka University
Osaka 565, Japan

INTRODUCTION

A novel type of protein which induced gelation of F-actin at low Ca^{++} concentrations was isolated from Ehrlich tumor cells and from rat hepatic cells. However, this gelation reactions can be halted by the addition of Ca^{++} with a half-maximal inhibition at about 2 µM. Solation is not due to severing of F-actin by this protein.

Intercellular distribution and intracellular localization of actinogelin were studied using anti-actinogelin antibody raised in rabbits. Free cells such as macrophages and lymphocytes, and tissue cells including epithelial cells and fibroblastic cells, contained actinogelin. Intracellularly, converge or focal points of microfilaments are preferentially stained with the antibody, followed by microfilaments localized in terminal web region of intestinal epithelial cells and other types of microfilaments. Based on these results, possible participation of actinogelin in regulation of microfilament organization is suggested.

Microfilaments in non-muscle cells differ significantly from those in muscle cells in at least one respect, that is, variability of intracellular distribution of microfilaments. In muscle, microfilaments are orderly organized in parallel to each other, whereas microfilament bundles in non-muscle cells are not arranged in parallel, but rather criss-cross randomly in the cytoplasm.

---

*Present address: Nippon Zoki Seiyaku, K.K., Osaka, Japan

Furthermore, distribution patterns of microfilament bundles are drastically changed depending on the state of cells; for example, those in cell movement, cell adhesion, cell division and so forth. In addition, presence of unbundles microfilaments and a large amount of non-filamentous actin in non-muscle cells (14) further complicate the regulation of microfilament organization in non-muscle cells.

For our earlier discovery of the possible participation of microfilaments in regulation of HVJ (Sendai virus)-induced cell fusion (1), and cooperation of microfilaments and microtubules in phagocytosis (10), we were interested in regulatory mechanism of the microfilament system. During this line of study, we found that actin related gelation of crude extracts of Ehrlich tumor cells was controlled by micromolar concentration of Ca^{++} (11). Since this concentration of Ca^{++} is well known as an intracellular regulator of diverse cellular functions, we started to study the molecular basis of this Ca^{++} sensitivity.

As will be described in this chapter, a novel protein, named actinogelin, was isolated first from Ehrlich tumor cell extracts and later from rat hepatic cells, and was found to induce gelation of F-actin at low Ca^{++} but not at Ca^{++} concentrations higher than several µM. Molecular properties and intracellular distribution of this protein will also be reported.

MATERIALS AND METHODS

Actiongelin was purified from Ehrlich tumor cells as described previously (N. Nimura and A. Asano, submitted for publication). Actin was isolated from rabbit skeletal muscle by the method of Spudich and Watt (15), and heavy meromyosin was prepared as described by Yagi and Yasawa (16).

Primary cultures of chicken embryo fibroblasts were prepared from 10-day-old embryos and immunofluorescent staining was performed on cells 16 hr after third passage. Mouse peritoneal cells, mostly consisting of macrophages, were obtained after thioglycolate activation.

Antibody to actinogelin was raised in rabbits, and purified by an affinity column (actinogelin-ultrogel AcA). Fluorescein isothiocyanate (FITC)-conjugated IgG (goat) against rabbit IgG was purified by DEAE-cellulose chromatography.

Indirect fluorescent labelling of cells by FITC-IgG was performed after formaldehyde fixation and 100% acetone treatment (4). Fluorescent labelling of heavy meromyosin with tetramethylrhodamine isothiocyanate (TRITC) was performed as described by Herman

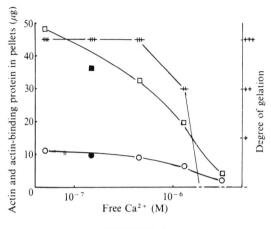

FIGURE 1

Inhibition of gelation with Ca^{++}. The degree of gelation is expressed as follows: +++, firm gel which could turnover easily without breaking the gel; ++, gel which showed no appreciable deformation when the tube containing it was brought to horizontal position; +, gel which deformed but could resist tilting of the tube; -, no gelation or gel easily broken down with slight disturbance. Open symbols, amounts of actin (□) and filamin (○) in the pellets. Calculation was based on the assumption that all protein was stained equally by Coomassie brilliant blue and amounts were expressed as µg protein. Solid symbols, amount of actin (■) and filamin (●) in a sample brought to a free Ca^{++} concentration of 3.1 µM, incubated for 30 min at 20°C and then brought back to 0.15 µM by the addition of EGTA. Duration of incubation for gelation at 20°C was 60 min. (Reprinted by permission from Nature 272:273, 1978).

and Pollard (7). Heavy meromyosin preparations containing about 0.5 mole of TRITC per mole of the protein was used for cell staining. Stained cells were observed using a Nikon epi-illumination instrument.

Flow birefringence was measured in an Edsal apparatus, and structural viscosity was estimated in a Low-Shear 100 Rheometer (Contraves, Zurich) (8).

RESULTS

<u>Characterization of gelation of crude cell free extracts and isolation of actinogelin</u>. Crude cell free extracts of Ehrlich tumor cells elicit gel when incubated 20-25°C provided that free Ca^{++} was kept below 10^{-6} M (Fig. 1). Gel formed by this system was compressed by centrifugation and the resultant pellets was

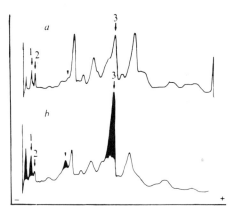

FIGURE 2

SDS-polyacrylamide gel electrophoretograms of crude extract and gel formed from the extract. (a), crude extract; (b), gel. Bands 1, 2, and 3 correspond to filamin, myosin heavy chain, and actin, respectively. Actinogelin is marked by an arrow. (Reprinted by permission from Nature 282:44, 1979).

analyzed by SDS-gel electrophoresis. As can be seen in Fig. 2, several proteins were concentrated in the gel fraction (compare Fig. 2a [crude extract] with Fig. 2b [gel fraction]). The most prominent protein peak in the gel fraction (peak 3) corresponded to authentic skeletal muscle actin. The peak marked as "1" showed an electrophoretic mobility that is identical with purified filamin of the Ehrlich tumor cell. Contents of the above two proteins in the pellets decreased as free Ca^{++} concentration increased from about 10^{-7} M to 10^{-5} M (Fig. 1). Beside these two, the other protein peak which migrate at about 100 kilodalton protein was also concentrated in the gel fraction (Fig. 2b).

To know what component is responsible for Ca^{++} sensitivity of the gelation reaction, we started to fractionate the crude extracts in hope of isolating the gelation-inducing factor and putative Ca^{++} sensitivity conferring factor using rabbit skeletal muscle actin as a main building block for the gel. As shown in Fig. 3, a fraction was obtained from a hydroxylapatite column eluate which induces gelation of F-actin only at low Ca^{++} condition (below 10^{-7} M free Ca^{++}). Since this fraction did not contain filamin, a protein which is known to induce Ca^{++} insensitive gelation of F-actin, we started to purify gelation-inducing factor in the fraction. As shown in Fig. 4, and which will be described in detail in a separate publication, the protein was purified to a homogeneity by combination of ammonium sulfate fractionation and several types of column

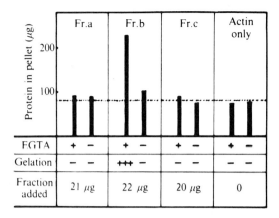

FIGURE 3

Purification of actinogelin by hydroxylapatite chromatography. Elution was carried out with a linear gradient prepared with an equal volume of 50 mM potassium phosphate and 0.25 M potassium phosphate (pH 6.8). Actinogelin was eluted as a major peak between 0.1 and 0.2 M potassium phosphate. Fractions with an appreciable actinogelin content on SDS-polyacrylamide gel electrophoresis were pooled as fraction b. Fractions just preceding the actinogelin peak were collected as fraction a, and those following the fraction b were used as fraction c. The gelation activity of the three fractions designated a, b, and c was measured as described in the text. (Reprinted by permission from Nature 282:44, 1979.)

chromatography (N. Mimura and A. Asano, submitted for publication).

The purified protein, named actinogelin, was found to induce gelation of F-actin only at low Ca^{++} concentration (Fig. 5), thus gelation-inducing activity and putative Ca^{++} sensitivity conferring activity seem to reside on the same molecule.

Molecular characterization of actinogelin. Subunit molecular weight of actinogelin estimated by SDS-gel electrophoresis is 110,000-115,000. Native molecular weight measured either by sedimentation equilibrium or by gel filtration is about twice that of denatured samples, thus the dimeric nature of the native actinogelin is evident (Table 1). Presence of a small amount of tetramer in actinogelin preparations was also found, but occurrence of higher oligomers in vivo is not clear at present since actinogelin tends to aggregate during storage.

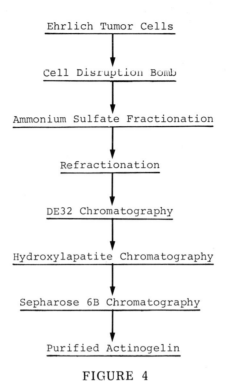

FIGURE 4

Flow diagram of purification of actinogelin.

Binding of actinogelin to F-actin was measured under different ionic conditions. High concentrations (≥ 0.3 M) of KCl completely inhibited gelation and at the same time inhibited the binding of actinogelin to F-actin, thus, this inhibition seems to be due to dissociation of the F-actin-actinogelin complex. On the other hand, only slight (0-20%) dissociation of actinogelin from F-actin occurred under high Ca^{++} (50 μM) condition, although this concentration of Ca^{++} completely inhibited the gelation reaction. Therefore we need some other explanation for Ca^{++} dependent solation.

One obvious possibility for the solation is dissociation of subunits of actinogelin under high Ca^{++} condition. But, as shown in Fig. 6, chemical cross-linking of actinogelin with dimethylsuberimidate produced identical results at either low or high Ca^{++} conditions. Native molecular weight estimated by gel filtration at low or high Ca^{++} concentrations did not reveal any change in subunit composition of the molecule (data not shown). Thus the above possibility is not substantiated.

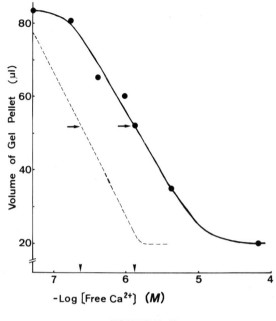

FIGURE 5

Effect of free Ca^{++} concentration on the actinogelin dependent gelation. Free Ca^{++} concentrations were manipulated by 0.5 mM EGTA and different amounts of $CaCl_2$ and calculated either using an apparent binding constant of Ca^{++} to EGTA, as published by Matsuda and Yagi (9)(——) or using a binding constant as determined by Schwarzenbach et al. (13)(----). The extent of gelation was estimated by measuring the volume of pellet after centrifugation at a defined condition. Arrows indicate half maximal inhibition of gelation.

A second possibility is that splitting (or severing) of F-actin occurs at high Ca^{++} condition as in the case of fragmin (6), gelsolin (17) and villin (2). But viscosity measurement of F-actin-actinogelin complex performed by K. Maruyama of Chiba University did not show such tendency (Figs. 7 and 8), instead, these experiments clearly showed that the viscosity of the complex measured at low shear condition is extremely high under low Ca^{++} conditions, whereas it is drastically decreased to almost the level of F-actin alone in the presence of 50 μM Ca^{++}. At higher shear stress, this difference decreased substantially indicating low rigidity of F-actin-actinogelin gel. An important finding is that in no case are flow birefringence values of F-actin-actinogelin complex lower than those of F-actin alone (Fig. 8). Therefore, it could be

TABLE 1: Properties of actinogelin.

I. Molecular Properties

Molecular weight (SDS-gel):	110,000–115,000
Molecular weight (native):	230,000–260,000
Subunit composition of native molecule:	dimer (tetramer)
Binding to F-actin($-Ca^{++}$):	actinogelin(dimer): actin(monomer) = 1:10–12
Effectors of F-actin-actinogelin binding:	KCl(0.3 M), complete dissociation Ca^{++}(50 μM), slight dissociation
Divalent cation sensitivity of actinogelin dependent gelation:	Ca^{++}; half maximal inhibition = 2 μM Sr^{++}; half maximal inhibition = 400 μM Ba^{++}; insensitive
Ca^{++} binding (equilibrium dialysis):	Yes
Antigenicity (ouchterlony):	Cross-reacted; Ehrlich tumor cell actiongelin vs rat liver actinogelin. No cross-reactivity; actinogelin vs chicken gizzard α-actinin

II. Inter- and Intracellular Distribution

Contents: > 0.3% of Ehrlich tumor cell extracts
> 0.1% of rat liver extracts

Immunologically detected in: chicken, embryo fibroblasts; mouse, 3T3 cells, intestinal epithelial cells, peritoneal macrophages, lymphocytes and embryo fibroblasts

Immunofluorescent staining:
Fibroblasts: stress fiber (focal points > periodical > continuous) no staining in nucleoplasm
Epithelial cells: terminal web > cytoplasm; no staining in microvilli and nucleoplasm

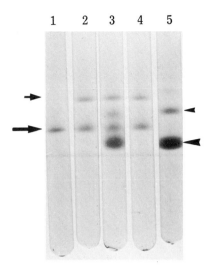

FIGURE 6

Chemical cross-linking of actinogelin under low or high Ca^{++} concentrations. Cross-linking was performed at pH 7.6 with 3 mg/ml of dimethylsuberimidate dihydrochloride, and subjected to SDS-disc gel electrophoresis at 4% acrylamide. Column 1: 10 μg of untreated actinogelin column 2: 16 μg of actinogelin cross-linked in the presence of 0.5 mM EGTA; column 3: the same as column 2 plus 20 μg of cross-linked hemocyamin (Sigma); column 4: 16 μg of actinogelin cross-linked in the presence of 50 μM $CaCl_2$; column 5: 20 μg of cross-linked hemocyanin. Large arrow = actinogelin monomer; small arrow = actinogelin dimer; large arrowhead = hemocyanin monomer (70,000 daltons); small arrowhead = hemocyanin dimer (140,000 daltons).

concluded that severing of F-actin is not the cause of Ca^{++} dependent solation. Further consideration on the mechanism of solation will be described in the Discussion.

Presence of actinogelin in different tissues and cells. Actinogelin is found to be present in a wide variety of cells and tissues. Among them, hepatic actinogelin could be purified to a homogeneity with a method modified from that employed for Ehrlich cell actinogelin (P.F. Kuo et al., in preparation). Using antibody raised in rabbits against Ehrlich tumor cell samples, immunological detection of actinogelin was performed. As summarized in Table 1, several types of cells contain actinogelin, at least in rodent.

Intracellular localization of actinogelin. Using indirect fluorescent antibody technique, intracellular localization of actinogelin was

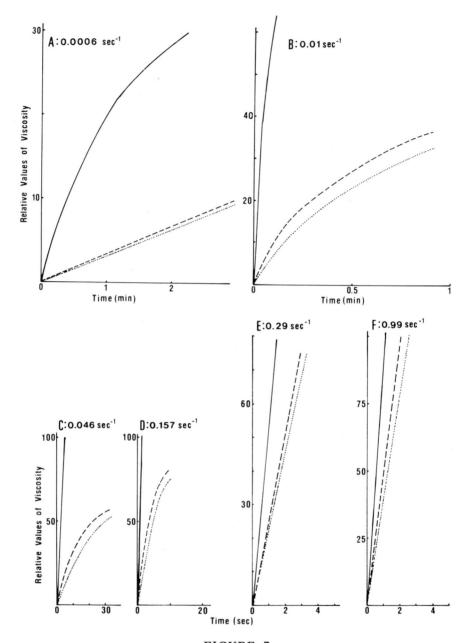

FIGURE 7

Increase of structural viscosity of F-actin-actinogelin mixture measured either in the presence or absence of Ca^{++} at different velocity gradients. Measurement of viscosity was started as soon as the cell of the apparatus was rotated. Temperature, 20°C; 0.5 mM EGTA = ——; 0.05 mM Ca^{++} = ---; control (F-actin alone, 0.5 mM EGTA = ···. (Reprinted by permission from J. Biochem. 89:317,1981).

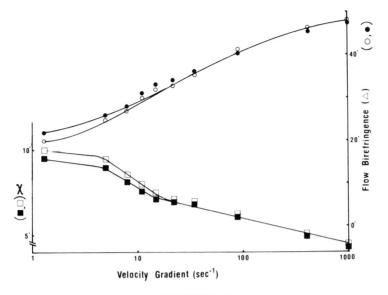

FIGURE 8

Extinction angle and flow birefringence of F-actin-actinogelin mixture. Medium contained either 0.5 mM EGTA (●,■) or 0.05 mM CaCl$_2$ (○,□). Extinction angle (X), –○–. Flow birefringence (△), –○–. (reprinted by permission from J. Biochem. 89:317, 1981.)

studied in several types of cells. In mouse peritoneal macrophages, actinogelin shows spot-like staining in more than half of the cells (Fig. 9b and 9c). Some cells of minor population exhibited more diffuse staining pattern (Fig. 9c). Staining of the same cell with TRITC-labelled heavy meromyosin resulted in similar staining but fluorescent spots were much less distinct than those stained for actinogelin (compared Fig. 9a to 9b). Entirely different staining pattern for actinogelin was observed after treatment of the cells with a Ca^{++} ionophore, A23187 (Fig. 9d), thus supporting Ca^{++} dependent regulation of microfilament distribution by actinogelin.

In fibroblastic cells, stress fibers were stained with anti-actinogelin (Fig. 10a), intense fluorescence was always observed on focal or converge points of stress fibers, and continuous staining of stress fibers was rather rare. Furthermore, adhesion plaques were also stained extensively (Fig. 10b). In intestinal epithelial cells, terminal web region is most intensively stained (Fig. 10c), together with some fluorescence in microfilaments in the other regions of cytoplasm. No staining seems to occur in microvilli (Fig 10c, and data not shown).

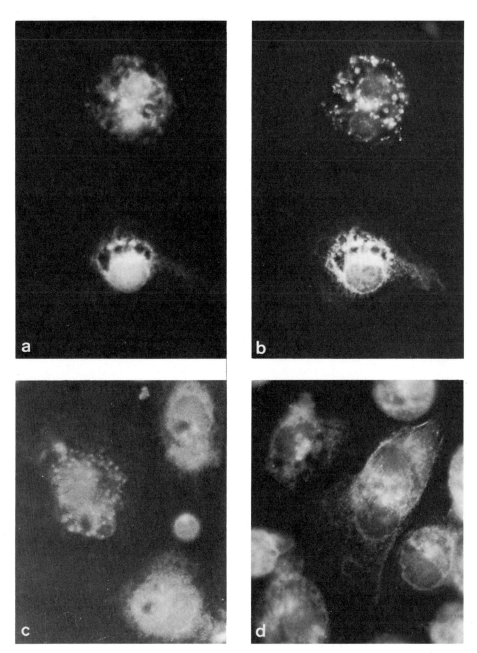

FIGURE 9

Fluorescent staining of mouse peritoneal cells mainly composed of macrophages. (a,c,d); Indirect immunofluorescent staining with anti-actinogelin IgG (rabbit) and FITC-labelled anti-rabbit IgG anti-body (goat). (b); Staining with TRITC-labelled heavy meromyosin. (d); Cells were incubated with 10 μg/ml of A23187 at 37°C for 20 min before fixation. (a,b); Control for (d) containing the same amount of dimethylsulfoxide (10 μg/ml).

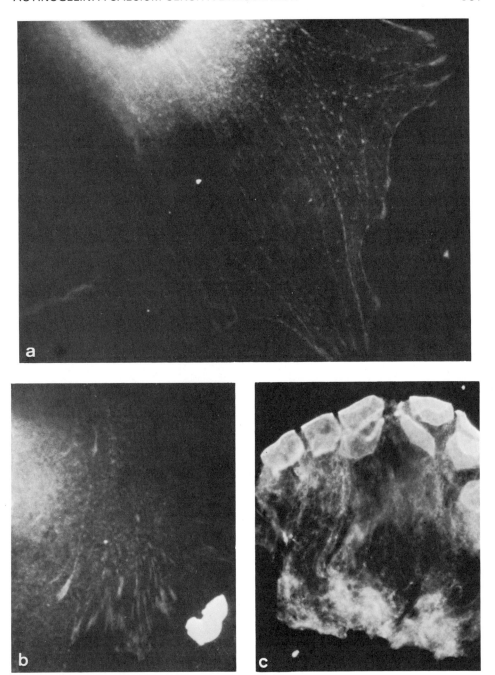

FIGURE 10
Indirect immunofluorescent staining of 3T3 cells (a,b) and mouse intestinal peritoneal cells using anti-actinogelin antibody.

DISCUSSION

From Ehrlich tumor cells and rat liver, we isolated a new type of Ca^{++} dependent regulatory protein, actinogelin, which seems to control intracellular distribution of microfilaments. In vitro, it can induce gelation of F-actin when added to several hundredth in mole per mole of actin, provided that free Ca^{++} concentration is kept around 10^{-7} M. However, at higher Ca^{++} (higher than several micromolar) concentrations, gelation was inhibited and solation of preformed gel could also be observed. Such gelation-solation interconversion was recently discovered to occur in F-actin-ABP (actin binding protein)-gelsolin system isolated from rabbit alveolar macrophages in Stossel's group (17), although gelation factor (ABP) and Ca^{++} dependent solation factor (gelsolin) are entirely different proteins. In addition, gelsolin was found to sever F-actin Ca^{++} dependently. Therefore, mechanism of regulation by Ca^{++} in both systems are quite different from each other.

The other microfilament-binding protein was isolated from microvilli of intestinal epithelial cells by Weber's group (2). This protein, villin, by itself cross-links F-actin at low Ca^{++} but severs F-actin filaments at higher Ca^{++}, thus it possesses ability of actinogelin and gelsolin, together. However, localization of villin is restricted only in microvilli of intestinal epithelial cells (3). Accordingly, possible cooperation of actinogelin and gelsolin in the same cell requires further consideration, since actinogelin and gelsolin have rather wide intercellular distribution (from this report, and from Stossel's presentation at the Cold Spring Harbor Symposium on Organization of the Cytoplasm in 1981).

If we have a cell with two types of Ca^{++} dependent regulators of microfilament system (i.e. actinogelin and gelsolin), the first problem to be considered is the difference in concentrations of Ca^{++} which affect these two regulators. Thinking in this direction we found that free Ca^{++} concentrations manipulated by EGTA-Ca^{++} mixture are not always calculated using the same binding constant by researchers of this field. Many people, including Stossel (17), calculated apparent binding constants at the pH employed using a binding constant of EGTA to Ca^{++} of $K = 10^{11.00}$ taken from the original measurement of Schwarzenbach et al. (13), whereas recent determination of the apparent binding constant around neutral pH gave slightly different values (5,9). Thus, we used the binding constant most recently determined (9) for the calculation, but for comparison we included in Fig. 5, by dashed line, the results of the same experiment calculated using a binding constant of $K = 10^{11.00}$. As can be seen in Fig. 5, more than five fold difference was evident for the Ca^{++} concentration required for half-maximal inhibition of the gelation reaction. Thus, detailed comparison of Ca^{++} sensitivity of different regulatory proteins is rather difficult.

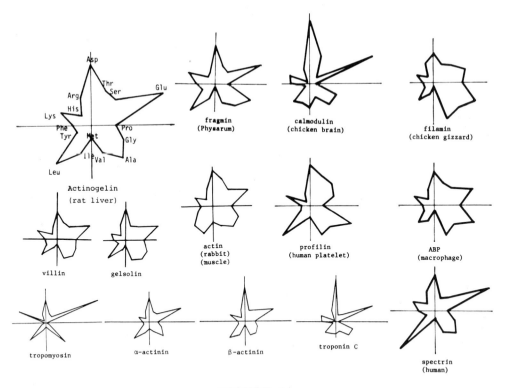

FIGURE 11

Star diagrams of amino acid compositions of actin related proteins.

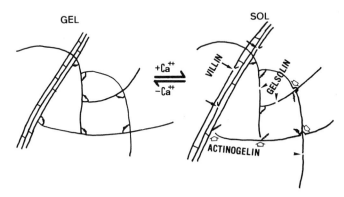

FIGURE 12

Schematic illustration of Ca^{++} dependent solation of F-actin related gel with three types of Ca^{++} sensitive actin-regulatory proteins.

The second problem to be considered is intracellular localization of these regulators. If such localization is not overlapping, presence of more than one Ca^{++} dependent regulator in a cell may be required. As reported here, actinogelin preferentially localized in converge or focal points of microfilaments, and therefore, differed from localization of gelsolin which did not show much punctuate distribution (from Stossel's presentation at the Cold Spring Harbor Symposium in 1981). Consequently, these Ca^{++} dependent regulators of microfilament system may take part in different functions of the filament system.

Mechanical weakness of actinogelin-F-actin gel to shear stress and punctuate distribution of actinogelin on stress fibers, if considered together, do not support participation of actinogelin in very rapid movement within cells, such as protoplasmic streaming. Rather, It may function as a regulator of microfilament redistribution.

Amino acid composition of hepatic actinogelin was determined and compared with those of the other actin-accessory proteins (Fig. 11). As can be seen, actinogelin is definitely different from Ca^{++} insensitive gelation factors, filamin and ABP of macrophages, in its amino acid composition. It also differed from Ca^{++} dependent modulator, calmodulin and troponin C. Amino acid compositions of villin, gelsolin and fragmin are rather similar to that of actinogelin, but α-actinin is the most similar in this respect, although muscle α-actinin is not a Ca^{++} sensitive protein. In this connection, it is interesting to note that one of the two α-actinins isolated from HeLa cell has very similar Ca^{++} sensitive gelation activity as actinogelin (K. Burridge's presentation at Cold Spring Harbor Symposium in 1981).

For elucidation of the mechanism of Ca^{++} dependent solation of F-actin-actinogelin gel, we need additional experiments. But one possibility is that actinogelin is an asymmetric protein having two different binding sites to F-actin, with one site insensitive to Ca^{++} whereas the other can be dissociated at high Ca^{++} concentrations. This mechanism may explain solation of the gel with Ca^{++} without severing of F-actin (Fig. 12). Solation induced by gelsolin or villin, although there is no evidence for the co-presence with actinogelin in the same cell, is also illustrated in Fig. 12.

ACKNOWLEDGEMENTS

The authors are grateful to Professor R. Sato of our Institute for discussions and encouragement; to Professor K. Maruyama of Chiba University for collaboration in rheological measurements. This work was supported in part by a Cancer Research Grant and a Grant-in-Aid for Special Project Research on Eucaryotic Cells from the Ministry of Education, Science and Culture of Japan and by the Naito Foundation.

REFERENCES

1. Asano, A. & Okada, Y. (1977): Life Sci. 20:117-122.
2. Bretscher, A. & Weber, K. (1980): Cell 20:839-847.
3. Bretscher, A. & Weber, K. (1980): J. Cell Biol. 86:335-340.
4. Dedman, J.R., Welsh, M.J. & Means, A.R. (1978): J. Biol. Chem. 253:7515-7521.
5. Godt, R.E. (1974): J. Gen. Physiol. 63:722-739.
6. Hasegawa, T., Takahashi, S., Hayashi, H. & Hatano, S. (1980): Biochemistry 19:2677-2683.
7. Herman, I.M. & Pollard, T.D. (1978): Exp. Cell Res. 114:15-25.
8. Marurama, K., Mimura, N. & Asano, A. (1981): J. Biochem. 89:317-319.
9. Matsuda, N. & Yagi, K. (1980): J. Biochem. 88:1515-1520.
10. Mimura, N. & Asano, A. (1976): Nature 261:319-321.
11. Mimura, N. & Asano, A. (1978): Nature 272:273-276.
12. Mimura, N. & Asano, A. (1979): Nature 282:44-48.
13. Schwarzenbach, G., Senn, H. & Anderegg, G. (1957): Helv. Chim. Acta 40:1886-1900.
14. Southwick, F.C. & Stossel, T.P. (1981): J. Biol. Chem. 256:3030-3036.
15. Spudich, J.A. & Watt, S. (1971): J. Biol. Chem. 246:4866-4871.
16. Yagi, K. & Yasawa, Y. (1966): J. Biochem. 60:450-456.
17. Yin, H.L., Zana, K.S. & Stossel, T.P. (1980): J. Biol. Chem. 255:9494-9500.

CALCIUM REGULATION OF ACTIN NETWORK STRUCTURE BY GELSOLIN

H.L. Yin and T.P. Stossel

Hematology-Oncology Unit, Massachusetts General Hospital, Department of Medicine, Harvard Medical School, Boston, Massachusetts 02114 USA

An increasing body of evidence implicates the protein actin as one of the most important constituents in the architecture and movement of cytoplasm in non-muscle as well as muscel cells (12). This chapter reviews a system which depends on calcium for the control of actin structure. The general principles behind this system is that calcium regulates the average length of actin filaments and thereby controls their viscosity and rigidity. The components of this system were first identified in rabbit lung macrophages which can be considered as a prototype of highly motile cells, and subsequently in a wide variety of cells and tissues. Therefore, this principle for the regulation of cell motility is likely to be valid for many motile eukaryotic cells.

The motor of the macrophage appears to reside in the peripheral cytoplasm beneath the plasma membrane. Actin, as well as a number of actin-associated proteins, are localized in this region (18,19). Under light microscopy, the cortical region excludes organelles contributing to the gel-like appearance of the cytoplasm. The thickness of cell cortex changes during various cell activities. It extends pseudopods when the cell spreads or moves on a surface, or engulfs objects by phagocytosis. These observations led early microscopists to infer that the formation and movement of the cortex arise from transformation between the gel and sol states of the peripheral cytoplasm (14). Some half a century later, the ideal of gel-sol transformation of the cytoplasm has been revived. and it may indeed be the basis by which cell movement is regulated.

The gel-like consistency of the peripheral cytoplasm is due to the interaction of two proteins, actin and actin binding protein. Actin, which is present in high concentration, possibly as much as 18 mg/ml (8,9), can assemble from monomers into long semi-flexible, double-helical polymers of variable lengths (17). The polymers thus formed are further organized into a three-dimensional network by a second protein, a Ca^{++} independent, high molecular weight acting binding protein (21,22). With such a high concentration of cross-linked actin polymers it is obvious that there should be sufficient rigidity to produce the gel-like consistency inferred to exist in the cortical cytoplasm and to maintain the shape of this region. However, it is necessary to explain how this structure can also be rendered more fluid to bring about changes in cell shape and directional movement of the peripheral cytoplasm.

To understand how the gel is isolated in the cell, we shall consider how it is assembled first. There is now conclusive evidence (3,10,16,27) that formation of actin network can be analyzed in terms of a network theory outlined by Flory (6). According to this theory, gelation is the cross-linking of long filaments to form a continuous three-dimensional network. One of the properties of a gel is rigidity. The rigidity of an actin solution increases as a function of actin-binding protein concentration, and it rises sharply when all of the filaments are cross-linked into a continuous network, indicating the formation of a gel. The concentration of cross-linking protein at which the state transition occurs is called the critical concentration for incipient gelation, a term defined mathematically as the ratio of the critical number of actin-binding protein cross-links (V_C) to the number of actin monomers in the polymer being cross-linked (N_O)(6). The abruptness of the sol-to-gel transition offers an excellent means for controlling cytoplasmic consistency.

Solation is the reverse of gelation. In principle, solation of an actin gel can be achieved by one of several mechanisms. First, by removal of the actin binding protein cross-links from the actin filaments. At present, there is little compelling evidence for regulation of gelation by this mechanism; second, by depolymerizing actin polymers. Although this seems reasonable intuitively, it is not very efficient, as will be shown below. Third, by decreasing the actin filament length distribution without altering the relative amounts of actin monomers and polymers. According to Flory's network theory, the critical actin cross-linker concentration is very sensitive to the length distribution of the polymers and is inversely proportional to the weight average degree of polymerization of the polymer (X_W).

$$V_c = N_o / X_w \qquad \text{[Equation 1]}$$

The second theoretical mechanism for solating gels, depolymerization of actin, can be predicted from Equation 1 to be relative ineffective because a decrease in the concentration of monomers in polymers (N_O) would also decrease the weight average degree of polymerization (X_W) simultaneously; the two parameters vary in such a way that they are counterproductive for changing the critical actin cross-linker concentration. The third mechanism, which is shortening of actin filaments without increasing actin monomer concentration, provides an extremely efficient means for altering V_c and appears to be the basis for regulation of actin gel structure by a Ca^{++} dependent regulatory protein isolated from rabbit lung macrophages.

This regulator is a 91,000 dalton, heat-labile globular protein. It shortens actin filaments when activated by Ca^{++} in the micromolar concentration range, contributing to the collapse of the three-dimensional lattice of actin filaments. When the ambient Ca^{++} concentration is decreased, the regulator is inactivated and the actin gel reforms. Since this protein causes irreversible gel-sol transformation of actin network, we have called it gelsolin (24).

The physicochemical properties of gelsolin are summarized in Table 1. Gelsolin binds 2 moles of Ca^{++} per mole of gelsolin with high affinity (Ka = 1.09×10^6 M^{-1})(26). It binds to both monomeric/oligomeric or polymeric actin (24,25). Ca^{++}-gelsolin shortens pre-existing actin filaments, as demonstrated directly by electron microscopy and indirectly by viscosity and flow birefringence measurements. Significant shortening of actin by Ca^{++}-gelsolin is observed without a comparable decrease in the sedimentability and turbidity of actin solutions, indicating that shortening is not primarily due to depolymerization of actin (27). When added directly to G-actin, Ca^{++}-gelsolin facilitates the nucleation step for the polymerization of actin prior to assembly, although the rate of elongation of the filaments following nucleation is decreased. Due to the increase in the number of nuclei, more filaments are formed and in the presence of a constant pool of actin molecules, this necessitates a decrease in average filament length (25). The ability of gelsolin to produce short filaments irrespective of the initial state of assembly of the actin offers flexibility for controlling the structure of the cytoplasm in which actin may exist as monomers or polymers, and actin filaments may or may not be cross-linked at different times in the cell.

Once the filaments are shortened by gelsolin, they remain short as long as the concentration of Ca^{++} is above the threshold level for the activation of gelsolin. This can be attributed to the capping to one end of the filament by gelsolin to prevent the actin fragments from reannealing. Binding of gelsolin to one end of the actin filament is demonstrated indirectly by electron microscopic analysis of the polarity of growth of actin filaments off of actin nuclei formed in the presence of gelsolin. Gelsolin blocks elongation

TABLE 1: Physicochemical properties of macrophage gelsolin.

Parameter	Value
Stokes radius, a	44 A
Sedimentation coefficient, $s_{20'}$	4.9 S
Partial specific volume, $\bar{V}$	0.73 cm g
Isoelectric point (calcium free)	6.1
Molecular weight	
gel-electrophoresis in presence of sodium dodecyl sulfate	91,000
From s and a	95,300
f/f_o (from M_r, $\bar{V}$ and a)	1.43
Calcium binding	
K_a	$1.09 \times 10^6 \, M^{-1}$
capacity	1.7 mol of Ca^{++} per mol of gelsolin

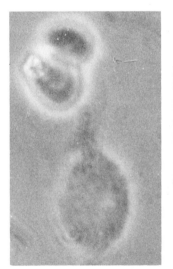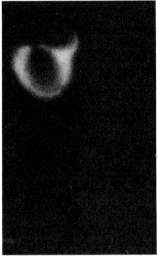

FIGURE 1

Immunofluorescent localization of gelsolin in rabbit granulocytes. Granulocytes from peripheral blood were plated on coverslips coated with heat-fixed yeast and allowed to phagocytize. The cells were fixed with paraformaldehyde and acetone sequentially, and labelled by indirect immunofluorescent staining technique as described by Yin et al. (28). The primary antibody was a specific goat anti-rabbit macrophage gelsolin, the secondary antibody was rhodamine-conjugated rabbit antigoat IgG. Phase contrast photomicrograph is presented on the left, fluorescent micrograph on the right. Magnification 2000 x.

of actin filaments from the normally fast-growing, or "barbed" end
of actin filaments, with reference to the direction of arrowheads
formed when the filaments are decorated with heavy meromysin (25).

Gelsolin shortens actin filaments without significantly increasing
the monomer concentration, and this is indeed the basis for the
solation of actin gels. As predicted by the gel theory, inhibition
of gelation by gelsolin can be overcome by increasing the concentration of the cross-linker, and the experimentally determined critical cross-linker concentration for gelation in the presence of gelsolin correlates well with the theoretical value derived from the
length of the shortened actin filament. In addition, the decrease
in gelation parallels the drop in viscosity of actin solutions in the
absence of cross-linking proteins (27), and gelsolin does not decrease the binding of actin cross-linking proteins to actin filaments
(24). Since solation of the gel is not due to a direct effect of gelsolin on the actin cross-linking protein, nor to depolymerization of
actin filaments, we conclude that gelsolin acts by reversible fragmentation of actin filaments.

The activity of gelsolin is dependent on the ambient calcium concentration. In the presence of gelsolin, the viscosity of actin decreases slightly as the free calcium concentration rises from 10^{-8}
to 10^{-7} M and falls sharply when the Ca^{++} concentration is above
10^{-7} M (27). Likewise, inhibition of actin gelation by gelsolin is
dependent on the calcium concentration (24). When the calcium
concentration is lowered by the addition of excess EGTA, both the
viscosity of actin and its gelation are restored, suggesting that the
effects of gelsolin are reversible.

Based on these studies, we propose the following model for the
action of gelsolin on actin filament. In the presence of Ca^{++} con
centration greater than 10^{-7} M, gelsolin binds Ca^{++} to form a gel-
solin-Ca^{++} complex. This complex is able to bind to both actin
filaments or actin monomers/oligomers with high affinity. When it
is added to preassembled actin filaments, it breaks the bond between adjacent actin monomers in a filament, shortening the filament. The point of breakage appears to be specified in such a way
that once the filament is broken, gelsolin remains attached to the
"barbed" end of the fragment. Capping of filament ends by gelsolin prevents them from re-annealing, so that the fragments remain short. When the gelsolin-Ca^{++} complex is added to G-actin,
it nucleates actin assembly and promotes formation of short, capped
filaments. Shortening of filaments prevent gelation. When the Ca^{++}
concentration is lowered, gelsolin dissociates from the fragments
which can then re-anneal to form long filaments and the actin network becomes cross-linked.

The mechanism by which gelsolin shortens actin filaments has
yet to be determined. Gelsolin-Ca^{++} acts rapidly on actin filaments

TABLE 2: Presence of gelsolin in cells and tissues.

Rabbit	lung macrophages splenic lymphocytes platelets intestinal epithelial cells brain thyroid taenia coli uterus bladder skeletal muscle cardiac muscle
Human	peripheral blood lymphocytes promyelocytic leukemic HL-60 line polymorphonuclear leukocytes platelets fibroblast lines (297, W-18-VA2) serum/plasma
Horse	serum/plasma
Rat	β-cells fibroblast line (W256)
Pig	kidney epithelial line (LLC-PK$_1$)

The presence of gelsolin was determined by the technique of Towbin et al. (23) utilizing a specific goat antigelsolin antibody. Whole cells (tissue) or high speed supernatants of cell (tissue) extracts were analyzed. In each of the systems listed a single polypeptide which comigrates with macrophage gelsolin was found to be cross-reactive with antigelsolin antibody.

and a maximal decrease in the viscosity of actin is observed at 30 sec, the earliest time point measured. It is equally effective at 25°C or 4°C. From the degree of shortening of actin filaments by gelsolin, we estimate that each of the added gelsolin breaks 0.8 bond between actin monomers in filament (27). All of these characteristics are consistent with a stoichiometric interaction between gelsolin and actin. The rapidity of shortening suggests that gelsolin may directly break the bond between adjacent actin molecules in a filament.

Several lines of evidence suggest that this Ca^{++} dependent system for the regulation of actin structure characterized in vitro is relevant for actin gel-sol transformation of living macrophages.

First, extracts of macrophage cytoplasm exhibits reversible Ca^{++} dependent gel-sol transformation. Actin binding protein, which is Ca^{++} independent, accounts for greater than 70% of the actin cross-linking activity, while gelsolin accounts for 75% of Ca^{++} dependent solation activity in macrophage extracts (26). Since these two proteins accounts for the bulk of gelation and solation activities in the extract, Ca^{++} dependent gel-sol transformation observed must be due to an effect of gelsolin on the gelation of actin by actin binding protein (26). Second, the expression of the regulatory function of gelsolin depends on variations in free calcium concentration likely to occur in living cells, and its effects are reversible. Third, by indirect immunofluorescent staining studies, we have shown that gelsolin is physically present in the cortical cytoplasm of phagocytic cells where the motor is believed to reside. During phagocytosis, the protein is further concentrated in pseudopodia (Fig. 1)(25), an area inferred to undergo rapid consistency change, in a distribution similar to that observed for actin, actin—binding protein, and myosin (19). These findings are compatible with the idea that gelsolin is an integral part of the motile apparatus of phagocytic cells, and it regulates the consistency of the cytoplasm. Through an effect on gel-sol transformation, gelsolin can be involved in the regulation of directional movement of the cytoplasm and intracellular organelles, as well as the maintenance of cell shape and remodelling of the cytoplasm. From the apparently random organization of the actin network in macrophages it is not immediately apparent how directional motion, crucial to many cell functions, is generated. We propose that if gelsolin can be activated differentially in the cell through a local change in Ca^{++} concentration, the difference in degree of gelation of the cytoplasm will orient its movement, as has been demonstrated in an in vitro motile system (20). Localized changes in cytosol Ca^{++} concentration may be achieved by different degrees of activation of a calmodulin, and Mg^{++}, ATP dependent calcium pump on the cytoplasmic surface of the cytoplasm (13).

The Ca^{++} dependent regulatory system of cytoplasmic consistency characterized in macrophages may also be applicable to a variety of other systems. First, extracts of many eukaryotic cells exhibit reversible, Ca^{++} dependent gel-sol transformation, and in many cases, gelation of actin is attributable to cross-linking of actin filaments by a high molecular weight actin binding protein. Second, gelsolin is present in a wide variety of cells and tissues from different species, as demonstrated by an immune SDS-polyacrylamide gel replica technique with a specific antibody against gelsolin (Table 2)(25). Third, other Ca^{++} sensitive proteins which regulate actin gel-sol transformation by a similar mechanism have also been described. Fragmin, a 43,000 dalton protein from Physarum polycephalum (11), will be discussed in a separate chapter. Villin, a 95,000 dalton protein from chicken intestinal epithelial cells, is structurally and functionally very similar to macrophage gelsolin (1,2,4,15). Interestingly, the most significant difference between these proteins is that unlike gelsolin which is present in a wide variety of cells, villin is reportedly restricted in its distribution to specialized cells containing brushborders (1). It is possible that villin represents a specialization of the Ca^{++} dependent regulator of actin gel-sol transformation, of which gelsolin is the prototype in eukaryotic cells.

ACKNOWLEDGEMENTS

This work was supported by USPHS grants HL-125183 and HL-19429.

REFERENCES

1. Bretscher, A. & Weber, K. (1979): Proc. Nat. Acad. Sci. 76: 2321-2325.
2. Bretscher, A. & Weber, K. (1980): Cell 20:839-847.
3. Brotschi, E.A., Hartwig, J.H. & Stossel, T.P. (1978): J. Biol. Chem. 253:8988-8993.
4. Craig, S.W. & Powell, L.D. (1980): Cell 22:739-746.
5. Davies, W.A. & Stossel, T.P. (1981): Arch. Biochem. Biophys. 206:190-197.
6. Flory, P.J. (1946): Chem. Rev. 39:137-197.
7. Glenney, J.R. Jr., Bretscher, A. & Weber, K. (1980): Proc. Nat. Acad. Sci. 77:6458-6462.
8. Hartwig, J.H., Davies, W.A. & Stossel, T.P. (1977): J. Cell Biol. 75:956-967.
9. Hartwig, J.H. & Stossel, T.P. (1975): J. Biol. Chem. 250: 5696-5705.
10. Hartwig, J.H. & Stossel, T.P. (1979): J. Molec. Biol. 134: 539-554.
11. Hasegawa, T., Takahashi, S., Hayashi, H. & Hatano, S. (1980): Biochemistry 19:2679-2683.
12. Korn, E.D. (1978): Proc. Nat. Acad. Sci. 75:588-599.
13. Lew, P.D. & Stossel, T.P. (1980): J. Biol. Chem. 255:5841-5846.
14. Lewis, W.H. (1939): Arch. Exp. Zellforsch. 23:8-26.
15. Mooseker, M.S., Graves, T.A., Whaton, K.A., Falco, N. & Howe, C.L. (1980): J. Cell Biol. 87:809-822.
16. Nunnally, M.H., Powell, L.D. & Craig, S.W. (1981): J. Biol. Chem. 256:2083-2086.
17. Oosawa, F. & Kasai, M. (1976): IN Subunits in Biological Systems, Part A. (eds) S.N. Timasheff & G.D. Fasman, Marcel Dekker, New York, pp. 261-322.
18. Reaven, E.P. & Axline, S.G. (1973): J. Cell Biol. 59:12-27.
19. Stendahl, O.I., Hartwig, J.H., Brotschi, E.A. & Stossel, T.P. (1980): J. Cell Biol. 84:215-224.
20. Stendahl, O.I. & Stossel, T.P. (1980): Biochem. Biophys. Res. Commun. 92:675-681.
21. Stossel, T.P. & Hartwig, J.H. (1975): J. Biol. Chem. 250: 5706-5712.

22. Stossel, T.P. & Hartwig, J.H. (1976): J. Cell Biol. 68:602-619.
23. Towbin, H., Staehelin, T. & Gordon, J. (1979): Proc. Nat. Acad. Sci. 76:4350-4354.
24. Yin, H.L. & Stossel, T.P. (1979): Nature 281:583-586.
25. Yin, H.L., Hartwig, J.H., Maruyama, K. & Stossel, T.P. (1981): J. Biol. Chem. 256:9693-9697.
26. Yin, H.L. & Stossel, T.P. (1980): J. Biol. Chem. 255:9490-9493.
27. Yin, H.L., Zaner, K.S. & Stossel, T.P. (1980): J. Biol. Chem. 255:9494-9500.
28. Yin, H.L., Labrecht, J.H. & Fattoum, A. (1981): J. Cell Biol. 91:901-906.

PHYSICAL PROPERTIES OF FRAGMIN, A Ca^{++} SENSITIVE REGULATORY PROTEIN OF ACTIN POLYMERIZATION ISOLATED FROM PHYSARUM PLASMODIUM

S. Hatano, T. Hasegawa*, H. Sugino and K. Ozaki

Institute of Molecular Biology, Faculty of Science,
Nagoya University, Chikusa-ku, Nagoya 464, Japan

Isolation of Fragmin from Physarum Plasmodium

In 1962 (13,14) we found that the water extract of acetone-dried powder of Physarum plasmodium contained an actin-like protein which bound to muscle myosin and formed actomyosin. This protein complex showed the typical properties of the actomyosin, for example, the viscosity dropped in a 0.5 M KCl solution and the superprecipitation was induced in a 0.1 M KCl solution on addition of ATP. However, no viscosity increase was observed in the water extract, even upon the addition of the neutral salts to the level of 0.1 M KCl (Table 1). It seemed to us that Physarum actin might not polymerize to F-actin. However, this was incorrect, because purified Physarum G-actin could polymerize to F-actin. We knew that non-muscle actin could exist in unpolymerized form in the crude extract, as shown by many investigators (1,3,6,10, 26,34,43,51,58).

In 1966 (15) we succeeded in insolating a partially purified G-actin preparation from the Physarum water extract. The addition of 0.1 M KCl caused G-actin to polymerize to F-actin, the structure of which was identical to muscle F-actin. However, when 2 mM $MgCl_2$ was added to this G-actin preparation, G-actin polymerized to another form of actin polymer, the Mg polymer. Mg polymer

---

*Present Address: Division of Cell Biology, National Center for Nervous, Mental and Muscular Disorders, 2620, Ogawa-Higashi-machi, Kodaira, Tokyo 187, Japan

TABLE 1: Purification of Physarum monitored by viscometry.

Old Method (48):

$$\text{Plasmodium} \xrightarrow{\text{acetone}} \text{Dry Plasmodium} \xrightarrow{\text{water}} \text{Extract I} \xrightarrow{\text{myosin}}$$

$$\text{Actomyosin} \xrightarrow{\text{acetone}} \text{Dry Actomyosin} \xrightarrow{\text{cysteine, ATP}} \text{Extract II}$$

$$(NH_4)_2SO_4 \xrightarrow{} \text{Actin I} \xrightarrow{\text{chromato.}} \text{Actin II} \xrightarrow{\text{1 M urea}} \text{Purified Actin}$$

	Extract I	Extract II	Actin I	Extract II	Actin II	Pure Actin
η_{sp}/C (dl/g) (0.1 M KCl added)	0	0	2.6	0	8.7	10.1
η_{sp}/C (dl/g) (2 mM MgCl$_2$ added)	0	0	1.1	0	4.7	9.5
η-Mg/η-K (percent)	—	—	33	—	54	94

showed an ATPase activity at room temperature and appeared to be a flexible polymer under the electron microscope (16,21,59).

To obtain a monospecific antibody to the Physarum actin, we improved our purification technique and, in 1976, finally obtained a pure actin preparation (22). The viscosity of purified Physarum F-actin was identical to that of purified rabbit striated muscle F-actin. We succeeded in producing the mono-specific antibody to actin which reacted with not only Physarum actin, but also actin distributing in other non-muscle cells (48,49). During the process of purification of Physarum actin, we found some factors which regulated actin polymerization were present in the crude G-actin preparation, so that when these factors were removed from the actin preparation, Mg polymer could not be induced (Table 1).

As a factor for inducing the formation of Mg polymer, β-actinin-like protein (Physarum actinin) was isolated from the water extract of plasmodium (22,24,36). Physarum actinin had a molecular weight of about 90,000 and $S_{20,w}$ of actinin was about 3.86S in a dilute buffer solution, and 4.77S in a 0.1 n KCl solution. No viscosity change of the actinin solution was observed on the addition of salts. The SDS-gel electrophoresis of Physarum actinin showed the single band which comigrated with Physarum and muscle G-actin, that is, Physarum actinin was the dimer of the subunit(s) of 43,000 daltons.

At that time we could not rule out the possibility that Physarum actinin might be a kind of partially denatured actin. Therefore we attempted further characterization of Physarum actinin. Hasegawa et al. (12) demonstrated that when the SDS-gel electrophoresis of Physarum actinin was carried out in the presence of 6 M urea, the single band of 43,000 daltons subunits did indeed split into two bands. The molar ratio of these bands was estimated as about 1:1, and the mobility of the lower band was just the same as actin. We assumed that the Physarum actinin was the protein complex which consisted of an actin and a new protein. From the preliminary analysis of the amino acid composition of Physarum actinin, this new protein appeared to have no cysteine residues. When the actinin was treated with 2-nitro-5-thiocyano benzoic acid (TNB-CN), which was known as a reagent which specifically split peptide bonds at the position of cysteine residues, the actin was broken into smaller peptides, whereas the new protein was not split by this reagent. Thus, the new protein, the fragmin, was purified from the TNB-CN treated fraction of Physarum actinin (12). The same protein, named as the actin modulating protein (AMP), was also isolated from Physarum by Hinssen (27,28).

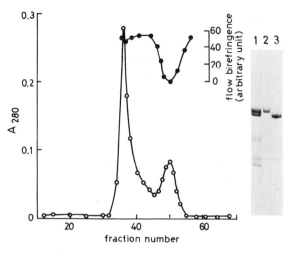

FIGURE 1

A simple method for the isolation of fragmin from Physarum actinin. Physarum actinin (the fragmin and actin complex) was prepared from acetone-dried powder of plasmodium or fresh plasmodium. The actinin was treated with 6 M urea solution containing 1 mM EGTA and 4 mM Tris-HCl buffer (pH 8.2) for 6 hr at 4°C, and was then dialyzed against 4 mM Tris-HCl buffer solution (pH 8.2) containing 1 mM EGTA for 6 hr to remove urea. The dialyzed solution was put on a column of sephadex G-100 buffered with 4 mM Tris-HCl buffer (pH 8.2) solution containing 1 mM EGTA and was then eluted with the same buffer. Fractions near the second peak showed the activity of fragmin, therefore these fractions were collected and pooled as fragmin preparation. The activity of fragmin of each fraction was determined as follows. A definite amount of each fraction was mixed with purified muscle G-actin soluton and the polymerization was then induced on addition of 2 mM $MgCl_2$. The flow birefringence of each solution was determined after 30 min (upper curve). Inserted photographs show SDS-gel electrophoretic patterns of: 1) actinin, 2) purified fragmin, and 3) Physarum actin. The SDS-gel electrophoresis was carried out in the presence of 6 M urea.

A Simple Method for the Isolation of Fragmin from Actinin (actin-fragmin complex)

The isolation method of fragmin described by Hasegawa et al. (12) was recently modified and improved in our laboratory. Actinin was treated with 6 M urea instead of TNB-CN for 6 hr at 4°C, and was then dialyzed against a buffer solution to remove urea. With this procedure actin was denatured and separated from the fragmin. The denatured actin formed new aggregates during the dialysis. No monomeric actin was present after the dialysis so that fragmin was easily separated by usual column chromatography (Fig. 1).

Physicochemical properties of fragmin. The molecular weight of fragmin was determined as 43,000 either by SDS-gel electrophoresis or by column chromatography with a dilute buffer solution. This suggested that the native state of fragmin was the monomer form in a dilute buffer solution.

Similar factors have been isolated from rabbit macrophage (gelsolin) (61,62) and chicken intestinal brush border cells (villin)(4). The molecular weights of these proteins were reported to be 91,000 and 95,000, respectively; which were nearly twice that of fragmin. There was a possibility that fragmin might be a product yielded by splitting a protein of 90,000 daltons into two halves. However, when we carried out the extraction of fragmin from acetone-dried powder of plasmodium, we found that the water extract showed very low proteinase activity. The eluted fractions of the chromatography of the water extract showed a single peak of the fragmin activity. These fractions contained fragmin with the molecular weight of 43,000 and no activity was found in the other fractions. So far there was no evidence to show that fragmin was produced by splitting the protein with 93,000 daltons by enzymatic action.

Although the molecular weight of fragmin was the same as that of Physarum actin, the amino acid composition of fragmin was clearly different from that of Physarum actin (Table 2). For example, fragmin contained no N^{τ}-methyl-histidine and no cysteine residues, while Physarum actin contained 1 N^{τ}-methyl-histidine and 3 cysteine residues. Fragmin contained more leusine and lysine residues, but less methionine and isoleusine residues, etc, than Physarum actin.

The most interesting property of the fragmin was its Ca^{++} sensitivity, that is, when Ca^{++} was absent (less than 10^{-7} M), fragmin showed no effects either on the polymerization of G-actin or on the fragmentation of F-actin. However, with the existence of fragmin and Ca^{++} (more than 10^{-6} M), G-actin could polymerize to short filaments, but F-actin was fragmented into shorter pieces. More fragmin was added to the actin solution, and the amount of short F-actin filaments was increased (Figs. 2-4).

Fragmin retained some common properties of β-actinin (37), as for instance: 1) fragmin accelerated the rate of polymerization of G-actin (Fig. 2B), probably promoting the nucleation of actin polymerization as β-actinin did, and 2) fragmin induced the formation of F-actin filaments holding the length shorter than usual and preventing possible association among these filaments. However, it should be emphasized that β-actinin had no Ca^{++} sensitivity, and had no ability to induce the fragmentation of F-actin.

TABLE 2: Amino acid composition of fragmin and Physarum actin (12).

Amino Acid	Fragmin	Actin
Asx	36	31
Thr	19	27
Ser	17	25
Glx	50	42
Pro	14	19
Gly	38	32
Ala	37	29
1/2Cys	0	3
Val	20	22
Met	3	15
Ile	15	25
Leu	36	27
Tyr	14	14
Phe	19	13
Lys	29	19
His	10	8
N^τ-Me-His	0	1
Arg	13	18
Total	370	370

Amino acid compositions of fragmin and actin were normalized to total amino acid residues of 370 for both proteins. Tryptophan was not determined.

Some differences existed between Physarum actinin and the reconstituted complex of actin and fragmin. Purified Physarum G-actin polymerized to Mg polymer on the addition of 2 mM $MgCl_2$ in the presence of Physarum actinin. The molecular configuration of Mg polymer was altered depending on the ATP concentration in the solution, as reported in previous papers (9,20). For example, when 1 mM ATP, together with 0.1 M KCl, was added to the solution of Mg polymer, the flow birefringence increased by 150% to 180% of the original level. On the other hand, purified G-actin polymerized to F-actin on the addition of 2 mM $MgCl_2$, even with the presence of fragmin. The F-actin, either long or short, showed no response on the addition of ATP. These facts suggested the existence of another factor in Physarum actinin preparation to form Mg polymer.

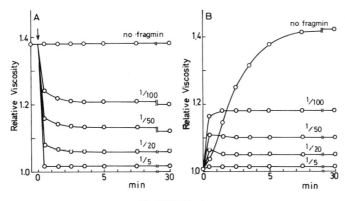

FIGURE 2

(A) Effect of fragmin on F-actin. Various amounts of fragmin were added to purified Physarum F-actin solutions. After adjusting free Ca^{++} concentration at pCa 5 with 2 mM EGTA-Ca buffer (as shown by the arrow), viscosity changes were monitored at 22°C. Solvent conditions: 0.1 M KCl, 2 mM $MgCl_2$, 1 mM ATP, 0.1 mM EGTA and 20 mM imidazole-HCl buffer (pH 7.0).
(B) Effect of fragmin on actin polymerization. After addition of various amounts of fragmin, Physarum G-actin was polymerized on addition of 0.1 M KCl and 2 mM $MgCl_2$. Polymerization was monitored by viscometry at 22°C. Solvent conditions: 1 mM ATP, 20 mM imidazole buffer (pH 7.0) and 1 mM Ca-EGTA buffer (pCa 5). Concentration of actin was 0.48 mg/ml. Molar ratios of fragmin to actin are shown in each figure (12).

Role of Fragmin in the Contraction and Relaxation
Cycle of Physarum Actomyosin

Physarum actomyosin showed a reversible superprecipitation (40). In the superprecipitated state, the actomyosin formed bundles of F-actin filaments decorated with myosin molecules. The bundle of F-actin dissembled to separate F-actin filaments again on the addition of ATP. This process was reversible and repeatable, suggesting a unique molecular feature of Physarum actomyosin. The state of actomyosin was regulated by the ATP concentration in the solution. However, the purified actomyosin showed no Ca^{++} sensitivity in our experiments. The Ca^{++} sensitivity of Physarum actomyosin was observed in the in vivo model of protoplasmic streaming of plasmodium, the "caffeine drops" (18,19,41,44,53).

When plasmodium was treated with 10 mM caffeine and 10 mM Tris-maleate buffer (pH 7.0) solution, some parts of the cortical gel layer of plasmodium were broken and the inner sol bulged out into the solution to form spherical plasmodial droplets (caffeine

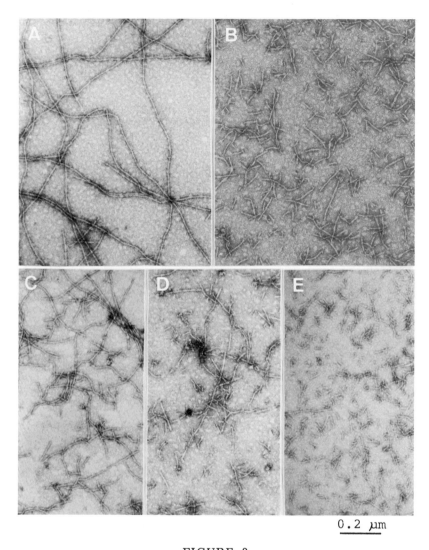

FIGURE 3

Electron micrographs of Physarum actin filaments. (A) F-actin; (B) F-actin fragments induced by addition of fragmin to F-actin solution (the molar ratio of fragmin to actin, 1:5); (C-E) Actin filaments polymerized in the presence of various amounts of fragmin. Molar ratios of fragmin to actin were 1:50 (C); 1:20 (D); and 1:5 (E) (12).

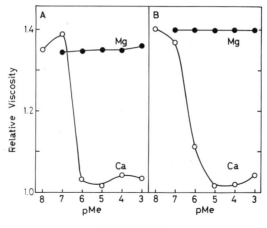

FIGURE 4

Effect of Ca^{++} and Mg^{++} on the activity of fragmin. (A) After addition of fragmin, Physarum G-actin was polymerized in the presence of 0.1 M KCl, 1 mM ATP, 20 mM imidazole-HCl buffer (pH 7.0) and various concentrations of free Ca^{++} and free Mg^{++}. Viscosity values 30 min after initiation of polymerization are plotted as a function of the divalent cation concentration. (B) Fragmin was added to Physarum F-actin solutions containing 0.1 M KCl, 1 mM ATP, 20 mM imidazole-HCl buffer (pH 7.0) and various concentrations of free Ca^{++} and free Mg^{++}. Viscosity values 30 min after addition of fragmin are plotted as a function of the divalent cation concentration. The concentration of actin and fragmin were 0.5 and 0.035 mg/ml, respectively. Concentrations of free Ca^{++} and free Mg^{++} were buffered with 1 mM Ca-EGTA and 1 mM Mg-EGTA, respectively. The Mg^{++} concentrations of pMg 3 and 4 were produced by addition of 0.1 mM $MgCl_2$ plus 0.01 mM EGTA and 1 mM $MgCl_2$ plus 0.1 mM EGTA, respectively (12).

drops). The "caffeine drops" retained the active proplasmic streaming and contractility of the cytoplasm. However, when Ca^{++} was removed from the solution by adding 1 mM ethylene glycol bis-(β-aminoethylether)-N-N'-tetraacetic acid (EGTA), the cytoplasm stopped the movement and expanded in the whole sphere of the "caffeine drop." When the excess amount of Ca^{++} was applied to these "caffeine drops," the granular cytoplasm contracted to form a small block in the middle of the sphere. After a while the inner sol was squeezed out into the hyaline zone. The contracted granular cytoplasm gradually transformed to the sol and finally disappeared. This rhythmic cycle of contraction-relaxation was repeated every few minutes in the solution containing Ca^{++}. The threshold concentration of Ca^{++} to maintain the movement was estimated as 10^{-6} M. If Ca^{++} induced only the contraction, as in the case of the striated muscle, the pattern of movement of the streaming cytoplasm would be temporary. No movement could be continued after the contraction. To repeat the continuous movement of the cytoplasm, the relaxation of the cytoplasm should be followed after the contraction. The factor like the native tropomyosin, has been isolated from Physarum by Kato and Tonomura (31). In the presence of the native tripomyosin, Physarum actomyosin could contract when Ca^{++} was added. A possible candidate for the factor which induces the relaxation of the actomyosin in the presence of Ca^{++} would be fragmin.

Sugino and Matsumura (personal communication) examined the effect of fragmin in vitro on the tension development of the artificial Physarum actomyosin thread. The actomyosin thread was prepared by mixing purified Physarum actin and myosin holding the 1 to 1 molar ratio. The actomyosin thread was immersed in a specially designed chamber filled with a buffer solution containing 0.03 M KCl, 5 mM $MgCl_2$, 20 mM imidazole buffer (pH 7.0) and 2 mM Ca-EGTA buffer. The tension of the thread was measured with a home-made tension meter. Exchange of the solutions in the chamber was performed by a controlled perfusion.

When the activation solution containing 0.1 mM ATP, 40 µg/ml of fragmin was applied to the thread at the Ca^{++} concentration of 10^{-8} M, the tension developed within 1 min. However, the tension development was markedly reduced when the Ca^{++} concentration of the activation solution was raised to 10^{-5} M. Sugino and Matsumura showed the marked reduction of oriented F-actin within the thread in the latter case by electronmicrographs.

Next, the isometric contraction of the thread was induced by applying the activation solution with the 10^{-8} M of Ca^{++}. After the tension reached a steady level, the Ca^{++} concentration of the solution was increased to 10^{-3} M by a stepwise procedure. It was confirmed that the fragmin markedly reduced the tension of the actomyosin thread at the Ca^{++} concentration of 10^{-5} M. Thus, fragmin induced the fragmentation of F-actin and eventually caused a decrease of the thread tension. In other words, fragmin could act as the relaxation factor of actomyosin at the given concentration of Ca^{++}.

It is believed that when the Ca^{++} concentration increases gradually in the living plasmodium the native tropomyosin mediates the contraction of actomyosin initially. As a result, bundles of F-actin filaments are composed. Fragmin then breaks the F-actin filaments within the bundle into short segments, so that the bundles are eliminated. This is the case for the relaxation of actomyosin.

According to this hypothesis: 1) the threshold concentration of Ca^{++} of the native tropomyosin should be lower than that of fragmin. According to Kato and Tonomura (31), the threshold value for the Physarum native tropomyosin was between 10^{-7} and 10^{-6} M. The threshold concentration of Ca^{++} for fragmin was between 10^{-7} and 10^{-6} M for the reduction of the tension of actomyosin thread. The threshold concentration of Ca^{++} for the native tropomyosin and fragmin appeared to be identical. However, rather than emphasize their resemblance for Ca^{++} sensitivity, the conclusion should be suspended until we carefully examine the chemical nature of these Ca^{++} regulatory proteins under the same conditions, using the same Ca^{++} buffer. Unfortunately we did not succeed in isolating the native tropomyosin-like protein from Physarum and 2) the effect of fragmin for the fragmentation of F-actin has to be the reversible process. It has been shown that the fragmentation of the F-actin initiated by fragmin occurs rather quickly within 30 sec when the Ca^{++} concentration in increased from 10^{-8} to 10^{-5} M (Fig. 2A). However, it took longer (several hours or more) for the reassociation of fragmented F-actin to normal F-actin, even when the Ca^{++} concentration was repidly reduced to the original level. If fragmin is directly involved in the cyclic changes of the contraction and relaxation of the actomyosin, the reassociation of fragmented F-actins should occur in a few minutes, because this cyclic change requires only a few minutes in the living systems wuch as "caffeine drop" and plasmodium. We have not succeeded in inducing such a rapid reassociation of fragmented F-actin under the physiological conditions of plasmodium in vitro.

Another possibility for the active role of fragmin for the physiology of plasmodium must be considered; fragmin bound to G-actin may form the protein complex of G-actin and fragmin in the presence of more than 10^{-6} M of Ca^{++}. The binding of G-actin and fragmin was so tight that these two molecules could not separate, even when the complex was treated with several mM of EGTA. As we reported, fragmin was isolated from plasmodium only in the state of the protein tightly bound with G-actin (Physarum actinin)(22,24). Therefore, there is a possibility that fragmin might exist as the complex with G-actin in plasmodium. G-actin can polymerize to the segmented F-actin filaments in the presence of fragmin. In fact, when we observed myosin B, which was directly isolated from plasmodium, with the electron microscope, we noted that the average length of F-actin filaments was much shorter than that of purified Physarum F-actin (see Figs. 6 and 7 in Ref. 17). This could be due to the presence of the fragmin and G-actin complex in the crude myosin B preparation. It is believed that the existence of these shorter F-actin filaments reflect the necessity for the assembly and disassembly of F-actin filaments which should accompany the cycle of the contraction and relaxation of the plasmodial gel layer (23,25).

TABLE 3: Classification of actin associated proteins (AAPs)

Group	Ca^{++} Sensitivity	Protein Factors	Remarks	
δ	–	DNase I (bovine pancreas: 31 K F* (34))	(porcine brain: 94 K: F* (46))	Inactivation
γ	–	γ-actinin (rabbit striated muscle: 35 K (32,33))	profilin (calf spleen: 16 K (6)) (hog thyroid gland: 13 K (8)) (human platelet: 16 K (39)) (calf thymus & brain (2)) (Acanthamoeba: 12 K (52))	Inhibition of nucleation
β	–	β-actinin (rabbit striated muscle: 37+ 34 K (37))	(porcine brain: 88 K (45))	

PHYSICAL PROPERTIES OF FRAGMIN 415

			promotion of nucleation, termination
+	gelsolin (rabbit macrophage: 91 K x 2 (61,62))	fragmin or AMP (Physarum: 43 K: F* (12,26,27)) Villin (chick brush border: 95K:F* (41))	
α	HMW; actin binding protein filamin (rabbit macrophage: 220 K x 2 (11,56))	(chicken gizzard: 250 K x 2 (55,60))	cross-link of F-actin; gelation, bundle formation
	LMW; α-actinin (rabbit striated muscle: 100 K x 2 (7,57)) (bovine brain: 100 K (54))	fascin (sea urchin: 58 K (47,58))	
+	actinogelin (Ehrlich tumor cell: 115 K (42))		

Materials from which factors were isolated, molecular weights and references numbers are described in parentheses. Factors marked with F* can induce fragmentation or depolymerization of F-actin. HMW, factors with high molecular weights; LMW, factors with low molecular weights.

Classification of actin associated proteins. Actin appears in various forms in a wide range of non-muscle cells, taking shapes such as unpolymerized form, spread microfilaments, bundles of microfilaments, etc. In general, the organization of actin in non-muscle cells should occur in the following processes:

G-actin ⇌ F-actin ⇌ filamentous network (gel) or the bundle of F-actin

Each process is a reversible reaction and individually regulated by specific factors. These controlling factors, including fragmin, have recently been isolated from many non-muscle cells as well as the striated muscle.

We believe that there is some confusion based on the uncertain description or the unclear characterization of the physicochemical nature of these factors. One of the reasons for the confusion may arise because of the lack of reasonable standards for the characterization and classification of these factors. It seems inadequate at present that the description of these factors is classified by following their physiological speculation in vitro and in vivo, or simply by the estimated molecular weights by SDS-gel electrophoresis and other techniques, because the role of most factors is not clearly known and similar molecular weight does not mean sharing the same regulatory function. At present we are attempting to classify these regulatory factors into four classes, as shown in Table 3.

Factors of the first group (δ) inhibit the polymerization of G-actin in a simple manner, that is, they bind to G-actin with the molar ratio 1:1 and inactivate G-actin as in the enzymatic inhibitors. The resulting complex shows no effect on the other active G-actin molecules, so that the overall concentration of the active G-actin is reduced. Therefore, the initial rate and the final extent of the polymerization are reduced. A typical case would be DNase (34). Recently a similar factor has been isolated from porcine brain (46). The polymerization process of G-actin in the presence of the sub-molar ratio of the factor to G-actin is shown in the upper figure in Table 3. Two states of actin appear after the polymerization, that is, the complex of G-actin and the factor, and F-actin. The complex and F-actin can be separated by column chromatgraphy or ultracentrifugation.

Factors of the second group (γ) seem to inhibit the rate of the polymerization of G-actin. This is believed to inhibit the formation of required nuclei for the actin polymerization. This is supported by the fact that when the nuclei (for example, sonicated F-actin fragments) are added to the solution in the early stage of the actin polymerization, the rate of polymerization is greatly accelerated. Profilin seems to be the representative factor of this group and seems to distribute widely in eukalyotic cells (2,6,8,39,52). A factor with similar properties is isolated from rabbit striated muscle (32,33). It is not certain whether or not the factors continue associating with the actin filaments after the polymerization. Considering that all G-actins polmyerize to normal F-actin at the final stage of the polymerization, the factor may separate from the nucleus in the process of the elongation of actin polymer. A typical case of polymerization of G-actin in the presence of a submolar ratio of the factor to G-actin is shown in the middle figure in Table 3.

Factors in the third group (β) accelerate the rate of the polymerization of G-actin. Factors bind to G-actin in the molar ratio of 1:1, and the produced complex of G-actin and the factor itself turns to the nucleus for the F-actin elongation. As a result of the formation of numerous nuclei, short F-actin filaments are composed. There is some evidence that the factor remains on the F-actin filament attaching to one of the terminal ends of the short filament (29,37). This inhibits the reassociation of short F-actin filaments. Some of the factors (gelsolin, fragmin) are Ca^{++} sensitive, but others are not. Some of them can induce the fragmentation or depolymerization of F-actin filaments. A typical polymerization process of G-actin in the presence of a submolar ratio of the factor to G-actin is shown in the lower figure in Table 3.

Factors of the fourth group (α) have at least two binding sites to F-actin. They can cross-link F-actin to form network structures of actin filaments (gel) or bundles of actin filaments. Regular arrangements of the factor within the bundle is reported (30). The Ca-sensitive factor for low molecular weight (actinogelin)(42) is isolated from Ehrlich tumor cells.

If, as we anticipate, some factors which regulate the actin organization can be isolated from non-muscle cells, the authors will try to classify the factors according to this criteria. It will be helpful for us and other investigators in allied areas to understand the important role of the regulatory factors in cell motility, and at the same time to avoid any further confusion in the description of the factors.

ACKNOWLEDGEMENTS

We thank Professor H. Sata for his help in preparing this manuscript. We also thank Professor F. Oosawa for his helpful discussion in preparing Table 3 for this manuscript. We are deeply indebted to the American Chemical Society for permitting us to reproduce Table 2, and Figs. 2-4, from Biochemistry 19:2677-2683 (1980). This work was supported in part by a Grant-in-Aid for Scientific Research from the Ministry of Education, Science and Culture, Japan (Projects 438037 and 511205).

REFERENCES

1. Abramowitz, J.W., Strucher, A. & Detwiller, T.C. (1975): Arch. Biochem. Biophys. 167:230-237.
2. Blikstad, I., Sundkvist, I. & Eriksson, S. (1980): Eur. J. Biochem. 105:425-433.
3. Bray, D. & Thomas, C. (1976): J. Molec. Biol. 105:527-544.
4. Bretscher, A. & Weber, K. (1980): Cell 20:839-847.
5. Bryan, J. & Kane, R.E. (1978): J. Molec. Biol. 125:207-224.
6. Carlsson, L., Nystrom, L.-E., Sundkvist, I., Markey, F. & Lindberg, U. (1977): J. Molec. Biol. 115:465-483.
7. Ebashi, S., Ebashi, F. & Maruyama, K. (1964): Nature 203:645.
8. Fattoum, A., Roustan, C., Feinberg, J. & Pradel, L.-A. (1980): FEBS Lett. 118:237-240.
9. Fujime, S. & Hatano, S. (1972): J. Mechanochem. Cell Motil. 1:81-90.
10. Gordon, D.J., Boyer, J.L. & Korn, E.D. (1977): J. Biol. Chem. 252:8300-8309.
11. Hartwig, J.H. & Stossel, T.P. (1975): J. Biol. Chem. 250:5696-5705.
12. Hasegawa, T., Takahashi, S., Hayashi, H. & Hatano, S. (1980): Biochemistry 19:2677-2683.
13. Hatano, S. & Oosawa, F. (1962): Ann. Rept. Res. Group Biophys. Japan 2:29.
14. Hatano, S. & Oosawa, F. (1966): J. Cell Physiol. 68:197-202.
15. Hatano, S. & Oosawa, F. (1966): Biochim. Biophys. Acta 127:488-498.
16. Hatano, S., Totsuka, T. & Oosawa, F. (1967): Biochim. Biophys. Acta 140:109-122.
17. Hatano, S. & Tazawa, M. (1968): Biochim. Biophys. Acta 154:507-519.
18. Hatano, S. (1970): Exp. Cell Res. 61:199-203.
19. Hatano, S. & Oosawa, F. (1971): J. Physiol. Soc. Japan 33:589-590.
20. Hatano, S. (1972): J. Mechanochem. Cell Motil. 1:75-80.
21. Hatano, S. & Totsuka, T. (1972): J. Mechanochem. Cell Motil. 1:67-74.
22. Hatano, S. & Owaribe, K. (1976): In Cold Spring Harbor Conferences on Cell Proliferation and Cell Motility. (eds) R. Goldman, T. Pollard & J. Rosenbaum, Cold Spring Harbor Laboratory, pp. 499-511.
23. Hatano, S., Matsumura, F., Hasegawa, T., Takahashi, S., Sato, S. & Ishikawa, H. (1979): In Cell Motility: Molecules and Organization. (eds) S. Hatano, H. Ishikawa & H. Sato, University of Tokyo Press, Tokyo, pp. 87-104.
24. Hatano, S. & Owaribe, K. (1979): Biochim. Biophys. Acta. 579:200-215.
25. Hatano, S., Owaribe, K., Matsumura, F., Hasegawa, T. & Takahashi, S. (1980): Can. J. Bot. 58:750-759.

26. Hinssen, H. (1972): Cytobiology 5:146-164.
27. Hinssen, H. (1981): Eur. J. Cell Biol. 23:225-233.
28. Hinssen, H. (1981): Eur. J. Cell Biol. 23:234-240.
29. Isenberg, G., Aebi, U. & Pollard, T.D. (1980): Nature 288: 455-459.
30. Kane, R.E. (1976): J. Cell Biol. 71:704-714.
31. Kato, T. & Tonomura, Y. (1977): J. Biochem. 81:207-213.
32. Kuroda, M. & Maruyama, K. (1976): J. Biochem. 80:315-322.
33. Kuroda, M. & Maruyama, K. (1976): J. Biochem. 80:323-332.
34. Lazarides, E. & Lindberg, U. (1974): Proc. Nat. Acad. Sci. 71:4742-4746.
35. Lindberg, U. (1977): J. Molec. Biol. 115:465-483.
36. Maruyama, K., Kamiya, R., Kimura, S. & Hatano, S. (1976): J. Biochem. 79:709-715.
37. Maruyama, K., Kimura, S., Ishii, T., Kuroda, M., Ohashi, K. & Muramatsu, S. (1977): J. Biochem. 81:215-232.
38. Maruyama, K. & Sakai, H. (1981): J. Biochem. 89:1337-1340.
39. Markey, F., Lindberg, U. & Eriksson, L. (1978): FEBS Lett. 88:75-79.
40. Matsumura, F. & Hatano, S. (1978): Biochim. Biophys. Acta 533:511-523.
41. Matthews, L.M. (1977): J. Cell Biol. 72:502-505.
42. Mimura, N. & Asano, A. (1979): Nature 282:44-48.
43. Nachmias, V.T. (1976): J. Molec. Biol. 107:623-629.
44. Nachmias, V.T. & Meyers, C.H. (1980): Exp. Cell Res. 128: 121-126.
45. Nishida, E., Kuwaki, T., Maekawa, S. & Sakai, H. (1981): J. Biochem. 89:1655-1658.
46. Nishida, E. (1981): J. Biochem. 89:1197-1203.
47. Otto, J.J., Kane, R.E. & Bryan, J. (1979): Cell 17:285-293.
48. Owaribe, K. & Hatano, S. (1975): Biochemistry 14:3024-3029.
49. Owaribe, K., Izutsu, K. & Hatano, S. (1979): Cell Struct. Funct. 4:117-126.
50. Pastan, I. (1976): J. Biol. Chem. 251:6562-6567.
51. Probst, E. & Luscher, F. (1972): Biochim. Biophys. Acta 278:577-584.
52. Reichstein, E. & Korn, E.D. (1979): J. Biol. Chem. 254: 6174-6179.
53. Sato, H., Hatano, S. & Sato, Y. (1981): Protoplasma 109:187-208.
54. Schook, W., Ores, C. & Puszkin, S. (1978): Biochem. J. 175: 63-72.
55. Shizuta, Y., Shizuta, H., Gallo, M., Davies, P. & Pasten, I. (1976): J. Biol. Chem. 251:6562-6567.
56. Stossel, T.P. & Hartwig, J.H. (1975): J. Biol. Chem. 250: 5706-5712.
57. Suzuki, A., Goll, D.E., Singh, I., Allen, R.E., Robson, R.M. & Stromer, M.H. (1976): J. Biol. Chem. 251:6860-6870.

58. Tilney, L.G. (1976): J. Cell Biol. 69:73-89.
59. Totsuka, T. & Hatano, S. (1970): Biochim. Biophys. Acta 223:189-197.
60. Wang, K. (1977): Biochemistry 16:1857-1865.
61. Yin, H.L. & Stossel, T.P. (1979): Nature 281:583-586.
62. Yin, H.L., Zaner, K.S. & Stossel, T.P. (1980): J. Biol. Chem. 255:9490-9493.

CALCIUM ION AND CALCIUM BINDING PROTEINS

S. Ebashi

Department of Pharmacology, Faculty of Medicine
University of Tokyo, Hongo, Tokyo 113, Japan

As indicated by the monumental work of Ringer on frog cardiac muscle in the late 19th century, physiologists working on muscle other than skeletal, already knew that Ca^{++} was necessary for muscle contraction. However, the fact that Ca^{++} would act on skeletal muscle in an apparently opposite way (i.e. its increase in the surrounding medium would reduce the contractibility and its decrease would induce repetitive twitches), prevented physiologists from pursuing the role of Ca^{++} in contraction. The level of biological science at that time was too immature to distinguish between the electrical phenomena at the surface membrane and the contractile processes in the protoplasm. In these circumstances, physiologists naturally gave priority to skeletal muscle and did not take seriously the events in other muscles. The essential role of Ca^{++} in contraction has thus been left unrecognized for nearly a hundred years.

Heilbrunn (4) has been recognized as the first person to state that Ca^{++} should be the factor to induce the contractile responses (4). He soaked the frog skeletal muscle bundle, of which the both ends were cut, in the isotonic $CaCl_2$. It shortened rather slowly, but intensely, becoming one-fifth or less of its initial length; this never happened with other solutions without Ca^{++}. From the present point of view, this "contraction" is very complex in its nature and cannot simply be accepted as direct evidence for the role of Ca^{++} in physiological contraction. However, his intuition has undoubtedly introduced a new concept into muscle science.

Kamada was keenly interested in this report, but not satisfied with it, and so started a new experiment in collaboration with Kinosita (10). Prior to this study, Kamada explored a unique technique, injecting chemical agents into protoplasm by a glass micropipette. He tried to apply this technique to muscle (10). It was described in their paper (10) as follows (underlines by present author):

Microinjection Experiments*

However, it is more desirable to <u>bring the solution directly to the protoplasm</u>, leaving all the other part of the fibre in the normal state as far as possible. For this reason, injection of solution into the fibre was done with <u>micropipettes</u> having the perforated tip of <u>2-5 microns in external diameters</u>. The technique of the microinjection was fundamentally the same as that used in the case of Paramecium (Kamada, 1938) and of sea urchin eggs (Kamada, 1941).

The experiment was successful:

The injection causes around the tip of the micropipette <u>a localized slow contraction</u> (a hump-like swelling due to local shortening and thickening of the fibre, associated with a localized movement of protoplasm toward the tip of the micropipette).

The very important fruit of this experiment was the discovery of the reversible nature of contraction induced by Ca^{++}.

...the cross-striations, which have been condensed at about the injected spot, are soon made to scatter again, and the fibre can restore the original state once more so as to <u>respond to a second injection</u> as before. Hence the reaction induced by the microinjection may be considered as fundamentally of <u>a reversible nature</u>.

Owing to the unfavorable circumstances that the paper was published in a Japanese journal, Japanese Journal of Zoology (10), during World War II, it could not make an impact on muscle research in the world, even on Japanese muscle scientists. In 1947 Heilbrunn

---

* Footnote:
Prior to this, Chambers and Hale (1) used a micropipette filled with 0.1 M $CaCl_2$ to stimulate a muscle fiber. However, they applied it on the outer surface of the muscle fiber under nearly frozen conditions. Thus their experiment was clearly different from Kamada's, but it is interesting that they could induce local contracture by such a procedure.

also reported a similar experiment (5) in collaboration with Wiercinski.

After World War II, most muscle scientists were attracted by the revolutionary discovery of Szent-Gyorgyi and his colleagues. The fact that the actomyosin-ATP system was not apparently influenced by Ca^{++} had compelled them to ignore the Heilbrunn's concept (and, consequently, Kamada's one) for many years. The history after that (i.e. from the discovery of the "relaxing factor" [1951] to that of troponin [1965], the first Ca^{++} binding protein of biological importance) was described in previous review articles (2,3,7).

Establishment of the role of Ca^{++} in muscle contraction naturally led us to look for another site on which Ca^{++} would perform its regulatory action. One of the fruits of this pursuit was the discovery, in 1967, of Ca^{++}-dependent nature of phosphorylase b kinase by Ozawa et al. (11). Based on this finding, in 1968 we stated (2):

> We do not know whether the phosphorylase b kinase is the only enzyme sensitive to Ca ion at concentrations as low as 10^{-6} M. It is, however, interesting that Ca ion assumes a controlling effect on certain crucial steps of various metabolic processes, i.e. <u>Ca ion is a common mediator between function and metabolism</u>.

At that time, however, Ca^{++} was not popular among most biochemists outside the muscle field. No one could imagine that the "calcium age" would visit the biochemical field within a very short time. Even those biochemists who had somehow accepted the Ca^{++} concept in muscle contraction were apt to think that Ca^{++} would be the factor specific to only muscle. Perhaps Kakiuchi was the sole exception, a biochemist at that time who could sense the importance of Ca^{++}, he did not overlook a phenomenon (8), i.e. the activation of phosphodiesterase by a minute amount of Ca^{++}, which might have appeared to be a trifle to other persons. His further inquiry into its mechanism has reached a great discovery (9), i.e. that of "modulator protein," now commonly called calmodulin.

We now have various Ca binding proteins. Among them, parvalbumin has the longest history, i.e. found in 1955 (6) and soon crystallized, but its Ca^{++} binding was recognized much later and its function has not yet been established. Usually Ca binding proteins are rather small in their molecular weight, but in the case of molluscan myosin its heavy chain seems directly involved in its Ca binding. Thus it is difficult to depict the common characteristics of Ca binding proteins.

As is well known, calmodulin, or modulator protein, is an incredibly conservative protein and, therefore, must be the ancestor of some Ca binding proteins. It is a common feature of all living organisms in quiescent state that they contain virtually no Ca^{++} in their cytoplasm; this is the real basis on which Ca^{++} can exert its regulatory role in intracellular processes. I believe that this feature was deeply associated with the origin of life and, therefore, the investigation into calmodulin will shed a crucial light on the question about the origin of life.

On the other hand, the mode of action of Ca^{++} is characterized by its great diversity. We can find not only various kinds of Ca binding proteins, but also many different mechanisms in utilizing Ca^{++}; there is no general pattern for Ca^{++} mechanisms. Therefore, we must be very careful in discriminating truly Ca^{++} dependent physiological mechanism from a mere in vitro phenomenon. It should be emphasized that Ca^{++} can exercise its physiological functions only under very restricted conditions and that this is the very reason why Ca^{++} can act on various aspects of biological functions.

REFERENCES

1. Chambers, R. & Hale, H.P. (1932): Proc. Roy. Soc. Lond, B 110:336-352.
2. Ebashi, S. & Endo, M. (1968): In Progress in Biophysics and Molecular Biology, Vol. 5. (eds) J.A.V. Butler & D. Noble, Pergamon Press, Oxford & New York, pp. 123-183.
3. Ebashi, S. (1980): In The Croonian Lecture, 1979, Proc. Roy. Soc. Lond. B 207:259-286.
4. Heilbrunn, L.V. (1940): Physiol. Zool. 13:88-94.
5. Heilbrunn, L.V. & Wiercinski, F.J. (1947): J. Cell. Comp. Physiol. 29:15-32.
6. Henrotte, J.G. (1955): Nature 176:1221.
7. Huxley, A.F. (1980): In Muscle Contraction, Its Regulatory Mechanisms. (eds) S. Ebashi, K. Maruyama & M. Endo, Japan Science Societies Press, Tokyo; Springer-Verlag, New York, pp. 3-18.
8. Kakiuchi, S. & Yamazaki, R. (1970): Proc. Japan Acad. 46: 387-392.
9. Kakiuchi, S., Yamazaki, R. & Nakajima, H. (1970): Proc. Japan Acad. 46:587-592.
10. Kamada, T. & Kinoshita, H. (1943): Japan. J. Zool. 10:469-493.
11. Ozawa, E., Hosoi, J. & Ebashi, S. (1967): J. Biochem. 61: 531-533.

APPENDIX

POSTER PRESENTATIONS
(ABSTRACTS)

FATTY ACID SENSITIVE Ca^{++} DEPENDENT GUANYLATE CYCLASE ACTIVITY IN SYNAPTIC PLASMA MEMBRANES FROM RAT BRAINS. T. Asakawa, K. Hayama & K. Enomoto, Department of Pharmacology, Saga Medical School, Saga 840-01, Japan.

Guanylate cyclase activity in synaptic plasma membranes purified from rat brain was assayed with Mg^{++} as the major bivalent cation. The cyclase activity with Mg^{++} was activated by seven- to eight-fold in the presence of an unsaturated fatty acid. The activity induced by linoleate was greatly stimulated by Ca^{++} at concentrations that were in the physiological range. The concentration of free Ca^{++} required for the half maximal stimulation was estimated to be 0.17 µM by using an EGTA buffering system. The addition of $CaCl_2$ at the concentration higher than 0.2 mM caused great inhibition. When the activity was assayed with Mn^{++} as the major bivalent cation, Ca^{++} at a low concentration had no effect on the activity, and higher concentration caused inhibition.

Among phenothiazine derivatives, chlorpromazine and fluphenazine at 100 µM inhibited both basal and Ca^{++} stimulated activity, but the inhibitory effect of trifluoperazine was much less. Extensive washing of the membrane preparation with 1 mM EGTA did not diminish Ca^{++} dependency of the activity. The addition of calmodulin purified from pig brain to the assay mixtures with the membranes washed with EGTA had no effect on the Ca^{++} stimulated activity. These results suggest that guanylate cyclase activity in vivo in brains is regulated by free Ca^{++} levels in cells and that calmodulin may not participate in the stimulation of the activity by Ca^{++}.

MODULATION OF Ca^{++} DEPENDENT K^+-GATING OF ERYTHROCYTE GHOSTS BY EXTERNAL Ca-EGTA. A.M. Benjamin & D.M.J. Quastel, Department of Pharmacology, Faculty of Medicine, The University of British Columbia, Vancouver, B.C. V6T 1W5 Canada.

It is now well known that calcium acts at the inner surface of cell membranes to cause an enhanced permeability of the membrane to K^+. Using $^{86}Rb^+$ as a marker for K^+ under K^+ exchange conditions, we find that the presence of EGTA and Ca^{++} together in the incubation medium affects the rate of efflux of $^{86}Rb^+$ from red cell ghosts preloaded with $^{86}Rb^+$ and with Ca^{++} buffered with EGTA. At an interval Ca^{++} of about 50 nM (3 mM EGTA, 1 mM Ca^{++}, pH 7.1) the rate of $^{86}Rb^+$ efflux is little, if at all, increased from when internal Ca^{++} is 0; either external EGTA (up to 6 mM) or Ca^{++} (up to 4 mM) alone has no effect on the flux rate. However, the presence of 0.2-0.5 mM Ca-EGTA in the external medium raises flux rate as much as ten-fold. Higher concentrations (up to 1-2 mM) diminish the rate. The peak rate is insensitive to an excess of Ca^{++} or of EGTA externally, but depends on internal Ca^{++}. It is diminished by 4-aminopyridine (1 mM), trichloroethanol (2 mM), chlorpromazine (10 µM-0.1 mM), pentobarbital (0.1 mM) and 1-propranolol (5 mM), unaffected by ethanol (40 mM) and enhanced by 1-propranolol at

25 µM. External Ca-EDTA or Mg-EDTA, but not Ca-citrate (or Mg^{++} in the presence of EGTA) can substitute for Ca-EGTA in enhancing and suppressing flux rate. At concentrations of Ca^{++}-EGTA above 1-2 mM, $^{86}Rb^+$ flux is also enhanced. This occurs independently of internal Ca^{++}, is not suppressed by chlorpromazine and is associated with an enhanced permeability to inulin.

Supported by the Muscular Dystrophy Association of Canada.

DO "CALCIUM ANTAGONISTS" CAUSE VASODILATION BY INTERFERING WITH A CALCIUM BINDING PROTEIN RATHER THAN WITH TRANS-MEMBRANE CALCIUM INFLUX? S.-L. Bostrom, B. Ljung, S. Mardh*, S. Forsen** & E. Thulin**, AB Hassle Research Laboratories, S-48183 Molndal, Sweden; *Institute of Medical & Physiological Chemistry, Biomedical Center, Uppsala University, S-75123, Uppsala, Sweden; Physical Chemistry 2, Chemical Center, S-22007 Lund, Sweden.

Felodipine, 4-(2,3-dichlorophenyl)-1,4-dihydropyridine-2,6-dimethyl-3,5-dicarboxylic 3-ethylester and 5-methylester is a new antihypertensive agent which specifically reduces peripheral resistance by inhibition of vascular smooth muscle. The present experiments were designed to elucidate the mechanism of action. Effects on K^+ contractures and on transmembrane $^{45}Ca^{++}$ influx in K^+-rich solution were studied in isolated rat portal vein. Furthermore, the interaction between ^{14}C-felodipine and a Ca-sensitive actomyosin preparation from porcine aorta was studied and so were effects of felodipine on Ca^{++} binding to calmodulin and on the ^{113}Cd-NMR spectrum. It was found that contractile responses to graded Ca^{++} readmission in K^+ depolarized Ca^{++} depleted portal veins were insurmountable inhibited by 0.1 pM felodipine (1 hr exposure). Felodipine in a concentration of 1 nM failed to reduce the cellular $^{45}Ca^{++}$ uptake after pulse labelling with 1.2 mM Ca^{++}, while 0.7 µM felodipine or 2 mM $LaCl_3$ reduced the uptake by 25 and 90%, respectively. Felodipine was found to accumulate in the tissue. When incubated with a crude actomyosin preparation, ^{14}C-felodipine was detected by polyacrylamide electrophoresis to occur in the molecular weight region of 15,000 to 20,000 daltons, corresponding to calcium binding proteins. When calmodulin was incubated with ^{14}C-felodipine, the radioactivity was found at the site of protein staining. Potentiometric titration of calmodulin resulted in an uptake of 4 mol of Ca^{++}/mol protein. In the presence of 4 equivalents of drug the Ca^{++} binding was halved. Addition of drug to calmodulin produced changes in the ^{113}Cd-NMR spectrum indicating an effect on at least one binding site through conformational changes. A binding constant for the drug of 10^5-10^6 M^{-1} was obtained. Overall, these results indicate that felodipine exerts its vasodilating effects by interaction with a Ca^{++} binding protein rather than inhibition of Ca^{++} influx.

THE OCCURRENCE OF Ca^{++} DEPENDENT PROTEIN KINASES IN THE BRAIN. K. Fukunga, K. Matsui & E. Miyamoto, Department of Pharmacology, Kumanoto University Medical School, kumamoto 860, Japan.

Cyclic AMP, cGMP and Ca^{++} are intracellular mediators in diverse biological activities. The effects of these regulators can be seen largely through activation of protein kinases. Cyclic AMP and cGMP dependent protein kinases have been extensively studied over a wide range of the animal kingdom. During the course of studies on the significance of Ca^{++} in the central nervous system, we found that the level of a Ca^{++} dependent and calmodulin-independent protein kinase (peak I), in contrast to that of cAMP dependent protien kinase, increases at a later stage of development in the brain after birth. Subsequently, we have found another Ca^{++} dependent protein kinase (peak II), which was stimulated with a small amount of calmodulin in the cytosol of the brain. These two types of Ca^{++} dependent protein kinases were partially purified from the cytosol fraction of rat brain by DEAE-cellulose, sephadex G-200, and calmodulin affinity column chromatography and compared. Endogenous proteins and chicken gizzard myosin light chains were used for peaks I and II, respectively, as substrates. The molecular weights of the enzymes differed as determined by gel filtration analysis. Peak I had little affinity for calmodulin, whereas peak II had good affinity. The activities of both enzymes were stimulated in the presence of physiological concentrations of Ca^{++}. These results indicate that different types of Ca^{++} dependent protein kinases occur in the cytosol fraction of the brain. Similar enzymes were also found in other muscle and non-muscle tissues.

R-24571: A NEW POWERFUL INHIBITOR OF PLASMA MEMBRANE Ca^{++} TRANSPORT ATPase AND OF CALMODULIN REGULATED FUNCTIONS. K. Gietzen, A. Wuthrich* & H. Bader. Department of Pharmacology & Toxicology, University of Ulm, Oberer Eselsberg, D-7900 Ulm, German Federal Republic, and *Department of Veterinary Pharmacology, University of Bern, Langgass Str. 124, CH-3000 Bern, Switzerland.

The Ca^{++} transport ATPase of red blood cells (RBCs) is regulated by the ubiquitous Ca^{++} binding protein, calmodulin, and is dependent on the cytoplasmic Ca^{++} concentration. Beside RBC Ca^{++} transport ATPase calmodulin mediates control of a number of important Ca^{++} dependent cellular functions. Compound R-24571 (1-[bis(p-chlorophenyl)methyl]-3-[2,4-dichloro-β-(2,4-dichlorobenzyloxy)phenethyl]imidazoliniumchloride) is found to be a powerful inhibitor of RBC Ca^{++}-ATPase as well as Ca^{++} transport into inside-out RBC vesicles with an IC_{50} value of 0.5 and 2 μM, respectively. The inhibitory action of R-24571 is more specific on the calmodulin dependent fraction of Ca^{++} transport ATPase as compared to the basal Ca^{++} transport ATPase (determined in the absence of calmodulin) and can be antagonized by increasing concentrations of calmodulin in an apparently competitive manner. With respect to other ATPases, the action of R-24571 is relatively specific for RBC Ca^{++} transport

ATPase. Mg^{++} ATPase requires a 40 times higher concentration for half-maximal inhibition (IC_{50} = 20 µM) whereas (Na^+ + K^+) transport ATPase is only slightly affected in the investigated concentration range ($\leq$ 20 µM). Thus, R-24571, which also was shown to inhibit the activation of brain phosphodiesterase by calmodulin with a much higher potency than trifluoperazine (VanBelle, Biochem. Soc. Trans. 9:133P, 1981), seems to be the most powerful inhibitor of calmodulin regulated processes that has been described hitherto.

LOCALIZATION OF FLUORESCENTLY LABELLED CALMODULIN IN LIVING SEA URCHIN EGGS DURING EARLY DEVELOPMENT. Y. Hamaguchi & F. Iwasa*. Biological Laboratory, Tokyo Institute of Technology, Okayama, Meguro-ku, Tokyo 152, Japan, and *Department of Biology, College of General Education, University of Tokyo, Komaba, Meguro-ku, Tokyo 153, Japan.

Changes in the localization of calmodulin in the living eggs, of the sand-dollar Clypeaster japonicus during fertilization and mitosis were investigated by observing the fluorescence of porcine brain calmodulin labelled with N-(7-dimethylamino-4-methylcoumarinyl)-maleimide (DACM) which was microinjected into the eggs. Calmodulin fluorescence was localized at the sperm aster in fertilized eggs. During mitosis, the fluorescence was associated with the mitotic apparatus, in which the spindle poles at metaphase were most intensely fluorescent. This fact was confirmed with mitotic apparatus isolated from the eggs preloaded with DACM-calmodulin. During anaphase, the fluorescence around the poles spread to the astral rays and toward the central region of the spindle. The interzone became distinctly fluorescent at telophase. Calmodulin was distributed over the entire cell cortex, but no difference was found in fluorescence at the region of the cleavage furrow from the rest of the cell cortex. These results suggest that calmodulin plays significant roles in the motility of the cell during fertilization and mitosis.

Ca^{++} BINDING PROTEIN(S) IN MITOCHONDRIAL MATRIX. O. Hatase, M. Tokuda, T. Itano, H. Matsui & A. Doi, Department of Physiology, Kagawa Medical School, Ikenobe, Miki Town, Kagawa 761-07, Japan.

The change in Ca^{++} concentration in cytoplasm plays an important role in cellular regulatory mechanism. Ca^{++} binding protein, calmodulin (CDR), activates many enzymatic activities, such as phosphodiesterase, Ca^{++}-ATPase, myosin light chain kinase, etc. In the mitochondria the concentration of free Ca^{++} is of a micromolar order, but it is still ambiguous as to how it is regulated in the matrix (MX) space of mitochondria. We postulated that it would perform a regulatory role there, modulating the changes in configuration, enzymatic activities, and the permeability of the inner mitochondrial membrane. The MX fraction was prepared by mild sonication in hypotonic medium from rat liver mitochondria. It was fractioned by salting out with ammonium sulfate (40%). After intensive dialyzation, the residual fraction was gel filtrated by sephadex G-100 (G-150). Further purification of the Ca^{++} binding protein(s) was performed by affinity chromatography. Partially purified Ca^{++} binding protein(s) appeared to be smaller in molecular weight than CDR, and the K_{diss} for Ca^{++} was about 10^{-5} M. It probably forms complexes with a certain protein(s) in the presence of Ca^{++}. We consider that this protein(s) might be an intramitochondrial free Ca^{++} regulator (like CDR, cytoplasmic Ca^{++} regulator). Some enzymatic activities were studied in the presence and absence of this protein.

PAROTID GLAND AS A TARGET ORGAN OF VITAMIN D ACTION: CONTRIBUTION OF PHOSPHOLIPID REGULATION TO CALCIUM TRANSPORT. M. Hayakawa, H. Aoki, N. Terao, Y. Abiko & H. Takiguchi. Department of Biochemistry, Nihon University School of Dentistry, Matsudo, Chiba 271, Japan.

Rat parotid glands contain a high concentration of Ca^{++}, which plays an important role in the secretory process. It has become clear that rat parotid gland is one of the target organs for vitamin D (VD). The objective of the study was to clarify the physiological role of VD in Ca^{++} metabolism of rat parotid gland. Male Donryu rats, weighing 80-100 g were used. The dissociation constant value of $1\alpha,25$-dihydroxycholecalciferol binding protein in the cytosol of parotid gland ($1.88 \pm 0.15 \times 10^{-9}$ M) was significantly higher than that of intestinal mucosa ($p < 0.001$). Oral administration of VD_3 to VD-deficient rats caused the level of both Ca^{++} stimulated adenosine triphosphatase (Ca^{++}-ATPase) activity and ^{45}Ca uptake in parotid gland microsomes to decrease about 30%, and the uptake of ^{45}Ca was closely related with Ca^{++}-ATPase activity ($r = 0.98$). Administration of VD_3 to VD-deficient rats caused the level of free Ca^{++} concentration in the cytosol of parotid gland to decrease about 58%. The level of uptake of $[^3H]$-methyl group in microsomal phosphatidyl-N-monomethylethanolamine of VD-deficient rats from S-adenosyl [methyl-^3H]-L-methionine was decreased by the administration of VD_3. Moreover, Ca^{++}-ATPase activity in the microsomes decreased with the addition of phospholipase C in vitro. Our findings suggest that Ca^{++} transport control by VD may be associated with methylation of phosphatidylethanolamine in membrane of rat parotid gland.

CALMODULIN IN STIMULUS-ENZYME SECRETION COUPLING IN EXOCRINE PANCREAS. S. Heisler, C. Noel, L. Chauvelot, H. Lambert, L. Desy-Audet & A. Fortin, Department of Pharmacology & Toxicology, Faculty of Medicine, Laval University, Quebec G1K 7P4, Canada.

Increases in cytoplasmic calcium concentrations are often associated with the transfer of biological information across receptive, mammalian cell membranes. Calmodulin, a putative intracellular calcium receptor, may have a functionally important part in this transfer process since calcium-calmodulin complexes appear to activate several different enzymes directly involved in the expression of biochemical responsiveness to excitation. The object of this study was to investigate the role of calmodulin in secretion of digestive enzymes from exocrine pancreas. In the current study, calmodulin was identified in the cytoplasm of dispersed rat pancreatic acinar cells by affinity chromatography on 2-Cl-10 (3-aminopropyl)-phenothiazine-sepharose and gel electrophoresis. Five calmodulin binding proteins were subsequently identified in purified zymogen granule membranes using calmodulin-sepharose affinity columns and gel electrophoresis. The apparent molecular weights of these proteins was 76,500, 70,500, 57,000, 48,500 and 27,000; their importance in the exocytotic secretion of stored enzymes is still unresolved. Various drugs, especially neuroleptics, bind to calmodulin and inhibit enzyme activation dependent on the formation of calcium-calmodulin complexes. The involvement of calmodulin in enzyme secretion was assessed by evaluating the effects of several of these drugs on the enzyme secretory process from dispersed acinar cell suspensions. Trifluoperazine, thioridazine, chlorpromazine, chlorprothixene, amitriptyline, and W-7 all inhibited carbachol stimulated amylase secretion in a dose dependent fashion. Haloperidol, sulpiride, and phenobarbital were without similar effect. The data support the possibility that calmodulin is involved in stimulus-enzyme secretion coupling in the exocrine pancease. (Supported by MRC Canada).

EFFECT OF CALMODULIN AND CYCLIC AMP-DEPENDENT PROTEIN KINASE ON THE SARCOPLASMIC RETICULUM FROM CANINE HEART. M. Hirata, T. Inamitsu & I. Ohtsuki, Department of Pharmacology, Faculty of Medicine, Kyushu University, Fukuoka 812, Japan.

Sarcoplasmic reticulum fraction (SR) from canine cardiac ventricle bound 50-60 nmol Ca/mg protein in the presence of Mg-ATP. Calmodulin prepared from canine brain increased the initial rate of Ca binding (30%) and the maximum capacity (20%), but did not change the ATPase activity. It increased the affinity of SR for Ca ions by 0.2 pCa unit. It also phosphorylated two protein bands on the SDS-gel pattern of SR at the positions of 22,000 and 6,000 daltons with the same time course as the Ca binding. Cyclic AMP dependent protein kinase showed no significant effect on the Ca binding of SR in the absence and presence of calmodulin. But it phosphorylated two protein bands at the same position as in the case of calmodulin. In the presence of 5 mM oxalate, both Ca uptake and ATPase were potentiated by the protein kinase. This enhancement, however, could only be observed 2-3 min after the start of reaction, although the phosphorylation of SR was all but complete within 1 min. These findings suggest that calmodulin rather than the protein kinase plays some essential role in the physiological condition in muscle. It also seems probable that calmodulin and protein kinase stimulate the different Ca binding pathways of SR. (Present address of M. Hirata: Department of Biochemistry, Faculty of Dentistry, Kyushu University.)

FUNCTIONS OF CALCIUM IN THE DEGENERATION OF SKELETAL MUSCLE. S. Ishiura & H. Sugita*, Division of Neuromuscular Research, National Center for Nervous, Mental & Muscular Disorders, Kodaira, Tokyo 187, Japan, and *Institute of Brain Research, Faculty of Medicine, University of Tokyo, Tokyo 113, Japan.

The mechanism and control of intracellular protein turnover are not fully understood. There are many reports indicating a possible role for calcium in the regulation of muscle protein degradation. A novel calcium dependent proteinase (E.C.3.4.22.-) has been identified in chicken skeletal muscle. The enzyme is composed of a single polypeptide chain having a molecular weight of 80,000. It can not cleave any synthetic substrates except for Z-Try-ONp (k_m= 0.03 mM, K_{cat}= 0.18 sec^{-1} in 7.8% DMF, pH 7.5, at 30°C), but digest muscle structural proteins and cytoskeletal proteins, i.e. troponin, desmin, tubulin, etc. It selectively removes Z-line from myofibril. Immunofluorescent study shows that chicken (human) enzyme is localized in the myofibril, especially at the Z-line. This myofibril-bound proteinase has been isolated from chicken skeletal muscle and characterized. The half-maximal concentrations of calcium for activation is 0.18 mM, which is more sensitive than that of the soluble one. When calcium ions are introduced into the intact muscle with ionophore A23187 in vitro, Z-line loss is observed under electron microscope with concomitant release of α-actinin into the medium. Since the Z-line loss is not seen in the absence of calcium, this response can be attributed to an increase in cytosolic calcium influxed from external medium. A specific thiol-protease inhibitor E-64-c clearly suppresses the calcium induced release of α-actinin. Thus, calcium dependent proteinase may participate in the first step of myofibrillar degeneration. REFS.: 1) Ishiura et al., J. Biochem. 87:343, 1980; 2) Sugita et al., Muscle Nerve 3:335, 1980.

CALMODULIN AND PROPERTIES OF CALMODULIN RESPONSIVE Ca^{++} ATPase FROM RAT BRAIN SYNAPTIC PLASMA MEMBRANE. T. Itano*,+, O. Hatase* & J.T. Penniston+, *Department of Physiology, Kagawa Medical College, Kagawa 761-07, Japan, and +Section of Biochemistry, Department of Cell Biology, Mayo Clinic/Foundation, Rochester, MN 55905, USA.

Ca^{++} plays several important physiological roles in the nervous tissue, but the mechanisms by which the intracellular Ca^{++} concentration is controlled are still poorly understood. To analyze the control of intracellular Ca^{++}, it is very important to understand the plasma membrane Ca^{++}-ATPase which uses the energy supplied by ATP hydrolysis to extrude Ca^{++} from the cell. Recently we purified and characterized Ca^{++}-ATPase from red blood cell membranes (Niggli et al, J. Biol. Chem. 254:9955, 1979; Graf & Penniston, J. Biol. Chem. 256:1587, 1981). In this abstract we will describe the purification of Ca^{++}-ATPase from rat brain synaptic plasma membranes. Isolation of synaptic plasma membrane was done according to the method of Jones & Matus (Biochim. Biophys. Acta 356:276, 1974). To solubilize Ca^{++}-ATPase, synaptic plasma membrane was resuspended in a medium containing 1 mg Triton X-100/mg protein and 100 μM CaCl$_2$.

Purification of Ca^{++}-ATPase was by calmodulin affinity chromatography (Niggli et al., J. Biol. Chem. 254:9955, 1979). The purified enzyme shows a single band and has a molecular weight of about 130,000 as determined by SDS polyacrylamide gel electrophoresis. This enzyme has a high affinity for Ca^{++} and is stimulated by calmodulin. A phosphorylated intermediate of the Ca^{++}-ATPase was also detected in the presence of 0.1 µM Mg^{++} and 10 µM Ca^{++}. In all these properties this enzyme closely resembles the erythrocyte Ca^{++} pump, indicating that this may be the pump which expels Ca^{++} from the synapse of rat brain. (Supported by NIH Grant AM-21820).

A CALMODULIN BINDING PROTEIN FROM ESCHERICHIA COLI. T. Iwasa, Y. Iwasa, K. Matsui, K. Fukunaga & E. Miyamoto, Department of Pharmacology, Kumamoto University Medical School, Kumamoto 860, Japan.

Calmodulin, originally discovered as a protein modulator for Ca^{++} dependent cyclic nucleotide phosphodiesterase in mammalian brains (Kakiuchi & Yamazaki, 1970; Cheung, 1970), is not believed to be ubiquitous in eukaryotic cells. Calmodulin reacts with many known enzymes and proteins, whose physiological functions should be elucidated, in a Ca^{++} dependent fashion. As compared with the distribution of calmodulin, calmodulin binding proteins have been reported only in mammalian and avian tissues. Recently we have demonstrated that calmodulin-like activity in the soluble fraction of Escherichia Coli, and suggested that calmodulin may also occur in prokaryotic cells (Y. Iwasa et al., 1981). We will report the occurrence of a calmodulin binding protein in E. coli. A calmodulin binding protein was purified to apparent homogeneity from the soluble fraction of E. coli (0-143, K-X1) by ammonium sulfate fractionation followed by DEAE-cellulose, sephadex G-100 and calmodulin-agarose affinity column chromatography. The calmodulin binding protein was heat-stable and showed an M_r value of about 30,000. The calmodulin binding protein inhibited the activity of cAMP phosphodiesterase from bovine brain supported by Ca^{++} and calmodulin in a dose dependent fashion, but not the basal activity of the enzyme. Varying the concentration of Ca^{++} in the reaction medium did not affect the inhibition of the enzyme induced by the calmodulin binding protein. However, high concentrations of calmodulin could completely reverse the inhibition of phosphodiesterase by the calmodulin binding protein. It was demonstrated by calmodulin-agarose affinity column chromatography that the calmodulin binding protein bound to calmodulin in a Ca^{++} dependent fashion. The calmodulin binding protein had no activity of phosphodiesterase or myosin light chain kinase.

CALMODULIN BINDING PROTEINS IN SEA URCHIN EGGS. F. Iwasa & H. Mohri, Department of Biology, University of Tokyo, Komaba, Meguro-ku, Tokyo 153, Japan.

Calmodulin has been studied in relation to various Ca^{++} connected processes and is considered to be a multifunctional protein. Sea urchin egg exhibits several Ca^{++} regulated reactions and phenomena triggered by Ca^{++} during its early development and contains this protein. We examined calmodulin and calmodulin binding proteins in sea urchin eggs, Hemicentrotus pulcherrimus and Strongylocentrotus intermedius, in order to elucidate the physiological roles of calmodulin in differentiating cells. The cytosol of sea urchin eggs, to which fluorescein-labelled porcine brain calmodulin was added as a tracer, was applied to a sephacryl S-300 column and fractions with fluorescence were chromatographed on a DEAE-sepharose CL-6B and further applied to a calmodulin-sepharose 4B affinity column to obtain calmodulin binding proteins. Calmodulin in the cytosol of the eggs appeared to exist in a free form and in forms of complexes with other proteins either in a Ca^{++} dependent or in a Ca^{++} independent manner. As many as eleven Ca^{++} dependent calmodulin binding proteins were found. The apparent molecular weights estimated by sodium dodecyl sulfate-polyacrylamide gel electrophoresis (SDS-PAGE) were 94K, 81K, 77K, 71K, 67K, 56K, 49K, 46K, 44K, 40K and 38K daltons. The species of calmolulin binding proteins did not change after fertilization as far as analyzed by SDS-PAGE. Comparing them with calmodulin binding proteins from mammalian tissues, we could see the diversity of such proteins in sea urchin eggs. Although calmodulin is multifunctional, the roles might be limited to almost monofunctional in differentiated cells. On the other hand, participation of calmodulin in many Ca^{++} connected processes may be the case in undifferentiated eggs. The limiting factor for Ca^{++} dependent changes occurring at fertilization may not be such proteins but rather the concentration of free Ca^{++}.

INTERACTION OF CALMODULIN WITH BASIC PROTEINS. Y. Iwasa, T. Iwasa, K. Matsui, K. Higashi & E. Miyamoto, Department of Pharmacology, Kumamoto University Medical School, Kumamoto 860, Japan.

Calmodulin, originally reported as an activator of cyclic nucleotide phosphodiesterase (Kukiuchi & Yamazaki, 1970; Cheung, 1970), has been shown to be a multifunctional regulator for Ca^{++} dependent reactions. Calmodulin reacts with many enzymes and proteins whose physiological functions should be elucidated. It has been reported that hydrophobic agents such as antipsychotic drugs and 2-p-toluidinylnaphthalene-6-sulfonate have serious effects on calmodulin-enzymes interactions (Tanaka & Hidaka, 1980). We report that naturally occurring basic proteins such as myelin basic protein, histone and protamine also affect the interaction of calmodulin with calmodulin dependent enzymes. As well-known calmodulin dependent enzymes, bovine brain cyclic nucleotide phosphodiesterase and rabbit myometrium myosin light chain kinase were tested. The basic proteins inhibited the activities of the enzymes supported by Ca^{++} and calmodulin in a dose-dependent fashion. The inhibition of the enzymes could be reversed by high concentration of calmodulin but not with Ca^{++}. Kinetic analysis (Dixon plot) of the inhibition of activation of phosphodiesterase induced by the basic proteins revealed that the proteins inhibit the enzyme activity in a competitive fashion with calmodulin. K_i values for calmodulin of myelin basic protein, histone and protamine were 0.28 µM, 2.2 µg/ml and 6.2 µg/ml, respectively. The interaction of the basic proteins with calmodulin was Ca^{++} dependent. The basic proteins were retained by a calmodulin-agarose affinity column until the proteins were eluted by an ethylene gylcol bis(β-aminoethyl ether)-N,N,N',N'-tetraacetic acid (EGTA) containing medium. Thereby the basic proteins were released from the column. It should be pointed out that basic proteins, other than hydrophobic substances, also interact with calmodulin.

INSULIN AND SOMATOSTATIN SECRETION FROM PANCREATIC ISLETS: STUDIES ON THE EFFECT OF A23187 AND TRIFLUOPERAZINE. A. Kanatsuka, H. Makino, M. Osegawa, J. Kasanuki & A. Kumagai, Second Department of Internal Medicine, Chiba University School of Medicine, Chiba 280, Japan.

We investigated the role of calcium and calmodulin in insulin and somatostatin secretion from isolated rat pancreatic islets. Insulin and somatostatin released from 10 islets during 60 or 120 min incubation of 1 ml of KRB buffer were determined by radioimmunoassay. The addition of Ca ionophore A23187 (20 µM) to a medium containing 0.9 mM calcium and 5.5 mM glucose significantly increased the insulin and somatostatin secretion from islets. At low calcium levels (0.25, 0.9, 2.5 mM), changes in calcium level did not affect insulin or somatostatin secretion, but an increase to 7.5 mM calcium did enhance secretion. In the presence of 16.7 mM glucose, the addition of A23187 had different effects on B and D cells; insulin release was suppressed by A23187 while somatostatin secretion was enhanced. With 0.25 mM calcium, an increase in glucose to 16.7 mM enhanced insulin release but not somatostatin release. Insulin release induced by 16.7 mM glucose in a medium containing 0.9 mM calcium was suppressed by addition of 5 or 10 µM trifluoperazine during 120 min incubation in the medium, but somatostatin secretion was significantly enhanced by addition of the same doses of trifluoperazine. These findings indicate that an increase of calcium uptake may be an important process of both insulin and somatostatin secretion and calmodulin in the pancreatic islets may be involved in the secretory process. Sensitivity to calcium ions and calmodulin inhibitors may differ between B and D cells.

EFFECTS OF N-(6-AMINOHEXYL)-5-CHLORO-1-NAPHTHALENE SULPHONAMIDE (W-7) AND CHLORPROMAZINE (CPZ) ON CALCIUM MOVEMENTS IN VASCULAR SMOOTH MUSCLE. H. Karaki, H. Ozaki, T. Suzuki & N. Urakawa, Department of Veterinary Pharmacology, Faculty of Agriculture, University of Tokyo, Bunkyo-ku, Tokyo 113, Japan.

Effects of W-7 (2×10^{-4} M) and CPZ (2×10^{-5} M) on contractile tension and ^{45}Ca movements in adventitia-free rabbit aorta were studied. W-7 slightly decreased the resting tone and strongly inhibited the muscle contraction induced by 65.4 mM KCl. CPZ showed similar effects on muscle tension. Cellular ^{45}Ca retention was measured with a modified La^{+++} method (Karaki & Weiss, J. Pharmacol. Exp. Ther. 211:86, 1979). In normal solution with 1.5 mM Ca, the muscle strips took up 133.0 ± 6.0 nmol/g tissue (n = 9) during a 30 min ^{45}Ca loading period. During the high K-induced contraction, the ^{45}Ca retention increased by 59%. W-7 increased the resting ^{45}Ca retention by 48% and did not affect the K-induced increase in ^{45}Ca retention. CPZ did not change the resting ^{45}Ca retention but inhibited the K-induced increase. Since it has been reported that ^{45}Ca taken up by the muscle in high K solution is mainly retained by mitochondria (Karaki & Weiss, Blood Vessels 18:28, 1981), effects of anoxia on the increase in ^{45}Ca retention induced by high K and W-7 were examined. Anoxia did not change the K-induced contraction although it inhibited the K-induced increase in the ^{45}Ca retention. W-7 induced increase in the resting ^{45}Ca retention was also inhibited by anoxia. These data suggest that anoxia inhibits mitochondrial Ca uptake but not the K-induced Ca influx; and W-7 stimulates mitochondrial Ca uptake and does not inhibit the K-induced Ca influx. These data also support the suggestion (Hidaka et al., J. Pharmacol. Exp. Ther. 207:8, 1978) that W-7 produces relaxation of isolated vascular strips by inhibiting actin and myosin interaction.

REGULATION OF CALCIUM TRANSPORT IN MICROSOMAL PREPARATIONS ENRICHED IN SARCOPLASMIC RETICULUM FROM RABBIT SKELETAL MUSCLE. S. Katz & A. Wong, Division of Pharmacology, Faculty of Pharmaceutical Sciences, University of British Columbia, Vancouver, B.C. V6T 1W5, Canada

Previous studies in our laboratory have indicated that calmodulin significantly stimulates calcium transport in dog microsomal preparations enriched in sarcoplasmic reticulum (S.R.)(Katz & Remtulla, Biochem. Biophys. Res. Commun. 83:5603, 1979; Lopaschuk et al., Biochemistry 19:5603, 1980). This stimulation was additive to that produced by maximal stimulatory concentrations of cyclic AMP dependent protein kinase. Monovalent cations (KCl and NaCl) significantly stimulated calcium transport in this preparation but reduced the degree of stimulation observed in the presence of calmodulin. The control of excitation-contraction and relaxation in skeletal muscle is markedly different from that of cardiac muscle. The possible mechanism of regulation of skeletal muscle S.R. Ca^{++} transport was therefore investigated. ATP dependent oxalate facilitated calcium transport in S.R. preparations obtained from rabbit vastus lateral muscle (fast skeletal muscle) was significantly stimulated ($p < 0.005$) by KCl (10-110 mM). Calmodulin (0.3-12.0 μg/ml) did not stimulate calcium transport in this preparation. Cyclic AMP dependent protein kinase (5-200 μg/ml) in the presence of 110 mM KCl also had no stimulatory effect. In rabbit soleus (slow) S.R. preparations calcium transport was significantly lower than that observed in fast S.R. preparations. Calmodulin (0.3-12.0 μg/ml) had no significant effect on this activity. KCl produced a similar degree of stimulation in this preparation compared to that observed in fast skeletal muscle S.R. Cyclic AMP dependent protein kinase, in the presence of 110 mM KCl, had no significant effect. These data suggest that the mechanism of regulation of S.R. calcium transport in skeletal muscle preparations is different from that previously observed in cardiac muscle. (Supported by the British Columbia [Canada] Heart Foundation.)

A POSSIBLE INVOLVEMENT OF CALMODULIN-SPECTRIN INTERACTION IN THE REGULATION OF ERYTHROCYTE SHAPE: A STUDY WITH CALMODULIN ANTAGONISTS. K. Kidoguchi, A. Hayashi, K. Sobue*, S. Kakiuchi* & H. Hidaka**, Third Department of Internal Medicine, and *Institute of Higher Nervous Activity, Osaka University Medical School, Kita-ku, Osaka 530, Japan; and **Department of Pharmacology, Mie University School of Medicine, Tsu, Mie 514, Japan.

Recently, calmodulin has been shown to bind to spectrin in the presence of Ca^{++} (Sobue et al., Biochem. Int. 1:561, 1980; Sobue et al., Biochem. Biophys. Res. Commun. 100:1063, 1981). Since there is general agreement that spectrin is intimately implicated in the regulation of the erythrocyte shape and its deformability, the possibility arose that calmodulin-spectrin interaction may influence the erythrocyte shape change. Thus, with the aid of calmodulin antagonists, we examined this possiblity. Human erythrocytes, washed with saline, were incubated with drugs for 30 min at room temperature and fixed with 1% glutaraldehyde, then examined with a scanning electron microscope; 100 µM trifluoperazine (TF), chlorpromazine (CP), and promethazine (PM) produced spherocytes, a spherocyte-stomatocyte mixture, and a stomatocyte-discocyte mixture, respectively. The change induced by W-7 was similar to that with CP. These orders agree with their relative potencies in terms of the inhibition on the calmodulin dependent phosphodiesterase activity; the concentrations of TF, CP, W-7 and PM required for the half-maximum inhibition were 8.5, 35, 50 and 400 µM, respectively. The results are consistent with the view that calmodulin-spectrin interaction may be involved in the regulation of erythrocyte shape. This does not exclude the possibility that the observed shape changes may be due in part to the inhibition of calmodulin dependent Ca^{++} pump ATPase, with the consequent increase in the intracellular Ca^{++} level. However, when the incubation was carried out in a medium containing 2 µM Ca^{++}, 100 µM TF still produced similar shape change, indicating that, other than the Ca^{++} pump ATPase, there is a calmodulin-dependent process which is responsible for the shape change. A most likely candidate for this process is the calmodulin-spectrin interaction.

PURIFICATION OF CALMODULIN FROM HUMAN BRAIN. N. Kobayashi, Y. Nakeo, M. Kishihara, Y. Baba, T. Fujita & K. Hayashi*, 3rd Division, Department of Medicine, Kobe University School of Medicine, Kobe 650, Japan, and *Department of Biological Chemistry, Faculty of Pharmaceutical Science, Kyoto University, Kyoto 606, Japan.

Although various calmodulins have been isolated from both plants and animals, there are no reports on biological and physiological properties of human calmodulin in a homologous state. Since human calmodulin may help to elucidate the mechanism of the many Ca^{++} regulated biological processes in man, we made an attempt to isolate pure human brain calmodulin by six steps, as follows: Step 1: Homogenization; Step 2: CM sephadex C-50 batch filtration; Step 3: DEAE-sephadex A-25 column chromatography; Step 4: Ammonium sulfate, pH 4.0, precipitation; Step 5: DEAE-sephadex A-50 column chromatography; Step 6: Sephadex G-100 gel filtration. The purified protein migrated as a single band on SDS polyacrylamide gel electrophoresis and also comigrated with bovine brain calmodulin. Amino acid analysis of this protein was performed by Hitachi 835-50 single column system automatic acid analyzer, according to the method of Spackman et al., after hydrolysis with 6 N HCl at 110°C for 24 hr in an evacuated sealed tube. Several characteristic features of the amino acid composition of vertebrate calmodulin were similarly observed in this protein. These included a high aspartic and glutamic acid content, and the absence of tryptophan and cystein. The ability of this human brain calmodulin to stimulate calmodulin deficient bovine heart phosphodiesterase (PDE) was compared to that of bovine brain calmodulin. Human brain calmodulin had the same activity on PDE activation as did bovine brain calmodulin in the presence of Ca^{++}, but did not active PDE in the presence of Ca^{++} chelator, EGTA. Human calmodulin was purified from the brain by ammonium sulfate precipitation, gel filtration, and anion exchange chromatography. The biological and physicochemical properties of human brain calmodulin such as the ability to activate PDE, molecular weight, and amino acid composition were almost the same as bovine brain calmodulin.

PHYSIOLOGICAL ROLE OF CALMODULIN ON THE REGULATION OF HEPATIC GLYCOGENOSIS BY PHENYLEPHRINE. Y. Koide, S. Kimura* & K. Yamashita, Institute of Clinical Medicine, University of Tsukuba, Ibaraki 305, Japan, and *Endocrinology Division, National Cancer Center Research Institute, Tokyo 104, Japan.

The physiological role of calmodulin in the regulation of hepatic glycogenolysis by glucagon (G) and phenylephrine (PL) was studied. Effects of G, PL, dibutyryl cAMP (DBCA) and cAMP on the following parameters were examined in the rat liver perfused with trifluoperazine (ST) which blocks biological effects of calmodulin: 1) glucose output from perfused liver; 2) tissue level of cAMP and cGMP; 3) tissue level of phosphorylase a activity (Ph-a); 4) ^{45}Ca efflux from ^{45}Ca preloaded liver. While ST inhibited the effect of supermaximal concentration of G on cAMP level, it affected

Neither the increase in glucose output nor Ph-a activation induced by G. Dose-response curves on the net increase in glucose output by G, with and without ST, were identical. The activation of glucose output by DBCA and cAMP was not affected by ST. G, with and without ST, or ST alone, did not cause any significant changes in cGMP level. In contrast to G, PL caused rapid increase in Ph-a and glucose output without changes in cAMP level, and 0.01-0.1 mM ST significantly inhibited both effects of PL in a dose related manner. The inhibitory effect of ST on glucose output by PL was observed when ST perfusion was initiated either before (10 min) or after (5 min) PL administration. PL alone or with ST was again without effects on cGMP level. ST lowered the peak of PL-induced efflux of ^{45}Ca from the perfused liver. However, the magnitude of the inhibition was much smaller than that observed in glucose output or Ph-a. These effects of ST were independent of the extracellular calcium concentration. The results indicate that calmodulin is involved in the physiological regulation of hepatic glycogenolysis by PL, but not by G. It appears that the role of calmodulin is exerted through the regulation of phosphorylase kinase or calmodulin dependent protein kinase activity by binding to cytosolic calcium, the concentration of which is raised by PL.

THE EFFECTS OF CALMODULIN AND CALMODULIN INTERACTING AGENTS ON CALCIUM TRANSPORT IN SARCOPLASMIC RETICULUM OF SKELETAL MUSCLE.
M. Koshita & K. Hotta, Department of Physiology, Nagoya City University Medical School, Nagoya, Aichi 467, Japan.

It has been reported that calmodulin stimulates Ca^{++} transport in cardiac and smooth muscle microsomal fractions enriched in sarcoplasmic reticulum (SR). To clarify the involvement of calmodulin in the Ca^{++} transport system, the effects of calmodulin and calmodulin interacting agents on Ca^{++} uptake and ATPase activity in fragmented SR (FSR) of skeletal muscle were investigated. FSR was prepared from the leg muscle of bullfrog by differential centrifugation. Ca^{++} uptake was measured with the millipore filter method. Reaction mixture contained 50 µg FSR protein/ml, 20 mM Tris-maleate (pH 6.8), 100 mM KCl, 1 mM $MgCl_2$, 1 mM ATP and 0.1 mM $^{45}CaCl_2$. Ca ATPase activity was calculated by subtracting Mg-ATPase activity from Ca Mg-ATPase activity. N-(6-aminohexyl)-5-chloro-1-naphthalenesulfonamide (W-7) and trifluoperazine (TFP), calmodulin interacting agents, reduced the maximum Ca^{++} uptake and the initial rate of Ca^{++} uptake by FSR. The concentration of TFP producing 50% inhibition for Ca^{++} uptake was about 40 µM. Complete inhibition of Ca^{++} uptake was achieved at 100 µM of TFP. TFP (100 µM) also inhibited Ca Mg-ATPase activity, Mg-ATPase activity, and Ca ATPase activity to 45%, 45% and 43% of original level, respectively. Calmodulin did not alter the amount of Ca^{++} taken up by FSR significantly at steady level, however, it enhanced the initial rate of Ca^{++} uptake. Calmodulin (2 µg/ml) stimulated Ca Mg-ATPase activity, Mg-ATPase activity, and Ca ATPase activity by 19, 7 and 74% respectively.

These results suggest that calmodulin may have an important role in Ca^{++} transport of skeletal muscle SR. We thank Professor Yagi and Dr. Yazawa (Hokkaido University) for the generous gift of calmodulin from pig brain, and also Professor Hidaka (Mie University) for W-7. TFP was a gift from Yoshitomi Pharmaceutical Industry.

GOOD CORRELATION OF THE AMOUNT OF CYTOPLASMIC CALCIUM WITH THE CONTRACTILITY OF PHYSARUM POLYCEPHALUM. R. Kuroda & H. Kuroda, Department of Physiology, School of Dentistry, Aichi-Gakuin University, Nagoya 464, Japan.

The shuttle streaming of endoplasm in the plasmodium of Physarum polycephalum results from the alternating periodic contraction and relaxation of the actomyosin system in the ectoplasm. The activity of the actomyosin system may be modulated by the change of the local cytoplasmic Ca ion concentration. Ca in a dumb-bell shaped plasmodium showing active shuttle streaming was visualized with the electron microscopy by precipitation with K pyroantimonate, and the distribution of Ca between the cytoplasm and cellular organelles, especially vacuoles, was examined. The electron opaque precipitates were verified to be a Ca pyroantimonate by its susceptibility to removal by chelation with EGTA and x-ray microprobe analysis. Small Ca precipitates located in the cytoplasm were more abundant in the contracting half mass (3.6/μm^2) than in the relaxing one (1.9/μm^2). Proportion of the Ca containing vacuoles was smaller in the contracting half mass (41%) than in the relaxing one (71%) and vice versa as for the empty vacuoles. Ca ionophore X-537A was applied to one half mass as the motive force for endoplasmic streaming was being measured by the double chamber method of N. Kamiya. The motive force was increased by X-537A, suggesting that the contractility of the X-537A treated half mass increased. However, the period and phase of the shuttle streaming did not change, suggesting that Ca might not be a key substance for reversion of the direction of the endoplasmic streaming. The X-537A treated contracting plasmodium was significantly more abundant in the cytoplasmic Ca precipitates (8.7/μm^2) and empty vacuoles (62%) than the untreated contracting plasmodium. This ensemble of results indicates that Ca is released from the vacuoles into the cytoplasm in the contracting phase, and in the relaxing phase Ca is sequestered from the cytoplasm into the vacuoles.

EFFECT OF Ca^{++} ON INSULIN SENSITIVE PHOSPHODIESTERASE IN RAT FAT CELLS. H. Makino, A. Kanatsuka, M. Osegawa, A. Kumagai & S. Kakiuchi*, Second Department of Internal Medicine, Chiba University School of Medicine, Chiba 280, Japan, and *Institute of Higher Nervous Activity, Osaka University Medical School, Osaka 530, Japan.

In previous work (J. Biol. Chem. 255:7850, 1980) we observed that insulin activates membrane bound low Km cyclic AMP phosphodiesterase in fat cells. The molecular weight of the enzyme activated by insulin is 10^5 daltons larger than the basal enzyme. These observations prompted us to investigate whether calmodulin may be related to the activation mechanism of this enzyme. Isolated epididymal fat cells obtained from SD rats were incubated with or without 2 nM insulin for 7 min at 37°C and homogenized after washing. A crude microsomal fraction (fraction P-2) prepared by differential centrifugation was suspended in 0.25 M sucrose containing 10 mM Tes buffer, pH 7.5 (buffer A) at 0°C. Using fraction P-2 four different experiments were then performed: 1) Calmodulin was added to portions of fraction P-2. 2) Fraction P-2 was washed repeatedly with EGTA to remove endogenous calmodulin and resuspended in buffer A with 0.1 mM EGTA. Calmodulin and various concentrations of Ca^{++} were added to portions of washed fraction P-2. 3) Fraction P-2 was incubated with various concentrations of Ca^{++} for 3 hr. 4) Fraction P-2 was incubated with 10 mM Ca^{++} for various time intervals. All incubation was performed at 0°C and phosphodiesterase was assayed as described by Kono et al. (J. Biol. Chem. $\underline{250}$:7826, 1975). Calmodulin had no effect on the basal and plus insulin enzymes, both with and without endogenous calmodulin. Ca^{++} had a stimulatory effect on both basal and plus insulin enzymes. The effects of Ca^{++} were time and concentration dependent. These results suggest that calmodulin is not directly involved in the activation of the enzyme by insulin, but Ca^{++} is perhaps an activator of the enzyme but a yet unknown mechanism.

DETECTION OF CALCIUM WITH X-RAY MICROANALYSIS IN CYTOCHEMICAL REACTION PRODUCT OF ALKALINE AND ADENOSINE TRIPHOSPHATASE AT APICAL CELL MEMBRANE OF THE EPITHELIUM OF DUODENUM OF THE RAT AND HEN.
T. Makita, Veterinary Anatomy, Yamaguchi University, 1677-1 Yoshida, Yamaguchi City 753, Japan.

Calcium binding protein (Ca-BP) has been found in the epithelial cells, including goblet cells, of duodenum, kidney and avian oviduct. Alkaline (ALP) and adenosine triphosphatase (ATPase) have been localized along the plasma membrane of microvilli of epithelial cells. This interesting dual localization of Ca-BP and phosphatases and another well known fact that vitamin D3 administration can increase or induce both Ca-BP and phosphatase are the base for a working hypothesis that Ca-BP may be a sort of multi-enzyme complex containing both ALPase and ATPase. Additional evidence in support of that hypothesis we then surveyed Ca over apical cell membrane, including microvilli, goblet cell's mucous, and the cytochemical reaction product of phosphatase in the duodenum of chicken and rats, using either an energy dispersive (EDX) or wave length dispersive (WDX) x-ray microanalyzer attached to a scanning (SEM) or transmission (TEM) electron microscope. Ca was localized over microvilli and dense structure of terminal web which does not have phosphatase activity. There was too little Ca, if any, to be detected at the lateral cell wall where phosphatases were evident and mucous globules in goblet cell, which has been reported as a site of Ca-BP by immunocytochemists. Phosphatase reaction product, lead phosphate, at Gogli complex did not contain any significant amount of Ca either. Such discrepancy of localization of Ca, phosphatase reaction product and Ca-BP in the duodenal epithelium suggested that Ca-BP does not always contain ATPase and/or ALPase.

COMPARATIVE STUDIES ON THREE SPECIES OF CALCIUM DEPENDENT PROTEIN KINASES. R. Minakuchi, Y. Takai, U. Kikkawa, K. Kaibuchi, J. Takahashi & Y. Nishizuka. Department of Biochemistry, Kobe University School of Medicine, Kobe 650 Japan, and Department of Cell Biology, National Institute for Basic Biology, Okazaki 444, Japan.

Cellular activation and proliferation are often initiated by interaction of extracellular messengers with specific cell surface receptors, and Ca^{++} dependent protein phosphorylation seems to play a role in such activation processes. Muscle glycogen phosphorylase kinase (GP-kinase) and myosin light chain kinase (MLC-kinase) are both activated by Ca^{++} in a calmodulin dependent manner. Another species of Ca^{++} dependent protein kinase (C-kinase) which has been recently found in this laboratory is now purified from rat brain to nearly homogeneity. This enzyme is composed of a single polypeptide having an approximate molecular weight of 77,000. The enzyme absolutely requires micromolar concentrations of Ca^{++} and phospholipid for its activation, but contains no calmodulin nor is it activated by exogenous calmodulin. Instead, a small quantity of unsaturated diglyceride dramatically increases the affinity of enzyme for Ca^{++} as well as for phospholipid, and available evidence suggests that phosphatidylinositol hydrolysis induced by many extracellular messengers is directly coupled to the activation of C-kinase. It is also plausible that such a group of extracellular messengers may usually elevate Ca^{++} levels by causing the influx or intracellular movement of this divalent cation. Thus, the three species of Ca^{++} dependent protein kinases mentioned above are presumably activated simultaneously upon stimulation by single extracellular messengers. C-kinase thus activated does not react with myosin light chain but preferentially with some proteins associated with membranes. C-kinase is more susceptible to and profoundly inhibited by various phospholipid interacting drugs such as chlorpromazine, imipramine and dibucaine than GP-kinase and MLC-kinase. Some of the physical, kinetic and catalytic properties of the three species of Ca^{++} dependent protein kinases will be presented for comparison.

ACTIVATION OF CALCIUM AND MAGNESIUM STIMULATED ADENOSINE TRIPHOSPHATASE BY cAMP PROTEIN KINASE INHIBITOR AND CALMODULIN. A. Minocherhomjee & B.D. Roufogalis, Laboratory of Molecular Pharmacology, Faculty of Pharmaceutical Sciences, University of British Columbia, Vancouver, British Columbia V6T 1W5, Canada.

cAMP (1 μM) inhibits (Ca^{++} + Mg^{++}) ATPase activity in human erythrocyte membranes, probably by stimulating protein phosphorylation by an endogenous cAMP protein kinase (PK). cAMP protein kinase inhibitor (PKI) from bovine heart stimulated (Ca^{++} + Mg^{++}) ATPase activity over and above that expected from its antagonism of the PK dependent inhibition of the ATPase activity. Rabbit skeletal muscle PKI failed to stimulate the (Ca^{++} + MG^{++}) ATPase under the same conditions. The stimulation by bovine heart PKI was not due to contamination by calmodulin, which was removed by prior purification of PKI by DE-52 ion exchange chromatography. The activation by PKI was additive to that of saturating amounts of calmodulin at low Ca^{++} (0.58 µM) but not at high Ca^{++} (55 µM). This effect was seen both in erythrocyte "ghosts" and in Triton X-100 solubilized (Ca^{++} + Mg^{++}) ATPase preparations. In both cases PKI enhanced the apparent Ca^{++} affinity of the ATPase without increasing its V_{max}, in contrast to calmodulin which influenced both these parameters. The activation by PKI was blocked by 20 µM trifluoperazine but was not affected by modification of arginine groups by 1,2-cyclohexanedione. It was concluded that PKI from bovine heart stimulated (Ca^{++} + Mg^{++}) ATPase by a direct interaction with the enzyme complex, which increases its apparent affinity for Ca^{++}. Although both calmodulin and PKI are small, heat stable, acidic proteins, these properties alone do not appear sufficient for producing the activating effect observed. (Supported by the Canadian Heart Foundation and M.R.C. of Canada.)

REGULATION OF NICOTINAMIDE COENZYME LEVELS IN CHLOROPLASTS BY Ca^{++}.
S. Muto, S. Izawa* & S. Miyachi, Institute of Applied Microbiology, University of Tokyo, Bunkyo-ku, Tokyo 113, Japan, and *On leave from Department of Biology, Wayne State University, Detroit, Michigan 48202, USA.

Conversion of NAD to NADP was observed when Chlorella cells, which had been kept in the dark, were illuminated. The conversion reaction was inhibited by CCCP and CMU. Similar light induced conversion of NAD to NADP was observed in Chlamydomonas cells and higher plant leaves as well as intact chloroplasts isolated from leaves. In isolated chloroplasts, it is well known that illumination causes a rapid increase in stroma ATP level. NAD kinase is the only enzyme that is known to catalyze the formation of NADP from NAD. We have previously shown that this enzyme is localized mostly in the chloroplasts. These results suggest that the light induced conversion of NAD to NADP is the result of phosphorylation of NAD by photochemically produced ATP and is catalyzed by NAD kinase in the chloroplasts. We found that NAD kinase from pea seedlings requires an activator protein for its activity. The protein was subsequently identified as calmodulin by Anderson and Cormier, and by us as well. Illumination of intact chloroplasts is known to cause an H^+ uptake and Mg^{++} release by the thylakoids. These changes are associated with H^+ efflux from the envelope. We have found that intact chloroplasts isolated from wheat leaves take up Ca^{++} from suspending medium in the light, apparently in exchange with the released H^+. Since NAD kinase requires both Mg^{++} and Ca^{++} for its activity, illumination may thus activate and result in formation of NADP in the chloroplasts.

ROLE OF Ca^{++}, cGMP AND cAMP IN TASTE TRANSDUCTION MECHANISM. S. Nagahama & K. Kurihara, Faculty of Pharmaceutical Sciences, Hokkaido University, Sapporo 060, Japan.

The initial event of taste reception is adsorption of chemical stimuli on microvilli membranes of taste cells. The adsorption leads to a release of a chemical transmitter from taste cells which induces an increase in the gustatory nerve activities. It has been inferred that Ca and cyclic nucleotides may be involved in the taste transduction mechanism but no direct evidence has been obtained due to experimental difficulties; chemical applied to the tongue surface hardly penetrate taste cells, or below the tight junction between taste cells. In the present study, the lingual artery of the bullfrog was perfused with artificial solution and the effects of Ca, Ca-channel blockers, cGMP and cAMP added to the perfusing solution on the gustatory nerve responses were examined. The responses to chemical stimuli of Group 1 ($CaCl_2$, NaCl, distilled water, D-galactose and L-threonine) were highly dependent on Ca concentration in the perfusing solution; the responses were greatly decreased at low Ca concentration and increased with an increase in Ca concentration. The responses were greatly suppressed by addition of the Ca-channel blockers ($MnCl_2$ and verapamil). The responses were greatly enhanced by addition of cGMP to the perfusing solution and suppressed by cAMP. The responses to chemical stimuli of Group 2 (HCl, quinine and ethanol) were practically not affected by a decrease in Ca concentration in the perfusing solution, addition of the Ca-channel blockers, cGMP and cAMP. The responses of the stimuli of Group 1 seem to be induced by Ca influx into the taste cell which is triggered by depolarization and modulated by the cyclic nucleotides in the taste cell. The responses of Group 2 seem to be induced without accompanying Ca-influx.

TETRAHYMENA CALMODULIN: INTRACELLULAR DISTRIBUTION AND GUANYLATE CYCLASE ACTIVATION. S. Nagao, K. Sobue*, S. Kakiuchi* & Y. Nozawa, Department of Biochemistry, Gifu University School of Medicine, Tsukasamachi-40, Gifu 500, Japan, and *Institute of Higher Nervous Activity, Osaka University Medical School, Kita-ku, Osaka 530, Japan.

Tetrahymena calmodulin (M.W. 15,000) was purified by trichloroacetic acid precipitation, DEAE-cellulose column chromatography, and gel filtration on Ultragel AcA 44 column. Its identification as calmodulin was based on capability of activating the bovine brain phosphodiesterase, mobility on polyacrylamide gel electrophoresis, and similarity in amino acid composition. Nearly all (93%) of the total amount of calmodulin was localized in two soluble compartments, the ciliary and postmicrosomal supernatant fractions. Moreover, the calmodulin-binding activities were found to be localized mainly in microsomes and to some extent in cilia and plasma membrane. In the presence of calcium ion at low concentrations (pCa 6.0-4.6), the protein induced a marked enhancement of the activity of guanylate cyclase in this organism. However, Tetrahymena calmodulin did not exert any influence on guanylate cyclase activities in human platelet, rat brain and rat lung. The activation of Tetrahymena guanylate cyclase of the calmodulin was blocked by trifluoperazine. The concentration of the drug which produced 50% inhibition of the activated activity was 40 µM. Since no other calmodulins so far examined stimulated the activity of Tetrahymena guanylate cyclase, Tetrahymena calmodulin is unique in its potency to activate the Tetrahymena guanylate cyclase activity. Furthermore, amino acid sequence of the Tetrahymena calmodulin displayed profound differences from that of brain calmodulin; substitutions at eleven residues and one deletion (Yazawa et al., Biochem. Biophys. Res. Commun., in press). It is postulated that the structural difference between Tetrahymena calmodulin and other calmodulins from different tissues may be attributable to the specific activation of the guanylate cyclase.

EFFECTS OF Ca^{++} AND DEPOLARIZATION ON CYCLIC GMP CONTENT IN EXCISED SUPERIOR CERVICAL SYMPATHETIC GANGLIA OF THE RAT WITH AND WITHOUT DENERVATION AND AXOTOMY. Y. Nagata, M. Ando & T. Nanba, Department of Physiology, School of Medicine, Fujita-Gakuen University, Toyoake, Aichi 470-11, Japan.

The superior cervical sympathetic ganglion is used as a model system which makes it possible to define a physiological role of cyclic GMP in the nervous system. Either electrical stimulation of preganglionic nerve or addition of acetylcholine (ACh) to the ganglion of the rat caused substantial increase in cyclic GMP content in the tissue. This increase of cyclic GMP level is not inhibited by nicotinic ACh antagonists such as hexamethoniom or curare, but is mostly depressed by the muscarinic cholinergic antagonist, atropine. Thus, the increase of cyclic GMP by ACh, released from nerve terminals on postganglionic neurons, may result in the receptor coupled activation of guanylate cyclase at the postsynaptic sites. When cervical ganglion is incubated in high K^+ solution, a remarkable increase in cyclic GMP content is found. In the absence of Ca^{++} in the medium, K^+-induced stimulation of cyclic GMP level was not detected. Neither atropine nor hexamethonium had any effect on K^+-induced increase of cyclic GMP. Veratridine-induced depolarization also caused the increase of cyclic GMP level in the ganglion, and this effect was completely antagonized by tetrodotoxine. Unlike intact ganglia, the ganglia denervated for 1 to 7 days did not show increases of cyclic GMP content by raising K^+ and by depolarization with the addition of veratridine. Ganglia axotomized for 1 to 4 days responded to elevated K^+, but failed to respond to ACh in cyclic GMP accumulation. These results suggest that depolarization of preganglionic nerve terminals by elevated K^+ or by veratridine in the presence of Ca^{++} in the medium stimulates cyclic GMP accumulation, which in turn may trigger the release of the neurotransmitter, ACh, but not at the postganglionic receptor sites in the ganglion.

A SIMPLE PROCEDURE FOR PURIFICATION OF MEMBRANE BOUND CALMODULIN. K. Nakaya, S. Nakajo & Y. Nakamura, Laboratory of Biological Chemistry, School of Pharmaceutical Sciences, Showa University, Shinagawa-ku, Tokyo 142, Japan.

Studies from several laboratories have shown that calmodulin is almost equally distributed between soluble and membrane fractions in several tissues (1-4). Considerable amounts of calmodulin were not released from the membrane fraction even by treatment with 2 mM EGTA (3,4). However, little is known of the biological significance of the membrane bound calmodulin. This is at least partly because of the difficulties in preparing the membrane bound calmodulin. We report here a simple procedure for purifying calmodulin from the particulate fraction of rat liver. Rat liver was homogenized in the presence of 2 mM EGTA and the particulate fraction was sedimented by centrifugation at 100,000 x g for 1 hr. The particulate fraction was solubilized with 0.3 M lithium diiodosalyclate, extracted with phenol, and the aqueous layer was lyophilized. This fraction was further subjected to gel filtration. No significant differences between the purified protein and calmodulin from the soluble fraction of rat brain were detected. Both proteins were identical with respect to electrophoretic migration, chromatographic behavior, and molecular weight. In addition, they had quite similar amino acid compositions and were capable of activating phosphodiesterase in a calcium dependent manner. References: 1) Cheung et al., In Cyclic Nucleotides in Disease (ed) B. Weiss, 1975, pp.321, University Park Press, 2) Gnegy et al., Proc. Nat. Acad. Sci. 73: 352, 1976, 3) Teshima & Kakiuchi, J. Cyclic Nucleotide Res. 4:219, 1978, 4) Uenishi et al., J. Biochem. 87:601, 1980.

POSSIBLE ROLE OF CALMODULIN IN PANCREATIC ISLETS. H. Niki, A. Niki, T. Koide & H. Hidaka*, Department of Internal Medicine, Aichigakuin University School of Dentistry, Nagoya, Aichi 464, Japan, and *Department of Pharmacology, Mie University School of Medicine, Tsu, Mie 514, Japan.

A rise in the intracellular concentration of calcium is considered to be an important factor in the stimulation of insulin secretion. Calmodulin and a Ca^{++}-calmodulin dependent adenylate cyclase have been demonstrated in pancreatic islets. To elucidate the role of calmodulin in islets, we have studied the effects of calmodulin antagonists, N-(6-aminohexyl)-5-chloro-1-napthalenesulfonamide (W-7) and trifluoperazine (TFP), on islet-cell functions. Islets of Langerhans were isolated from rat pancreas by the collagenase digestion method. Both W-7 and TFP (25-100 µM) inhibited glucose (10 mM) induced insulin release from batch incubated islets. The maximal inhibition by W-7 (approximately 30%) was smaller than that by TFP (approximately 75%). W-7 did not affect the rate of formation of $^{14}CO_2$ from [U-^{14}C]glucose in islets. Total radioactivity of ^{3}H-leucine incorporated into the proinsulin and insulin fractions during the incubation with glucose (10 mM) was not modified but the radioactivity in the insulin fraction was decreased by W-7. Increase in islet cyclic AMP induced by glucose (10 mM), with or without 3-isobutyl-1-methylxanthine, a phsophodiesterase inhibitor, was inhibited by TFP but was not modified by W-7. The difference may be due to TFP having a higher affinity to lipid and being primarily accumulated in the cell membrane. In conclusion, calmodulin is involved in a part of stimulus secretion coupling in the pancreatic B-cell and probably in the conversion of proinsulin to insulin, but may not be involved in glucose induced accumulation of cyclic AMP in islets.

EFFECT OF CALMODULIN INHIBITOR ON PHOSPHOLIPID METABOLISM ALTERATION INDUCED BY A TUMOR PROMOTOR IN HeLa CELLS. H. Nishino, A. Iwashima & T. Sugimura*, Department of Biochemistry, Kyoto Prefectural University of Medicine, Kawaramachidori, Kamikyoku, Koyoto 602, Japan, and *National Cancer Center Research Institute, Tsukiji, Chuoku, Tokyo 104, Japan.

Although many of the biological and biochemical effects of tumor promotors were observed, it is not clear which of the effects induced by these compounds are related to the actual promotion of tumors. Within a few minutes or hours after exposure of cells to a promotor, some effects on cell membranes are observed, i.e. effects on phospholipid metabolism, transport of small membranes, and expression of fibronectin. Such early events may be related to a promoting action. In the present study, we investigated the effects of teleocidin, a new kind of tumor promotor, on membrane phospholipid metabolism, which seems to be affected by calmodulin, in HeLa cells. Stimulation of phospholipid metabolism, as measured by the incorporation of ^{32}P and labelled precursors of the polar head groups, was detected within a few hours after treatment with HeLa cells with teleocidin (20 ng/ml). Acceleration of ^{32}P incorporation into phosphatidylcholine is prominent, although many kinds of phospholipids were labelled with ^{32}P after treatment with the promotor. In the presence of trifluoperazine (25 μM), a calmodulin inhibitor, the stimulatory effect of teleocidin on phosphatidylcholine turnover was diminished. It has also been demonstrated that some biological effects of tumor promotors are expressed depending on Ca^{++} and calmodulin. Therefore, it might be possible that tumor promoting action is achieved in the presence of active calmodulin.

HYDROPHOBIC INTERACTION OF CALMODULIN IN ANTAGONISTS WITH Ca^{++} CALMODULIN COMPLEX. T. Ohmura, T. Tanaka & H. Hidaka, Department of Pharmacology, Mie University School of Medicine, Tsu, Mie 514 Japan

We reported previously that the function of the hydrophobic region is important in the calmodulin-enzyme(s) interaction (J. Biol. Chem. 255:11078, 1980). A calmodulin antagonist such as N-(6-aminohexyl)-5-chloro-1-naphthalene-sulfonamide, which binds to calmodulin in the presence of Ca^{++} and selectively inhibits calmodulin dependent enzyme activities, contains hydrophobic moiety in its molecule structure. It is likely that the hydrophobic moiety of the calmodulin antagonist interacts with calmodulin in the presence of Ca^{++}, followed by inactivation of calmodulin. We then synthesized dechlorinated analogs of naphthalenesulfonamide derivatives and investigated the change in hydrophobicity of these calmodulin antagonists. The hydrophobicity of chlorine deficient derivatives which was determined as octanol-water partition coefficients, was lower than that of these derivatives with the chlorine molecule. Moreover, these chlorine deficient naphthalenesulfonamides were less potent in suppressing the fluorescence of hydrophobic probe, 2-p-toluidinylnaphthalene-6-sulfonate in the presence of a Ca^{++}-calmodulin complex, indicating that these compounds without chlorine have a lower hydrophobicity. We then determined the affinity of naphthalenesulfonoamides by displacement of [^{3}H]N-6-aminohexyl)-5-chloro-1-naphthalenesulfonamide (W-7) from calmodulin using an equilibrium binding technique. It was found that dechlorination of these derivatives decreased the affinity for calmodulin. Furthermore, dechlorinated naphthalenesulfonamides were also less potent in inhibiting Ca^{++}-calmodulin dependent enzymes such as Ca^{++} dependent cyclic nucleotide phosphodiesterase and myosin light chain kinase. These results suggest that calmodulin antagonists such as napthalenesulfonamide derivatives may bind to Ca^{++}-calmodulin complex through hydrophobic interactions.

DECREASE OF ENDOGENOUS THROMBOXANE SYNTHESIS AND ITS REVERSAL BY α-BLOCKER, CALMODULIN AND LEUKOTRIENE C RECEPTOR-BLOCKING AGENT IN HYPOTHALAMUS OF SHR SR (NOVEL SUBSTRAIN C). E. Ohtsu & K. Abe, Department of Pediatrics, Kitasato University School of Medicine, Sagamihara, Kanagawa 228, Japan.

Some biosynthetic alterations for 6-keto-prostaglandin (PG)F/α, thromboxane (TX)B_2 and PGE_2 in the brain stem of spontaneously hypertensive rats (SHR; stroke resistant type)(Ohtsu & Matsuzawa, J. Nutri. Sci. Vitaminol. 26:343, 1980) have been investigated for their pharmacological receptor response. Endogenous synthesis of PGD_2 was not detected in hypothalamus or in pons-medulla. The possible genetic deficiency of thromboxane synthetase activity and its reversal (in vitro) by phentolamine (0.5 mg/ml), calmodulin (0.1 mg/ml; purified human brain calmodulin; donated by W. Schreiber, University of Washington) and FPL (5 μg/ml; a leukotriene C receptor blocking agent; given by Dr. Koshihara, Tokyo Metropolitan

Institute of Gerontology) were also confirmed in hypothalamus of SHR SR of 3-month-old (after establishment of hypertension). Ca^{++} (2.5 mM) increased each biosynthetic formation several-fold in hypothalamus of progenitor control rats (WKY; n = 10), but it did not do so in SHR SR (n = 20). For comparison, a novel chemokinetic and pro-aggregatory lipooxygenase product, leukotriene B_4 was identified as an elevation associated with Ca^{++} supersensitive platelets and leukocytes from Kawasaki disease (mucocutaneous lymphnode syndrome) using new thin-layer chromatography (m/e 217). (Ohtsu, Int. Symp. Leukotriene & Lipooxygenase Prod., Florence, 1981). TX synthesis in hypothalamus of SP substrain was elevated, insensitive to CaM and blocked by FPL, while TX decreased in pons-medulla of SR.

DES(Ala-Lys)CALMODULIN IN PORCINE BRAIN. T. Okuyama, N. Ishioka, T. Kadoya & T. Isobe, Department of Chemistry, Faculty of Science, Tokyo Metropolitan University, Setagaya, Tokyo 158, Japan.

From porcine brain extract, we have isolated a protein which is similar to calmodulin. This protein was designated as Des(Ala-Lys) calmodulin since it lacked the C-terminal Ala(147)-Lys(148). The content of Des(Ala-Lys)calmodulin was about one-tenth that of calmodulin. The protein was separated from calmodulin by isocratic elution chromatography on DEAE-sephadex A-50 using 0.1 M potassium phosphate (pH 7.1), containing 0.28 M NaCl and 1 mM EDTA as an eluant. The purified protein was distinguishable from calmodulin by its slightly larger mobility on 10% polyacrylamide gel electrophoresis without SDS. The protein gave an amino acid composition very similar to calmodulin, and contained one trimethyl-lysine residue. Comparative tryptic peptide mapping of both the purified protein and calmodulin was performed by high performance liquid chromatography on a Hitachi-Gel 3013N. This mapping result, and the following amino acid analyses of the isolated peptides, showed that the C-terminal fragment of the purified protein was 2-residue shorter than the equivalent fragment of calmodulin, which indicated a lack of the Ala-Lys sequence at the C-terminus of calmodulin. Des(Ala-Lys)calmodulin may be derived by enzymatic removal of the C-terminal dipeptide of calmodulin, or by a different type of cleavage of "pro-calmodulin" which is not detected yet. This would imply that several post-translational modifications, not only methylation of lysine residue but also specific peptide bond cleavages, are involved in the control mechanisms that regulate the biological activity of this "multi-functional" calcium mediator.

A NEW Ca^{++} SENSITIVITY OF THE ATPase AND SUPERPRECIPITATION ACTIVITIES OF ACTO-GIZZARD MYOSIN. H. Onishi & S. Watanabe*, Department of Chemistry, Faculty of Science, Tokyo Institute of Technology, Okayama, Meguro-ku, Tokyo 152, Japan, and *Department of Applied Chemistry, Faculty of Technology, Gumma University, Kiryu-shi 376, Japan.

We previously reported on the myosin light-chain kinase and phosphatase dependent Ca^{++} sensitivity of chicken gizzard actomyosin.

We now report a new Ca^{++} sensitivity independent of phosphorylation-dephosphorylation of myosin light chains: 1) The Ca^{++} sensitivity was practically unaffected by tropomyosin. 2) With 0.3 mg/ml acto-unphosphorylated myosin (reconstituted from actin and unphosphorylated myosin in a weight ratio of 1:2), the Ca^{++} sensitivity was hardly detectable when the KCl concentration was higher than 90 mM (e.g. the ATPase activity in 90 mM KCl was 0.5 nmol/min/mg myosin in the presence of 0.1 mM Ca^{++} and 0.4 nmol/min/mg myosin in its absence). 3) It was, however, easily detectable when the acto-unphosphorylated myosin concentration was increased to 1.5 mg/ml (e.g. the ATPase activity in 90 mM KCl was 11 nmol/min/mg/ myosin in the presence of 0.1 mM Ca^{++} and 1.1 nmol/min/mg myosin in its absence). We also reported that when the KCl concentration was reduced to a low level, a highly aggregated form of myosin was observed in the electron microscope and that the ATPase activity of myosin was increased again by reducing the KCl concentration to less than 0.15 mM. It is, therefore, suggested that the new Ca^{++} sensitivity is perhaps due to a highly aggretated form of myosin.

THE Ca^{++} DEPENDENT REGULATION IN CONTRACTION OF ACTOMYOSIN FROM PLATELET AND LEUCOCYTE. T. Onji, S. Nosaka & N. Shibata, Departments of Internal Medicine and Anesthesiology, Center for Adult Diseases, Osaka 537, Japan.

The contractile mechanisms of porcine platelet (PL) and polymorphonuclear leucocyte (PMN) has been studied. Methods: "Natural actomyosin (AM)" was prepared from PL and PMN by ammonium sulfate fractionation after extraction of the cell homogenate with isotonic-low ionic strength buffer in the presence of ATP and EDTA; "regulatory factor (RF)" which is so-called tropomyosin kinase (TMK) or tropomyosin leiotonin (TMLN) fraction was isolated from bovine carotid artery; tropomyosin was purified from the above RF by precipitation with 15-30 mM $MgCl_2$. Results: The Mg^{++}-ATPase activity of the natural AM from both PL and PMN was increased up to three-to five-fold in the presence of Ca^{++} by the addition of RF dose dependently; the activation was strictly dependent on Ca^{++}, i.e. no activation was observed in the absence of Ca^{++}; concomitantly, the superprecipitation which did not take in the absence of RF was manifested markedly in its presence; however, when purified TM was added to the natural AM, no activation was observed, even in the presence of Ca^{++}; in addition, when calmodulin purified from bovine brain was added to the natural AM, no activation was observed, even in the presence of Ca^{++}. Conclusion: The above findings show that arterial RF is capable of regulating the Ca^{++} dependent AM ATPase activity of PL and PMN. However, purified TM or calmodulin alone is not sufficient to activate the AM ATPase, even in the presence of Ca^{++}. Thus, although PL and PMN share common mechanisms of the regulation of AM-ATPase with arterial smooth muscle, the essential factor remains to be identified.

CALCIUM DEPENDENT ADENYLATE CYCLASE FROM RABBIT CEREBRAL CORTEX.
M. Sano & G.I. Drummond*, Institute for Developmental Research,
Aichi Colony, Kasugai, Aichi, 480-03, Japan, and *Biochemistry
Group, Department of Chemistry, University of Calgary, Calgary,
T2N 1N4, Canada.

Brostrom et al. proposed that the cerebral cortex contains two forms of adenylate cyclase, one which is calcium dependent and the other calmodulin independent, as determined from studies of adenylate cyclase in the particulate fraction. Westcott et al. reached similar conclusions from studies based on fractionation of detergent solubilized enzyme on calmodulin sepharose. In this study we resolved adenylate cyclase in detergent-dispersed preparations of rabbit cerebral cortex by two methods, gel filtration and anion exchange chromatography followed by ammonium sulfate fractionation, into two compartments. One component was stimulated by Ca^{++} and Ca^{++} plus calmodulin and was relatively insensitive to NaF and guanylylimidoliphosphate (Gpp(NH)p). A second component was sensitive to Gpp(NH)p and NaF, but relatively insensitive to Ca^{++} and calmodulin. Each form shows the opposite sensitivity to Gpp(NH)p and NaF, that is, the calmodulin sensitive activity was insensitive to these ligands while the calmodulin insensitive fraction responded to NaF and Gpp(NH)p. Our data support the possibility that two independent forms of adenylate cyclase exist in the cerebral cortex, one regulated by guanine nucleotide regulatory protein and another by Ca^{++} and calmodulin. The calmodulin sensitive enzyme fractions were stimulated two-fold by Ca^{++} and this was increased in the presence of calmodulin added to the assay. However, calmodulin which had been added to the gel filtration column was well resolved from adenylate cyclase. The fraction which contained Ca^{++} sensitive adenylate cyclase failed to stimulate phosphodiesterase. When calmodulin binding protein was added, stimulation of the adenylate cyclase by Ca^{++} was effectively eliminated. Therefore the Ca^{++} sensitivity appears to be mediated by a calmodulin tightly associated with adenylate cyclase.

CALMODULIN ANTAGONIST, W-7 INHIBITS THE INITIATION OF DNA SYNTHESIS AND CAUSES MORPHOLOGICAL CHANGE OF CHO-K1 CELLS IN CULTURE. Y. Sasaki, T. Tanaka & H. Hidaka, Department of Pharmacology, Mie University School of Medicine, Tsu, Mie 514, Japan.

W-7 (N-(6-aminohexyl)-5-chloro-1-naphthalenesulfonamide) is a putative calmodulin antagonist which binds calmodulin dependently to purified calmodulin and inhibits Ca^{++}-calmodulin related enzyme activities. The proliferation of CHO-K$_1$ cells which grew exponentially was suppressed dose dependently by the treatment with W-7 at concentrations ranging from 20 to 50 µM. The addition of 25 µM W-7 to the medium suppressed the proliferation of CHO-K$_1$ cells obtained with mitotic shake procedure from a monolayer in exponential growth. The inhibition was reversible. Immediately after the exclusion of W-7 from the culture medium, the incorporation of [^{3}H] thymidine into acid insoluble fraction was observed. The cell division was subsequently observed at 6 hr after the exclusion. When synchronous CHO-K$_1$ cells by the treatment with 2.5 mM thymidine were used for experiments, similar results were obtained: the cell division occurred at 6 hr after the exclusion of thymidine. On the other hand, W-7 produced morphological changes in CHO-K$_1$ cells to a more round and compact form and inhibited the reverse transformation of CHO-K$_1$ cells caused by the treatment with cholera toxin. Similar results were obtained with the treatment with colcemid, an inhibitor of the polymerization of microtubular protein. These data suggest that the inhibitory effect of W-7 on CHO-K$_1$ cell proliferation is mediated through a selective cell cycle block and that calmodulin may promote the initiation of DNA synthesis. Furthermore, it is proposed that W-7 acts by disrupting microtubules in plasma.

OCCURRENCE OF SOLUBLE CALCIUM DEPENDENT PROTEIN PHOSPHORYLATION IN DIVERSE TISSUES. H. Schulman, J. Luh, P. Nose & J.P. Whitlock Jr., Department of Pharmacology, Stanford University School of Medicine, Stanford, California 94305, USA.

Calcium dependent protein phosphorylation systems have been identified in cytosol fractions from rat liver, bovine brain, rat brain, and in HeLa cell nuclei. Calcium, at physiological concentrations (0.1-5.0 µM), activated protein kinases in each system that was examined. Cytosol fractions were prepared by homogenization of tissue and high-speed centrifugation. Rat liver cytosol contained a calcium dependent kinase that preferentially phosphorylated an endogenous substrate protein with apparent M_r of 95,000. Cytosol from bovine cerebral cortex was fractionated by DEAE-cellulose chromatography. Column fractions were incubated with 2.5 µM $\gamma[^{32}P]$ATP in the absence or presence of calcium and analyzed by SDS-polyacrylamide gel electrophoresis and autoradiography. A calcium and calmodulin dependent protein kinase and endogenous substrates were eluted with 0.05 N NaCl. Similar treatment of rat brain cytosol revealed the presence of three distinct calcium and calmodulin dependent protein kinases that were eluted at approximately 0.04 N, 0.11 N and 0.15 N NaCl. Each kinase appears to have a different domain of endogenous substrates. In addition, these kinases appear to be differentially regulated by calcium. Analysis of phosphorylation in HeLa cell nuclei revealed the presence of a calcium dependent protein kinase that selectively phosphorylated histone H3. The existence of soluble calcium dependent protein phosphorylation in such diverse tissues supports the hypothesis that some of the effects of calcium on soluble enzyme systems may be mediated by protein phosphorylation. (Supported by USPHS grants MH-32752 [NIMH] and CA-24580 [NIH]).

CALCIUM INDUCED CHANGE OF THE ACCESSIBILITY OF ANION TO TRIMETHYL-LYSINE-115 AS DETECTED BY ^{1}H-NMR SPECTROSCOPY. Y. Shibata, T. Miyazawa & S. Kakiuchi*, Department of Biophysics & Biochemistry, University of Tokyo, Bunkyo-ku, Tokyo 113, Japan, and *Institute of Higher Nervous Activity, Osaka University Medical School, Kita-ku, Osaka 530, Japan.

Calmodulin (bovine brain) has one prominent amino acid residue, trimethyl-lysine (Tml), at position 115 (Watterson et al., J.Biol. Chem. 255:962, 1980). To elucidate the role of this positively charged residue of calmodulin, it is important to investigate the microenvironment of this residue in the presence or absence of Ca^{++}. We have used a paramagnetic anion, hexacyanochromate ($[Cr(CN)_6]^{3-}$), and monitored the environmental change of Tml residue using ^{1}H-NMR spectroscopy. This paramagnetic anion enhances the longitudinal relaxation rates of the nuclei near its binding site (Inagaki et al., J. Biochem. 86:591, 1979). Thus, from the paramagnetic effect of this anion on the relaxation rate of methyl protons of Tml-115, the accessibility of this Tml residue could be evaluated. In the presence of excess amounts of Ca^{++}, the relaxation rate enhancement of methyl protons of Tml-115 induced by $[Cr(CN)_6]^{3-}$ is comparable to that of N-methyl protons of choline, which is used as a freely accessible model compound of Tml side chain. This indicates that, in the presence of Ca^{++}, Tml residue is exposed to the solvent and interacts with an anion almost freely. On the other hand, the relaxation rate enhancement in the absence of Ca^{++} is considerably weaker than that in the presence of excess Ca^{++}. This means that, in the Ca^{++}-free state, an anion in the solution is hardly accessible to Tml-115 residue. As calmodulin interacts with the target proteins only in the presence of Ca^{++}, this concomitant change of the accessibility to Tml residue possibly suggests that the Tml residue is involved in the binding with the target proteins.

Ca^{++} INDEPENDENT ACTIVATION OF Ca^{++} DEPENDENT CYCLIC NUCLEOTIDE PHOSPHODIESTERASE BY SYNTHETIC COMPOUNDS, QUINAZOLINESULFONAMIDE DERIVATIVES. T. Sone*, E. Yamada, T. Tanaka & H. Hidaka, Department of Pharmacology, Mie University School of Medicine, Tsu, Mie 514, Japan, and *Pharmaceutical Division Biochemical Research Center Asahi Chemical Industries, Nobeoka, Miyazaki 882, Japan.

Some synthetic compounds (quinazolinesulfonamides) have been found to activate calcium independently Ca^{++} calmodulin dependent cyclic nucleotide phosphodiesterase. The increase in the activity by the compounds was comparable to the activation observed with calcium and calmodulin. This activation was observed with 2,4-dipiperidino-6-quinazolinesulfonamides but not with 2-piperidino-6-quinazolinesulfonamide. Ca^{++} dependent phosphodiesterase was purified from bovine heart and brain by calmodulin affinity chromatography. cAMP and cGMP phosphodiesterase were purified from human platelets by DEAE-cellulose chromatography. 2,4-Dipiperidino-6-quinazolinesulfonamides activate Ca^{++} dependent phosphodiesterase in the absence of Ca^{++} with increase in V_{max} and decrease in K_m

value, similar to those observed with Ca^{++} calmodulin. These quinazolinesulfonamides did not activate Ca^{++} dependent phosphodiesterase in the presence of Ca^{++} calmodulin complex. Moreover, these compounds are stimulators of neither cAMP nor cGMP phosphodiesterase activities but are potent inhibitors. These results suggest that 2,4-dipiperidono-6-quinazolinesulfonamides may activate Ca^{++} dependent phosphodiesterase activity calcium independently through mechanisms similar to those seen with some lipids. On the other hand, calmodulin antagonists such as N-(6-aminohexyl)-5-chloro-1-naphthalenesulfonamide (W-7) inhibited quinazolinesulfonamides induced activation of the phosphodiesterase selectively. These results suggest that these quinazolinesulfonamides are calcium independent activators of Ca^{++} phosphodiesterase and that these compounds may be useful for in vitro studies of calmodulin.

CALMODULIN IN PARKINSON'S DISEASE AND LESCH-NYHAN SYNDROME. Y. Suzuki, M. Shibuya, N. Kageyama, H. Hidaka*, O. Hornykiewicz** & K.G. Lloyd***, Department of Neurosurgery, Nagoya University, Nagoya 466, Japan; *Department of Pharmacology, Mie University, Tsu 514, Japan; **Institute of Biochemical Pharmacology, University of Vienna, Austria; and ***Department of Neuropharmacology, L.E.R.S, Bagneux, France.

Despite numerous documentations on the physiological roles of calmodulin (CaM), little is known of its alterations in the basal ganglia in patients with Parkinson's disease or Lesch-Nyhan syndrome. We report here significant decreases in CaM contents in the striatum in patients with these diseases. Intrastriatal distribution of CaM and its possible alteration in a state of denervation supersensitivity was also examined in rats treated with intranigral injection of 6-hydroxydopamine (6-OHDA) or intrastriatal injection of kainic acid. Human brains were obtained at autopsy, dissected and kept below -40°C until assayed. The corpus striatum was homogenized, boiled and sonicated. CaM contents were measured from this homogenate on the basis of its ability to activate CaM deficient, Ca^{++} dependent fraction of cGMP phosphodiesterase (PDE) partially purified from human brain or aorta using 0.4 µM cGMP as a substrate. Results: 1) CaM content in the striatum from three Parkinsonian patients (152 ± 2 units/mg protein) was significantly lower than that of three controls (247 ± 11 (p 0.05). 2) CaM content in the caudate nucleus from three patients with Lesch-Nyhan syndrome (185 ± 18 units/mg protein) was also significantly lower than that of three age-matched controls (330 ± 2)(p 0.05).3) CaM content in the striatum was significantly decreased only in kainic acid treated (66 ± 7% of control side, p <0.05, 1 week after 2 µg, intrastriatal injection) on but not in 6-OHDA treated (91 ± 6% that of control side, 1 month after 8 µg, intranigral injection). Conclusions: Decrease of striatal CaM in Parkinsonian and Lesch-Nyhan patients indicates the presence of postsynaptic (to dopamine neurons) functional deteriorations in both of these diseases and CaM per se is not responsible for the denervation of supersensitivity seen in 6-OHDA treated rats.

EFFECTS OF PHOSPHOLAMBAN PHOSPHORYLATION CATALYZED BY CYCLIC AMP DEPENDENT AND CALMODULIN DEPENDENT PROTEIN KINASES ON CALCIUM TRANSPORT ATPase OF CARDIAC SARCOPLASMIC RETICULUM. M. Tada, M. Inui, M. Kadoma, M. Yamada, T. Kuzuya, H. Abe & S. Kakiuchi*, First Department of Medicine, and *Institute of Higher Nervous Activity, Osaka University School of Medicine, Osaka 553, Japan.

To elucidate the role of 22,000 dalton protein phospholamban (PN), a putative regulator of Ca^{++} dependent ATPase of cardiac sarcoplasmic reticulum (CSR), we examined the relationship between cyclic AMP (cAMP) and calmodulin (CaM) dependent phosphorylation of PN and their effects on ATPase activity and Ca transport of CSR. CSR was prepared from dog heart, treated with buffered 0.6 M KCl-0.3 M sucrose. CSR was incubated with [γ-^{32}P]ATP or unlabelled ATP, cAMP, cAMP dependent protein kinase and/or exogenous CaM, and subsequently assayed for ATPase activity and Ca uptake by CSR. cAMP dependent phosphorylation of PN was independent of Ca^{++}, whereas CaM dependent phosphorylation of PN was dependent on Ca^{++} within a range between 0.2 and 50 μM. cAMP and CaM dependent phosphorylation of PN occurred independently; when both kinases were operative, the amounts of phosphorylation were additive. Under these conditions, the phosphoproteins formed by cAMP and CaM dependent protein kinases electrophoretically migrated as 11,000 dalton components when sodium dodecyl sulphate-solubilized phosphoproteins were boiled prior to polyacrylamide gel electrophoresis. The ATPase activity was stimulated by either cAMP or CaM dependent phosphorylation of PN at Ca^{++} concentrations up to 2 μM. The extent of stimulation of ATPase activity was additive when both types of phosphorylation were functional. Ca uptake was similarly augmented by cAMP and/or CaM dependent phosphorylation of PN. These results indicate that Ca^{++} dependent ATPase and Ca transport of CSR are regulated by PN phosphorylation catalyzed by cAMP and CaM dependent protein kinases, thus suggesting a dual role of phospholamban in active Ca transport. It remains to be seen whether such a dual role is attributable to a specific structure of phospholamban that may exhibit a heterodimeric configuration (2 x 11,000 daltons).

CALCIUM DEPENDENT CONTROL OF CILIARY MOTILITY IN PARAMECIUM. K. Takahashi, A. Murakami, C. Shingyoji & Y. Mogami, Zoological Institute, Faculty of Science, University of Tokyo, Hongo, Tokyo 113, Japan.

Ciliary reversal in Paramecium is a classical example of calcium dependent regulation of cell motility. To investigate the mechanism whereby the normal ciliary beat is modified by an increase in the intracellular free calcium, we studied the effects of local iontophoretic application of calcium, magnesium, etc, to demembranated Paramecium cilia under various conditions. Paramecium caudatum were demembranated with Triton X-100 and subjected to mild sonication to obtain fragments of the cell surface structure bearing only a few cilia. Reactivation of beat in these cilia normally occurs in the presence of magnesium and ATP in the medium. ATP in a divalent cation-free medium did not induce any ciliary movement. A brief ciliary beat was induced in the presence of ATP, however, by local application of either calcium or magnesium. The response to calcium required a low concentration of magnesium in the medium. The response to calcium was similar in certain respects to the response to magnesium; in particular, it consisted of circular motion which was counterclockwise when viewed from above. These observations indicated that the role of calcium in ciliary reversal may actually be more complex than originally thought. The effects of phenothiazines and vanadate are also discussed.

INVOLVEMENT OF CALMODULIN IN PHAGOCYTOTIC RESPIRATORY BURST OF LEUKOCYTES. K. Takeshige & M. Minakami, Department of Biochemistry, Kyushu University School of Medicine, Fukuoka 812, Japan.

The burst in respiration which is observed in polymorphonuclear leukocytes during phagocytosis or by stimulation of the cell with reagents such as digitonin, cytochalasin D or calcium ionophore A23187 is accompanied by the formation of active oxygens such as superoxide radicals and hydrogen peroxide. A rise in the cytosol calcium concentrations is considered to be an important factor in the stimulation of the oxidative metabolism, but the mechanism by which calcium stimulates the metabolism is unknown. In this communication, we show effects of calmodulin inhibitors on superoxide release by stimulated leukocytes and suggest the involvement of calmodulin in stimulating the oxidative metabolism of leukocytes during phagocytosis. The suggestion is based on the following findings: 1) The superoxide release of guinea pig polymorphonuclear leukocytes induced by phagocytosis or by stimulation with cytochalasin D, digitonin or calcium ionophore A23187 was completely inhibited by the calmodulin inhibitors such as trifluoperazine or N-(6-aminohexyl)-5-chloro-1-naphthalene sulfonamide (W-7) at 10 µM. 2) The activity of a particulate NADPH dependent superoxide forming enzyme which is assumed to be responsible for the cellular superoxide release was also inhibited by the inhibitors, to the value obtained by the resting particles. 3) The activation of bovine heart phosphodiesterase by a boiled extract of guinea pig leukocytes which was dependent on calcium ions and abolished by trifluoperazine was observed. This finding suggests the presence of calmodulin in guinea pig leukocytes.

Ca^{++} AND PHOSPHOLIPID TURNOVER AS AN INITIATOR OF INSULIN RELEASE FROM LANGERHANS' ISLETS OF RAT PANCREAS. K. Tanigawa, H. Kuzuya, H. Imura, Y. Takai* & Y. Nishizuka*, Second Division of Internal Medicine, Kyoto University School of Medicine, Kyoto 606, Japan, and *Department of Biochemistry, Kobe University School of Medicine, Kobe 650, Japan.

Ca^{++} is thought to be an important intracellular messenger which regulates insulin release from the B-cells of Langerhan's islet of pancreas. On the other hand, phosphatidylinositol turnover is related to glucose stimulated insulin release. Recently, a Ca^{++}

activated, phospholipid dependent protein kinase originally described in rat brain (Takai et al., J. Biol. Chem. 254:3692, 1979) was found in Langerhans' islets of rat pancreas. The enzyme was present in the cytosol in an inactive form and was activated by reversible attachment to membranes in the presence of a micromolar concentration of Ca^{++}. Active components of membranes were identified as unsaturated diacylglycerol and phospholipid. Among various phospholipids, phosphatidylserine was the most active. This diacylglycerol appeared to be derived from phosphatidylinositol turnover which was stimulated by glucose, and to serve as a second messenger for the activation of this enzyme. The enzyme thus activated was inhibited selectively by phospholipid interacting drugs such as chlorpromazine and dibucaine. These drugs inhibited glucose-stimulated insulin release from the isolated islets of rat pancreas. Formation of diacylglycerol by bacterial phospholipase C (Clostridium perfingens) induced insulin release in a dose dependent manner. In the absence of Ca^{++}, the effect of phospholipase C on insulin release was abolished. The enzyme was independent of cyclic nucleotide and was clearly distinct from cyclic nucleotide dependent protein kinases. These results suggest that a different protein kinase system which is coupled with phosphatidylinositol turnover may be involved in the mechanism of glucose stimulated insulin release in the presence of Ca^{++}.

ENHANCEMENT OF VINCRISTINE CYTOTOXICITY BY CALMODULIN INHIBITORS AND OVERCOMING OF VINCRISTINE RESISTANCE IN VIVO AND IN VITRO.
T. Tsuruo, Cancer Chemotherapy Center, Japanese Foundation for Cancer Research, Toshima-ku, Tokyo 170, Japan.

Antitumor effect on anticancer agents can be enhanced if drug influx- and efflux-mechanisms can be controlled. In cancer chemotherapy, the drug resistance of tumor cells is also caused by the changes in the transport mechanisms of the drugs across plasma membrane. We found that calmodulin inhibitors greatly enhanced the cytotoxic activity of vincristine in tumor cells, especially in their vincristine resistant subline, through a possible inhibition of vincristine efflux. We used W-7, N^2-dansyl-L-arginine-4-t-butylpiperidine amide (No.233), verapamil, prenylamine, anafranil as calmodulin inhibitors. Mouse leukemia P388, human myelogenous leukemia K562, and their vincristine resistant sublines (P388/VCR and K562/VCR) were used. No. 233 showed most prominent inhibition of calmodulin. No. 233 and verapamil greatly enhanced the cytotoxic activity of vincristine in P388, K562 and, especially in resistant sublines of these cells. Perfect overcoming of vincristine resistance could be attained in vitro. Similar results were also obtained with other calmodulin inhibitors. We have reported that verapamil enhanced the amount of intracellular vincristine through a possible inhibition of the drug efflux (Tsuruo et al., Cancer Res. 41:1967, 1981). Calmodulin inhibitors reported here also enhanced the intracellular amount of vincristine. The results suggest that calmodulin may possibly be involved in anticancer agent-efflux mechanism of tumor cells. Calmodulin is a possible target in prospective cancer chemotherapy.

CALMODULIN OF COPRINUS MACRORHIZUS. I.Uno & T. Ishikawa, Institute of Applied Microbiology, University of Tokyo, Bunkyo-ku, Tokyo 113, Japan.

In Coprinus macrorhizus, a basidiomycete, cAMP plays important roles in growth and fruiting body formation (Uno et al., Proc. Nat. Acad. Sci. 71:479, 1974). In the course of study of cAMP dependent protein kinase in this fungus (Uno & Ishikawa, Biochim. Biophys. Acta 675:197, 1981), a calcium dependent protein kinase was found, and calmodulin was identified as a NAD kinase activator in the presence of calcium. Calmodulin was purified using DEAE-cellulose and fluphenazine-sepharose column chromatography. Membrane bound adenylate cyclase, phosphodiesterase, NAD kinase, and calcium dependent protein kinase were prepared from cell-free extract of this fungus. The activities of these enzymes were inhibited by EGTA, but activated by calmodulin in the presence of calcium. The enzyme activities elevated in the presence of calmodulin and calcium were inhibited by the addition of fluphenazine (FP), trifluoperazine (TP) or chlorpromazine (CP). Ki values of FP, TP and CP for NAD kinase activity were 30, 50 and 100 µM, respectively, and Ki values for other enzymes were similar. Further, growth and fruiting body formation of this fungus was inhibited. Ki value of growth was similar to those for the enzymes. Calmodulin of this fungus showed characteristics similar to that of pig brain, but the molecular weight was 13,000 less than that of pig brain. These results suggest that calmodulin control activities of various enzymes, especially those related to cAMP metabolism, and consequently regulates growth and differentiation in C. macrorhizus.

ANATOMICAL DISTRIBUTION OF CALMODULIN IN SEVERAL AQUATIC ORGANISMS AND SOME PROPERTIES OF SEA SQUIRT CALMODULIN. S. Watabe, H. Kumagai*, K. Masuda & K. Hashimoto, Laboratory of Marine Biochemistry, Faculty of Agriculture, and *Department of Biophysics & Biochemistry, Faculty of Science, University of Tokyo, Bunkyo-ku, Tokyo 113, Japan.

Although there are data on calmodulin as related to mammals, little is known of this compound in lower aquatic organisms. This is true especially with respect to its anatomical distribution. We therefore made an attempt to determine calmodulin concentration in various tissues of several aquatic organisms, and also to isolate and characterize sea squirt calmodulin. Various tissues were excised from the carp, two requiem sharks, sea squirt, sea cucumber and scallop. The calmodulin concentration in each tissue was determined by Butcher's method using Ca^{++} dependent cyclic nucleotide phosphodiesterase. The bryozon and three marine algae were also assayed for calmodulin. In the three fishes, the brain showed the highest calmodulin level (180-280 µg tissue), followed by the gill (52 µg), liver (39-56 µg), cardiac muscle (15-17 µg), ovary (9-12 µg), testis (1-3 µg), and ordinary muscle (about 1 µg). The invertebrate muscles examined generally gave higher levels; sea squirt mantle 12 µg, sea cucumber longitudinal 32 µg, and scallop adductor 13 µg. In the case of sea squirt, the content was highest in the liver (110 µg), followed by the gill sac (45 µg) and ovary (42 µg). The bryozon exhibited a level of 23 µg/g, and the sea lettuce 1.6 µg. Then calmodulin was isolated from the gill sac of sea squirt by the method featuring tubulin-sepharose affinity chromatography. When electrophoresed on SDS-slab gel in the presence of EGTA, sea squirt calmodulin thus obtained was separated into two bands, one of which had the same mobility as that of various calmodulins so far reported. The amino acid composition was comparable to those of various calmodulins, except for higher contents of serine (14 mol/mol) and glycine (20 mol/mol), and the lower content of methionine (3 mol/mol). A molecular weight of 16,500 was calculated for this calmodulin, assuming that it contains two tyrosine residues.

EFFECTS OF ANTIMICROTUBULE DRUGS AND ALPHA ADRENOCEPTOR BLOCKING AGENTS ON CALMODULIN.

K. Watanabe, J.S. Law, E.F. Williams, R. Nuwonkowala & W.L. West, Department of Pharmacology, College of Medicine, Howard University, Washington D.C. 20059, USA.

A number of studies indicate the involvement of calmodulin (CaM) in the regulation of microtubule polymerization and/or in the functions regulated by microtubules such as insulin release and catecholamine release. Thus CaM may serve as a receptor for drugs that affect microtubules. Among the antimicrotubule drugs (colchicine, podophyllotoxin, griseofulvin and vinca alkaloids), dimeric indole vinca alkaloids (vinblastine [VB], vincristine, desacetylvinblastine amide) were selective inhibitors of CaM activated PDE activity. This action of vincas resides in the catharanthine moiety of VB molecule. The other half of VB molecule, vindoline, which is an inactive compound as a mitotic inhibitor, was the weakest inhibitor of PDE action. This action was not CaM dependent. Maytansine which binds VB site of tubulin but structurally unrelated drug selectively inhibited the CaM activated PDE activity with a potency similar to vincristine. These results indicate that CaM can serve as a receptor for vinca type of antimicrotubule drugs. Since an $alpha_2$ inhibitor, yohimbine, which contains the indole moiety, also affects microtubules, a series of alpha adrenoceptor blockers were examined. A relative order of the potency is phenoxybenzamine (PBA) = dibenamine > phentolamine > yohimbine > prazosin > tolazoline and the first four drugs were selective inhibitors of calmodulin action. Inhibition (50%) was obtained by 2 µM of PBA. Since antineoplastic alkylating agents (up to 4 mM) were not strong inhibitors of CaM action, alkylating activity of PBA may not be a primary mechanism. Our preliminary results showed that ^{3}H-VB bound to CaM in a Ca^{++} dependent manner, however, binding to albumin is not. Trifluoperazine reduced VB binding but PBA had no effect. PBA enhanced the fluorescence of TNS to CaM Ca^{++} dependently. Although our working hypothesis led to the finding of new CaM inhibitors, whether these drugs also act on CaM in vivo remains to be elucidated. These results suggest the importance of the high lipophillic nature along with the certain configuration of the drugs in binding to CaM. These results also suggest the existence of more than one drug binding site in CaM. (Supported by 1-T02-GM05000-MARC, Bio-Medical Supporting Grant 5S07-RRo5361, and an Institutional Research Grant from the American Cancer Society, No. 132).

INTRACELLULAR CALCIUM LINKED TO PROTEIN PHOSPHORYLATION AND MEMBRANE POTASSIUM PERMEABILITY IN SYMPATHETIC NEURONS. F.F. Weight, H.C. Pant & S.M. McCort, Laboratory of Preclinical Studies, National Institute on Alcohol Abuse and Alcoholism, Rockville, Maryland 20852, USA.

The possible relationship of intracellular calcium, $[Ca^{++}]_i$, to protein phosphorylation and membrane permeability in vertebrate neurons was investigated in sympathetic ganglia of the bullfrog using both biochemical and electrophysiological methods. The permeability of the membrane was investigated using standard intracellular recording techniques. Protein phosphorylation was analyzed by incubating intact ganglia in vitro in a Ringer's solution containing ^{32}P inorganic phosphate, polyacrylamide SDS gel electrophoresis and liquid scintillation counting. Spontaneous rhythmic hyperpolarizations (SRH) were induced by the addition of 5 mM theophylline to the Ringer or by the intracellular injection of citrate ions (see also: J. Physiol. 298:251, 1980). Membrane resistance decreased during the SRH, they reversed near E_K and the reversal potential was sensitive to $[K]_o$. The frequency of the SRH was increased by the addition to the Ringer's of increased $[Ca]_o$ (8 mM), A23187 (10 µM), and increased $[K]_o$ (8 mM). The frequency of the SRH was decreased or they were abolished by dantrolene (43 µM). In the biochemical studies, the phosphorylation of proteins was decreased by addition to the Ringer's of theophylline (5 mM), increased $[Ca]_o$ (4, 8 and 20 mM), A23187 (1 and 10 µM), and increased $[K]_o$ (20 mM). On the other hand, dantrolene (43 µM) increased protein phosphorylation. The major change in the incorporation of ^{32}P was in protein bands at 18,000-20,000 daltons. The results suggest that in these vertebrate neurons, intracellular Ca^{++} is linked to membrane K^+ permeability and protein phosphorylation.

DYNAMICS OF CALMODULIN AND CYCLIC AMP-PHOSPHODIESTERASE IN PLASMA MEMBRANES OF RAT LIVERS AND ASCITES HEPATOMAS. K. Yamagami & H. Terayama, Zoological Institute, Faculty of Science, University of Tokyo, Hongo, Bunkyo-ku, Tokyo 113, Japan.

Calmodulin (CaM) in plasma membranes (PM) may be responsible for modulating Ca^{++} transport, cyclic nucleotide metabolism and phosphorylation of proteins, and henceforth for controlling cell proliferation. In the present study the contents and dynamics of CaM and cyclic AMP phosphodiesterase (PDE) in PM were compared between normal liver and hepatoma cells (rats). PM were prepared from liver and ascites hepatoma (AH-7974, AH-130) cells by the method

of Ray. CaM in PM was released by heating at 80°C for 5 min and assayed by the stimulation of CaM dependent PDE (bovine brain). PDE was assayed by the conventional method using 2 mM cyclic AMP as substrate. Snake venom 5'-nucleotidase was used for hydrolyzing 5'-AMP thus formed, and then Pi was assayed colorimetrically.
Results: 1) Hepatoma PM contained much smaller amounts of CaM (approximately ½) and PDE (approximately 1/3) compared to liver PM. 2) A part of CaM molecules in liver PM (but not hepatoma PM) was released by washing. The sufficiently washed liver PM had the specific binding sites for externally added CaM (bovine brain) (n = 140 pmol/mgprotein, Kd = 7.9×10^{-8} M), whereas only nonspecific binding with negative cooperativity was found in the washed hepatoma PM. 3) PM (liver and AH-7974) appeared to contain both CaM dependent and CaM independent PDE, but the stimulation by external Ca^{++} plus CaM was rather small. 4) Externally added CaM dependent PDE (bovine brain) was more bound to liver PM than hepatoma PM. Newly bound PDE was more sensitive to the stimulation by Ca^{++} plus CaM in hepatoma PM than in liver PM. Preincubation of PM with CaM did not affect the PDE binding to PM, but the stimulation by Ca^{++} plus CaM was lost. These results suggest that the CaM dependent regulatory system in hepatoma PM is somewhat different from that in liver PM.

COMPARISON OF PROPERTIES OF GLYCOGEN PHOSPHORYLASE KINASE IN VARIOUS RAT TISSUES. H. Yamamura, T. Taira, S. Nakamura, H. Tabuchi, A. Tsutou & E. Hashimoto, Department of Biochemistry, Fukui Medical School, Matsuoka-cho, Yoshida-gun, Fukui-ken 910-11, Japan.

Glycogen phosphorylase kinase, which seems to be a multifunctional protein kinase, requires Ca^{++} for its activity. Although skeletal muscle phosphorylase kinase has been extensively studied, no vigorous attempts have been made to characterize the nature of phosphorylase kinase from various tissues. The present studies are an attempt to clear the properties of phosphorylase kinase obtained from various tissues, and also to compare them with those of skeletal muscle enzymes. Properties of phosphorylase kinase, present in 100,000 x g supernatant from nine rat tissues (skeletal muscle, heart, brain, fat, lung, kidney, spleen, testis and liver) were examined of pH 6.8/8.5 ratio, Ca^{++} dependency, activation by cAMP dependent protein kinase (protein kinase A) and reactivity with an antiskeletal muscle phosphorylase kinase antiserum. If the enzyme has a low value of pH 6.8/8.5 ratio, then it has a high Ca^{++} dependency, highly activated by protein kinase A and strongly inhibited by antiserum. However, if the enzyme has a high value of pH 6.8/8.5 ratio, then it has a low Ca^{++} dependency, weakly activated by protein kinase A and weakly inhibited by antiserum. Heart enzyme has similar properties to those of muscle. However, enzymes from lung, kidney, spleen, testis and liver are quite different from skeletal muscle enzymes. Enzymes from brain and fat are different from both skeletal muscle and liver enzymes and there

PRIMARY STRUCTURE OF CALMODULINS: SEARCH FOR THE ENZYME BINDING SITE. M. Yazawa, Department of Chemistry, Faculty of Science, Hokkaido University, Sapporo 060, Japan.

Calmodulin is known to have two functional units in its structure, namely, putatively assigned Ca^{++} binding sites and a site for interaction with the calmodulin dependent enzymes. In an attempt to identify the latter site in the primary structure, several calmodulins with different affinities for calmodulin dependent enzymes were selected. Using the TCA precipitation method, calmodulins were isolated from rabbit, scallop, sea anemone, Tetrahymena and wheat germ, and their complete amino acid sequences were determined. The sequence study of wheat germ calmodulin is in progress. The C-terminal halves of the established structure are shown below, because the substitutions were localized in this region. In the proposed troponin I binding region, which may overlap the binding site, one important substitution was found at position 86 of the Tetrahymena protein. Substitutions occurred frequently at positions 143 and 147. Position 146 of the Tetrahymena protein was deleted. These differences in a sequence of C-terminal 6 residues are to be reflected to the apparent affinities for the enzymes. We conclude that the C-terminal sequence is involved in the enzyme binding site.

```
                70         80         90  X Y Z-Y-X -Z   110
Rabbit       TMMARKMKDTDSEEEIREAFRVFDKDGDGYISAAELRHVMTNL
Scallop      ------------------------------F-------------
Sea anemone  ------------------------------F-------------
Tetrahymena  SL-------------LI---K---R----L-T------------

                          120          X Y -Y-X -Z   145
             GEtmlLTDEEVDEMIREADIDGDGQVNYEEFVQMMTAK
             -- - --------------------------T---S-
             -- - --------------------------K---S-
             -- - --------------------HI------R-- A-
```

THE EFFECT OF MEMBRANE PHOSPHORYLATION ON CALCIUM UPTAKE IN THE MICROSOMES OF SEA URCHIN EGGS. T. Yoshioka & H. Inoue, Department of Physiology, School of Medicine, Yokohama City University, 2-33 Urafune-cho, Minami-ku, Yokohama 232, Japan.

In the eggs of Arbacia and other species, large fluxes of free calcium have been observed after fertilization and it has been suggested that this intracellular release of Ca^{++} may be the general factor promoting the activation of some metabolic factor in the eggs. Recently, we demonstrated that the degree of evoked stimulation was essentially undiminished by removal of extracellular Ca^{++}

during parthenogenesis. Therefore the concept of a release of Ca^{++} from intracellular store has been stressed. The present study was undertaken to describe in detail the kinetics of calcium exchange in microsomes of sea urchin eggs under phosphorylated state. Sea urchin egg microsomes were prepared by the method adopted in sarcoplasmic reticulum. Determination of Ca^{++} uptake was performed by filtration method using $^{45}CaCl_2$. The degree of membrane phosphorylation was expressed as a summation of protein phosphorylation and polyphosphoinositide phosphorylation. Polyphosphoinositide was phosphorylated maximally 5 min after microsomes were incubated in ATP containing solution, then decreased gradually. Protein phosphorylation, on the contrary, increased with incubation time monotonously until 30 min. Ca^{++} uptake by microsomes was markedly increased by 30 min preincubation in ATP solution (protein phosphorylation had occurred sufficiently), but no change in Ca^{++} uptake resulted by polyphosphoinositide phosphorylation. A great deal of difference in calcium binding to microsomes was found in fertilized and unfertilized eggs, but no difference in Ca^{++} permeability of microsomes in either case.

THYREOCALCITONIN EFFECT ON CALCIUM METABOLISM IN BONE TISSUE UNDER HYPOKINESIA. Y. Zorbas & P. Groza, Life Sciences Division, European Institute of Environmental Cybernetics, Athens 515/1, Greece.

The role of thyreocalcitonin (TCT) on calcium metabolism in different bones and the teeth of rats subjected to 60 days hypokinesia (HK) was investigated. Seventy days after each rat was administered 2 microcuries of Ca^{45}, they were divided into four groups. The first served as control, the second exposed to pure KH, the third received only TCT, and the fourth was given HK and TCT. The third and fourth groups were given daily s.c. injections of 5 µg of TCT and polyvinyl pyrrolidone (PVP). The rats in the first and second groups were administered 5 µg of PVP daily. Calcium concentration in the serum was determined. The molars, incisors, upper and lower jaws, parietal, scapular, shoulder, hip and tibial bones were reduced to ash, dissolved in HCl and the content of Ca^{45} and total calcium was measured. Radiometric measurements were made. The content of ash, total calcium and the quantity of Ca^{45} in the teeth and bones was evaluated. In comparison to the control, HK markedly reduced the increment in the weight of skeletal bones. The mineral content decreased in hindleg, scapular and mandibular bones, increased in parietal and maxillary bones, and remained unchanged in the shoulder bones. The rate of Ca^{45} resorption from the molars and nonreadily exchangeable bone fraction was impaired considerably. TCT and PVP decreased the Ca^{45} level in the calcified tissues of the third group rats. TCT and PVP increased the renovation rate of nonreadily exchangeable fraction in the fourth group rats as compared to that of the second group. Thus, TCT does not favorably affect the growth of bone tissues under HK.

PARTICIPANTS

ADELSTEIN, R.S.: Laboratory of Molecular Cardiology, National Heart, Lung, and Blood Institute, National Institutes of Health, Bethesda, Maryland 20205, USA.

ASANO, A.: Institute for Protein Research, Osaka University, Osaka 565, Japan.

BLUMENTHAL, D.K.: Departments of Pharmacology and Cell Biology, and Moss Heart Center, University of Texas Health Science Center at Dallas, Dallas, Texas 75235, USA.

BOTTERMAN, B.R.: Departments of Pharmacology and Cell Biology, and Moss Heart Center, University of Texas Health Science Center at Dallas, Dallas, Texas 75235, USA.

CHAFOULEAS, J.G.: Department of Cell Biology, Baylor College of Medicine, Houston, Texas 77030, USA.

CHARBONNEAU, H.: Department of Biochemistry, University of Georgia, Athens, Georgia 30602, USA.

CHAU, V.: Laboratory of Biochemistry, National Heart, Lung and Blood Institutes, National Institutes of Health, Bethesda, Maryland 20205, USA.

CHOCK, P.B.: Laboratory of Biochemistry, National Heart, Lung and Blood Institutes, National Institutes of Health, Bethesda, Maryland 20205, USA.

CONTI, M.A.: Laboratory of Molecular Cardiology, National Heart, Lung and Blood Institute, National Institutes of Health, Bethesda, Maryland 20205, USA.

CORMIER, M.J.: Department of Biochemistry, University of Georgia, Athens, Georgia 30602, USA.

DALY, J.W.: Laboratory of Bio-organic Chemistry, Institute of Arthritis, Diabetes, Digestive and Kidney Diseases, National Institutes of Health, Bethesda, Maryland 20205, USA.

DEMAILLE, J.G.: U-249 INSERM and Centre de Recherches de Biochimie Macromoleculaire du CNRS, BP 5015, 34033 Montpellier, France.

DE LANEROLLE, P.: Laboratory of Molecular Cardiology, Heart, Lung and Blood Institute, National Institutes of Health, Bethesda, Maryland 20205, USA.

HIRATA, M.: Department of Physiology, Faculty of Dentistry, Kyushu University, Fukuoka 812, Japan.

HUANG, C.Y.: Laboratory of Biochemistry, Heart, Lung and Blood Institutes, National Institutes of Health, Bethesda, Maryland 20205, USA.

JARRETT, H.W.: Department of Biochemistry, University of Georgia, Athens, Georgia 30602, USA.

KAIBUCHI, K.: Department of Biochemistry, Kobe University School of Medicine, Kobe 650, Japan, and Department of Cell Biology, National Institute for Basic Biology, Okazaki 444, Japan.

KAKIUCHI, S.: Institute of Higher Nervous Energy, Osaka University Medical School, Nakanoshima, Kita-ku, Osaka 530, Japan.

KANDA, K.: Institute of Higher Nervous Energy, Osaka University Medical School, Nakanoshima, Kita-ku, Osaka 530, Japan.

KILHOFFER, M.-C.: ERA CNRS 551, Laboratoire de Biophysique, Faculte de Pharmacie, Universite Louis Pasteur, BP 10, 67048 Strasbourg, France.

KLEE, C.B.: Laboratory of Biochemistry, National Cancer Institute, National Institutes of Health, Bethesda, Maryland 20205, USA.

KLUG, G.A.: Departments of Pharmacology and Cell Biology and Moss Heart Center, University of Texas Health Sciences Center at Dallas, Dallas, Texas 75235, USA.

KRINKS, M.H.: Laboratory of Biochemistry, National Cancer Institute, National Institutes of Health, Bethesda, Maryland 20205, USA.

KUMAGAI, H.: Department of Biophysics and Biochemistry, Faculty of Science, The University of Tokyo, Bunkyo-ku, Tokyo 113, Japan.

KUO, P.F.: Institute for Protein Research, Osaka University, Osaka 565, Japan.

MANNING, D.R.: Departments of Pharmacology and Cell Biology, and Moss Heart Center, University of Texas Health Science Center at Dallas, Dallas, Texas 75235, USA.

MARUYAMA, K.: Department of Biology, Faculty of Science, Chiba University, Chiba 260, Japan.

MARUYAMA, M.: Department of Pharmacology, Mitsubishi-Kasei Institute of Life Sciences, Machida, Tokyo 194, Japan.

PARTICIPANTS

EBASHI, S.: Department of Pharmacology, Faculty of Medicine, University of Tokyo, Hongo, Tokyo 113, Japan.

ENDO, T.: Department of Pharmacology, Mie University School of Medicine, Tsu 514, Japan.

FECHHEIMER, M.: Cell & Developmental Biology, Harvard University, 16 Divinity Avenue, Cambridge, Massachusetts 02138, USA.

FLOCKHART, D.A.: Howard Hughes Medical Institute, Department of Physiology, Vanderbilt University Medical School, Nashville, Tennessee 37232, USA.

FUJII, Y.: Department of Anatomy, Shinshu University School of Medicine, Matsumoto 390, Japan.

FUJISAWA, H.: Department of Biochemistry, Asahikawa Medical College, Asahikawa 078-11, Japan.

FUJITA, M.: Institute of Higher Nervous Activity, Osaka University Medical School, Nakanoshima, Kita-ku, Osaka 530, Japan.

FUKUNAGA, K.: Department of Pharmacology, Kumamoto University Medical School, Kumamoto-shi, Kumamoto 860, Japan.

GERARD, D.: ERA CNRS 551, Laboratoire de Biophysique, Faculte de Pharmacie, Universite Louis Pasteur, BP 10, 67048 Strasbourg, France.

HAIECH, J.: U-249 INSERM and Centre de Recherches de Biochimie Macromoleculaire du CNRS, BP 5015, 34033 Montpellier, France.

HARTSHORNE, D.J.: Muscle Biology Group, Departments of Biochemistry and Nutrition and Food Science, College of Agriculture, The University of Arizona, Tucson, Arizona 85721, USA.

HASEGAWA, T.: Institute of Molecular Biology, Faculty of Science, Nagoya University, Chikusa-ku, Nagoya 464, Japan.

HATANO, S.: Institute of Molecular Biology, Faculty of Science, Nagoya University, Chikusa-ku, Nagoya 464, Japan.

HATHAWAY, D.R.: Cardiology Branch, Heart, Lung and Blood Institute, National Institutes of Health, Bethesda, Maryland 20205, USA.

HIDAKA, H.: Department of Pharmacology, Mie University of Medicine, Tsu, Mie 514, Japan.

MATSUBARA, T.: Department of Orthopedic Surgery, Kobe University School of Medicine, Kobe 650, Japan.

MATSUDA, S.: Department of Chemistry, Faculty of Science, Hokkaido University, Sapporo 060, Japan.

MATSUI, K.: Department of Obstetrics and Gynecology, Kumamoto University Medical School, Kumamoto-shi, Kumamoto 860, Japan.

MEANS, A.R.: Department of Cell Biology, Baylor College of Medicine, Houston, Texas 77030, USA.

MIKUNI, T.: Department of Chemistry, Faculty of Science, Hokkaido University, Sapporo 060, Japan.

MIMURA, N.: Institute for Protein Research, Osaka University, Osaka 565, Japan.

MIYAMOTO, E.: Department of Pharmacology, Kumamoto University Medical School, Kumamoto-shi, Kumamoto 860, Japan.

MORIMOTO, K.: The Second Department of Surgery, Osaka University Medical School, Fukushima-ku, Osaka 553, Japan.

MURAMOTO, Y.: Institute of Higher Nervous Activity, Osaka University Medical School, Nakanoshima, Kita-ku, Osaka 530, Japan.

NAGAMOTO, H.: Department of Chemistry, Faculty of Science, Hokkaido University, Sapporo 060, Japan.

NAGATA, T.: Department of Anatomy, Shinshu University School of Medicine, Matsumoto 390, Japan.

NISHIDA, E.: Department of Biophysics and Biochemistry, Faculty of Science, The University of Tokyo, Bunkyo-ku, Tokyo 113, Japan.

NISHIZUKA, Y.: Department of Biochemistry, Kobe University School of Medicine, Kobe 650, Japan, and Department of Cell Biology, National Institute for Basic Biology, Okazaki 444, Japan.

NONOMURA, Y.: Department of Pharmacology, Faculty of Medicine, University of Tokyo, Hong, Tokyo 113, Japan.

OHNO, S.: Department of Anatomy, Shinshu University School of Medicine, Matsumoto 390, Japan.

PARTICIPANTS

OZAKI, K.: Institute of Molecular Biology, Faculty of Science, Nagoya University, Chikusa-ku, Nagoya 464, Japan.

PATO, M.D.: Laboratory of Molecular Cardiology, Heart, Lung, and Blood Institute, National Institutes of Health, Bethesda, Maryland 20205, USA.

PERSECHINI, A.: Muscle Biology Group, Departments of Biochemistry and Nutrition and Food Science, College of Agriculture, The University of Arizona, Tucson, Arizona 85721, USA.

PETRALI, E.H.: Department of Physiology, College of Medicine, University of Saskatchewan, Saskatoon, Canada S7N 0W0.

RANEY, B.L.: Department of Physiology, College of Medicine, University of Saskatchewan, Saskatoon, Canada S7N 0W0.

SAKAI, H.: Department of Biophysics and Biochemistry, Faculty of Science, The University of Tokyo, Bunkyo-ku, Tokyo 113, Japan.

SANO, K.: Department of Biochemistry, Kobe University School of Medicine, Kobe 650, Japan, and Department of Cell Biology, National Institute for Basic Biology, Okazaki 444, Japan.

SEAMON, K.: Laboratory of Bio-organic Chemistry, Institute of Arthritis, Diabetes, Digestive and Kidney Diseases, National Institutes of Health, Bethesda, Maryland 20205, USA.

SELLERS, J.R.: Laboratory of Molecular Cardiology, Heart, Lung and Blood Institute, National Institutes of Health, Bethesda, Maryland 20205, USA.

SHARMA, R.K.: Department of Biochemistry, University of Manitoba, Winnipeg, Manitobe, Canada R3E 0W3.

SHIBATA, S.: Department of Pharmacology, School of Medicine, University of Hawaii, Honolulu, Hawaii 96822, USA.

SILVER, P.J.: Departments of Pharmacology and Cell Biology, and Moss Heart Center, University of Texas Health Science Center at Dallas, Dallas, Texas 75235, USA.

SOBUE, K.: Institute of Higher Nervous Activity, Osaka University Medical School, Nakanoshima, Kita-ku, Osaka 530, Japan.

STULL, J.T.: Departments of Pharmacology and Cell Biology, and Moss Heart Center, University of Texas Health Science Center at Dallas, Dallas, Texas 75235, USA.

SUGINO, H.: Institute of Molecular Biology, Faculty of Science, Nagoya University, Chikusa-ku, Nagoya 464, Japan.

SULAKHE, P.V.: Department of Physiology, College of Medicine, University of Saskatchewan, Saskatoon, Canada S7N 0W0.

TAKAI, Y.: Department of Biochemistry, Kobe University School of Medicine, Kobe 650, Japan, and Department of Cell Biology, National Institute for Basic Biology, Okazaki 444, Japan.

TANAKA, T.: Department of Pharmacology, Mie University School of Medicine, Tsu, Mie 514, Japan.

TAYLOR, D.L.: Cell and Developmental Biology, Harvard University, 16 Divinity Avenue, Cambridge, Massachusetts 02138, USA.

USUDA, N.: Department of Anatomy, Shinshu University School of Medicine, Matsumoto 390, Japan.

VINCENZI, F.F.: Department of Pharmacology, University of Washington, Seattle, Washington 98195, USA.

WALSH, M.P.: Muscle Biology Group, Departments of Biochemistry and Nutrition and Food Science, College of Agriculture, The University of Arizona, Tucson, Arizona 85721, USA.

WANG, J.H.: Department of Biochemistry, University of Manitoba, Winnipeg, Manitoba, Canada R3E 0W3.

WEISS, G.B.: Department of Pharmacology, University of Texas Health Science Center at Dallas, Dallas, Texas 75235, USA.

YAGI, K.: Department of Chemistry, Faculty of Science, Hokkaido University, Sapporo 060, Japan.

YAMAUCHI, T.: Department of Biochemistry, Asahikawa Medical College, Asashikawa 078-11, Japan.

YAZAWA, M.: Department of Chemistry, Faculty of Science, Hokkaido University, Sapporo 060, Japan.

YIN, H.L.: Hematology-Oncology Unit, Massachusetts General Hospital, Boston, Massachusetts 02114, USA.

YU, B.: Department of Biochemistry, Chinese Medical College, Shinyan, China.

CONTRIBUTOR INDEX

Adelstein, R.S., 313
Asano, A., 375

Blumenthal, D.K., 219
Botterman, B.R., 219

Chafouleas, J.G., 141
Charbonneau, H., 125
Chau, V., 199
Chock, P.B., 199
Conti, M.A., 313
Cormier, M.J., 125

Daly, J.W., 93
de Lanerolle, P., 313
Demaille, J.G., 55

Ebashi, S., 189,421
Endo, T., 35

Fechheimer, M., 349
Flockhart, D.A., 303
Fujisawa, H., 267
Fujii, Y., 35
Fujita, M., 167
Fukunaga, K., 255

Gerard, D., 55

Haiech, J., 55
Hartshorne, D.J., 239
Hasegawa, T., 403
Hatano, S., 403
Hathaway, D.R., 303
Hidaka, H., 19,35
Hirata, M., 189
Huang, C.Y., 199

Jarrett, H.W., 125

Kaibuchi, K., 167,183,333
Kanda, K., 167
Kilhoffer, M.-C., 55
Klee, C.B., 303
Klug, G.A., 219
Krinks, M.H., 303
Kumagai, H., 153
Kuo, P.F., 375

Manning, D.R., 219
Maruyama, K., 183
Maruyama, M., 49
Matsubara, T., 333
Matsuda, S., 75
Matsui, K., 255
Means, A.R., 141
Mikuni, T., 75
Mimura, N., 375
Miyamoto, E., 255
Morimoto, K., 167
Muramoto, Y., 167

Nagamoto, H., 75
Nagato, T., 35
Nishida, E., 153
Nishizuka, Y., 333
Nonomura, Y., 189

Ohno, S., 35
Ozaki, K., 403

Pato, M.D., 313
Persechini, A., 239
Petrali, E.H., 281

Raney, B.L., 281

Sakai, H., 153
Sano, K., 333
Seamon, K., 93
Sellers, J.R., 313
Sharma, R.K., 199
Shibata, S., 49
Silver, P.J., 219
Sobue, K., 167,183
Stossel, T.P., 393
Stull, J.T., 219
Sugino, H., 403
Sulakhe, P.V., 281

Takai, Y., 333
Tanaka, T., 19,35
Taylor, D.L., 349

Usuda, N., 35

Vincenzi, F.F., 1

Walsh, M.P., 239
Wang, J.H., 199
Weiss, G.B., 111

Yagi, K., 75
Yamauchi, T., 267
Yazawa, M., 75
Yin, H.L., 393
Yu, B., 333

SUBJECT INDEX

A 23187, 282
Actin binding
   proteins, 174, 304
β-Actinin, 191
Actinogelin,
   chemical, 380
   cross-linking of, 380
   distribution,
      intercellular, 382
      intracellular, 382
   molecular,
      properties, 382
      weight
         SDS-gel, 382
         native, 382
   F-actin,
      binding to, 382
         effectors of, 382
      staining, spot-like, 385
Activator protein, 268
Activity-structure,
   relationship, 21
Actomyosin,
   Ca binding to natural, 193
   from muscle system, 256
   thread,
      artificial Physarum, 412
Adenylate cyclase, 93
Aequorin,
   cells injected with, 354
Affinity chromatography,
   W-7 coupled, 24
Alemethicin, 282
Amino acid,
   composition, 126, 390
Aminoglycoside,
   antibiotics, 111
Antimycin A, 111
Antipsychotic,
   activity, 3
ATPase, Ca^{++} pump, 2, 169

Brain A, 257
Brain regions, 94

Ca^{++}, 55, 303, 421
   -calmodulin-tubulin
      complex, 158
   cellular, 111
   channel,
      voltage sensitive, 116
   intracellular
      mobilization of, 112
   receptor, 141
   sensitivity, 407
      conferring factor, 378
   uptake,
      effects of,
         K^+, 115
         norepinephrine, 115
      mitochondrial, 120
      Scatachard-coordinate
         plots, of equilibrius, 113
   within muscle cell, 239
^{45}Ca,
   uptake, 113
   washout, 113
Caffeine drops, 411
Calcineurin, 304
Calcium, 35
   binding proteins, 55
   contractile process, 313
   ions, 155
Caldesmon, 174, 183
Calmodulin,
   binding protein, 168
      spectrin-like, 171
   fungal, 126
      peanut seed, 128
      sepharose, 130
   -tubulin complex, 164
      K_d for the, 164
Catecholamines,
   biosynthesis of, 267
Cell, 146
   cycle, 146
   non-muscle, 313
   proliferation, 19
   transformation, 144
Channel inhibitors,
   voltage sensitive, 111

Chinese hamster,
    ovary cells, 36
Chloroplast, 136
Colchicine-tubulin,
    complex, 163
Conformational changes, 66
Contraction,
    effect of calcium on, 352
Contraction-relaxation,
    cycle of, 411
Cyclic AMP, 93, 267, 333
    phosphodiesterase, 304
Cyclic nucleotide,
    phosphodiesterase, 154, 199
Cytoskeleton, 146, 167, 183
Cytosol fractions, 264

D-600, 111
Diaacylglycerol, 334
Dibucaine, 6
Difference UV absorption,
    spectrum, 77
DNA replication, 150
Drugs,
    anti-calmodulin, 1, 147

EGTA,
    binding constant of,
        to Ca^{++}, 388
Endocytosis,
    receptor mediated, 143
Enzymes,
    Ca^{++} dependent, 255
    calmodulin dependent, 255
Excitation-contraction,
    coupling, 55

F-actin, 174, 183, 380, 393, 403
    cross-link of, 415
Flip-flop mechanism, 178, 180
Fluorescein,
    fluorescence,
        excitation, spectrum of, 357
        intense,
            on focal points, 385
            on converge points, 385
Fluorescent analog,
    cytochemistry, 359
Fragmin, 405
Free fatty acids, 10

G & P complex, 93
G-actin,
    inactivation of, 414
Gelation, 350
    inducing factor, 378
    reaction
        Ca^{++} sensitivity of , 378
Gel-sol transformation, 393
Gelsolin, 191, 395, 407
Gentamicin, 118
Glutaraldehyde, 37
Gray matter,
    cortex, rat cerebral, 282
    microsomes, 282
Guanine nucleotides, 93

Hormone receptor, 93
Human platelets,
    thrombin induced serotonin
        release in, 337
Hydralazine, 111
Hydrophobic regions, 25

Imipramine, 5
Insulin secretion, 19
Interactions, cooperative, 246

Kinase,
    II, brain distribution, 268
    Ca^{++} stimulated, 281

Leiotonin,
    A, 191
    C, 191
Leukocytes,
    polymorphonuclear, 337
Lymphocytes,
    human peripheral, 337

Macrophages,
    rabbit lung, 393
Magnesium, 60
Mechanisms, regulatory, 239
Membranes, 334
Meromyosin, heavy, 318
Microinjection, 354
Micropipette, glass, 422
Microtubule,
    assembly, 155
    associated proteins, 153, 155

complex, cytoplasmic, 145
polymerization, 144
protein, 154
Modulator protein, 423
Molluscan myosin, 423
Myelin, kinase,
  phospholipid requiring, 282
Myosin,
  aorta, 193
  phosphorylation of, 241
  scallop, 193
  kinase, 194,319
    light chain, 194,242,263
  light chain,
    calmodulin deficient, 256
    phosphorylation of, 20,241
  molecule,
    dephosphorylation of, 319
    phosphorylation of, 313
    rephosphorylation of, 319
  phosphatase,
    light chain, 242

N-acetylimidazole, 89
NAD kinase, 129
Naphthalenesulfonamides, 19
Neomycin, 118
Nitrendipine, 118
Nitroprusside, 111
NN=2-hydroxy-1-(2-hydroxy-4-sulfo-1-naphthylazo)-3-naphthoic acid, 76
N-(2,2,5,5-tetramethyl-3-carbonyl-pyrroline-1-oxyl)-imidazole, 75
N-(2,2,6,6-tetramethyl-4-piperidine-1-oxyl)-maleimide, 75
Nucleation,
  inhibition of, 414
  promotion of, 415

Oleate, 10
Oligomycin, 120
Ovary cells,
  Chinese hamster, 36

Parvalbumin, 423
  preparation of, 76

pH in,
  amoeboid cells, 356
Phenothiazines, 19,282
Phosphatase,
  inhibition of, 319
Phosphatidyllinositol,
  turnover, 334
Phospholipids,
  acidic, 10
Phosphorylation, 20,241,263,313
  -b kinase, 267
  cyclic AMP stimulated, 281
  protein, 344
Photosynthesis, 136
Physarum,
  actin, 403
  actinin, 405
  actomyosin, 409
Phytochrome, 136
Platelet,
  aggregation inhibition, 30
  function, 19
Prostaglandin E1, 337
Protein,
  actin binding, 174,304
  contractile,
    actin, 313
    myosin, 313
  240K, 173
  kinase,
    Ca^{++} dependent, 255
    calmodulin independent, 255
    cAMP dependent, 244,334
    cGMP dependent, 334
    C-kinase, 334
    endogenous, 281
  phosphatase, 305
  regulatory, 239
Proteolysis, 263

Radioautography,
  technique of, 36
Radioimmunoassay, 144
Receptors, 14
  mechanisms, 333
Relaxing factor, 423

Second messengers, 303
  intracellular,

cAMP, 333
cGMP, 333
Serotonin,
    biosynthesis of, 267
[^{14}C]-Serotonin release, 30
Sites, Ca^{++}, 111
Smooth Muscle, 49,111,174,183,189,241
Sodium nitroprusside,
    muscle relaxation, 342
Solation-contraction,
    coupling hypothesis, 364
Spectrin, 168
Sr^{++}, 115
Sr binding, 193
Staining,
    adhesion plaques, 385
    terminal web region, 385
Superprecipitation,
    reversible, 409
Synaptic,
    membrane, 264
    vesicle, 264

Tb^{+++} luminescence, 62
Tissues,
    non-muscle, 255
Transmembrane control, 335
Trifluoperazine, 2,159
Trimethyllysine, 127
Triton X-100, 281
Tropomyosin, 186
    skeletal, 191
    smooth muscle, 191
Troponin, 423
Tryptophan 5-
    mono-oxygenase, 267
Tubulin, 144,153
    -sepharose 4B, 156
Tyr, 83,86
    99; 83,86
    138; 77,83,86
Tyrosine 3-mono-
    oxygenase, 267

Vascular,
    contraction, 19
    smooth muscle, 111
Villin, 407
Viruses,
    oncogenic, 150